Continuum and Computational Mechanics for Geomechanical Engineers

ISRM Book Series
Series editor: Xia-Ting Feng[1] and Reşat Ulusay[2]
[1] *Institute of Rock and Soil Mechanics, Chinese Academy of Sciences, Wuhan, China*
[2] *Hacettepe University, Department of Geological Engineering, Ankara, Turkey*

ISSN: 2326–6872
eISSN: 2326–778X

Volume 7

International Society for Rock Mechanics

Continuum and Computational Mechanics for Geomechanical Engineers

Ömer Aydan

Department of Civil Engineering, University of the Ryukyus, Nishihara, Okinawa, Japan

CRC Press
Taylor & Francis Group
Boca Raton London New York

CRC Press is an imprint of the
Taylor & Francis Group, an **informa** business

CRC Press/Balkema is an imprint of the Taylor & Francis Group, an informa business

Typeset by codeMantra

Library of Congress Cataloging-in-Publication Data
Names: Aydan, Ömer, author.
Title: Continuum and computational mechanics for geomechanical engineers / Ömer Aydan, Department of Civil Engineering, University of the Ryukyus, Nishihara, Okinawa, Japan.
Description: Boca Raton : CRC Press, [2021] | Series: ISRM book series, 2326-6872 ; volume 7 | Includes bibliographical references and index.
Identifiers: LCCN 2020046075 (print) | LCCN 2020046076 (ebook)
Subjects: LCSH: Geotechnical engineering—Mathematics. | Continuum mechanics.
Classification: LCC TA705 .A95 2021 (print) | LCC TA705 (ebook) | DDC 624.1/510151352—dc23
LC record available at https://lccn.loc.gov/2020046075
LC ebook record available at https://lccn.loc.gov/2020046076

Published by: CRC Press/Balkema
Schipholweg 107C, 2316 XC Leiden, The Netherlands
e-mail: Pub.NL@taylorandfrancis.com
www.crcpress.com – www.taylorandfrancis.com

ISBN: 978-0-367-68053-4 (Hbk)
ISBN: 978-0-367-68054-1 (Pbk)
ISBN: 978-1-003-13399-5 (eBook)

DOI: 10.1201/9781003133995
https://doi.org/10.1201/9781003133995

Contents

About the author xi
Acknowledgments xiii
Preface xv

1 Fundamental operations **1**

1.1 Scalar 1
1.2 Vector 1
1.3 Vector operations 1
1.3.1 Addition and subtraction 1
1.3.2 Dot product 2
1.3.3 Cross product 2
1.3.4 Unit vector 2
1.3.5 Coordinate systems and base vectors 3
1.3.6 Vector operations on a Cartesian coordinate system 3
1.4 Tensors of rank n *4*
1.4.1 Definition of tensors of rank *n* 4
1.4.2 Tensor operations 6
1.5 Matrix representation of tensors 7
1.5.1 Matrix representation of vectors 7
1.5.2 Matrix representation of tensors 7
1.6 Coordinate transformation 8
1.7 Derivation of tensorial quantities 9
1.7.1 Derivative of a scalar function 9
1.7.2 Divergence 9
1.7.3 Rotation 9
1.7.4 Gradient of a vector: second-order tensor 9
1.7.5 Divergence of a tensor (second-order tensor) 10
References 10

2 Stress analysis **11**

2.1 Definition of a stress vector 11
2.2 Stress tensor 11
2.3 Relationship between a stress vector and a stress tensor: Cauchy's law 12

2.4 Stress transformation 14
2.5 Principal stresses and stress invariants 15
2.6 Geometrical representation of stress tensor on the Mohr circle for the 2D condition 15
References 17

3 Deformation and strain **18**
3.1 Preliminaries 18
3.2 Derivation of a strain tensor using the Lagrangian description 19
3.3 Derivation of a train tensor using the Eulerian description 21
3.4 Relationship between the small strain theory and the finite strain theory 23
3.5 Geometrical interpretations of a strain tensor 23
3.5.1 Uniaxial deformation 23
3.5.2 Simple shear deformation 24
References 25

4 Fundamental conservation laws **26**
4.1 Fundamental conservation laws for one-dimensional cases 26
4.1.1 Mass conservation law 26
4.1.2 Momentum conservation law 27
4.1.3 Energy conservation laws 28
4.1.4 Fundamental governing equations for coupled hydro-mechanical phenomena 30
4.2 Multi-dimensional conservation laws 35
4.2.1 Mass conservation laws for seepage and diffusion phenomena 35
4.2.2 Momentum conservation law 36
4.2.3 Angular momentum conservation law 36
4.2.4 Energy conservation law 36
4.3 Derivation of governing equations in the integral form 36
4.3.1 Mass conservation law 36
4.3.2 Momentum conservation law 38
4.3.3 Angular momentum conservation law 39
4.3.4 Energy conservation law 40
References 41

5 Constitutive laws **43**
5.1 One dimensional (1D) constitutive laws 43
5.1.1 1D linear constitutive laws 43
5.1.2 1D non-linear constitutive laws for solids 49
5.2 Multi-dimensional constitutive laws 57
5.2.1 Fourier's law 57
5.2.2 Fick's law 57
5.2.3 Darcy's law 57
5.2.4 Hooke's law 58
5.2.5 Newton's law 58

5.2.6 Kelvin–Voigt's law 58
5.2.7 Navier-Stokes law 58
5.3 Non-linear behavior (elasto-plasticity and elasto-visco-plasticity) for solids 59
5.3.1 Elasto-plastic law 59
5.3.2 Elasto-visco-plasticity 62
5.3.3 Yield/failure criteria 65
5.4 Equivalent models for discontinua 70
5.4.1 Equivalent elastic compliance model (Singh's model) 71
5.4.2 Crack tensor model (CTM) 73
5.4.3 Damage model 73
5.4.4 Microstructure models 74
5.4.5 Homogenization technique 76
References 77

6 Laboratory tests **80**
6.1 Laboratory tests on mechanical properties 80
6.1.1 Uniaxial compression tests 81
6.1.2 Direct and indirect tensile strength tests (Brazilian tests) 82
6.1.3 Triaxial compression tests 85
6.1.4 Post-failure behavior in uniaxial and triaxial compression tests 85
6.1.5 Direct shear tests 89
6.1.6 Tilting tests 93
6.1.7 Experimental techniques for creep tests 97
6.2 Thermal properties of rocks and their measurements 101
6.3 Tests for seepage parameters 107
6.3.1 Falling head tests 108
6.3.2 Transient pulse test method 109
6.4 Tests for diffusion parameters 114
References 118

7 Methods for exact (closed-form) solutions **121**
7.1 Basic approaches 121
7.1.1 Intuitive function methods 121
7.1.2 Solution by separating variables 121
7.1.3 Complex variable method 122
7.2 Closed-form solutions for solids 122
7.2.1 Visco-elastic rock sample subjected to uniaxial loading 122
7.2.2 Visco-elastic layer on an incline 127
7.2.3 One-dimensional bar embedded in rock 132
7.2.4 Circular cavity in the elastic rock under a far-field hydrostatic stress 134
7.2.5 Unified analytical solutions for circular/spherical cavity in an elasto-plastic rock 137
7.2.6 Foundations-bearing capacity 153
7.2.7 Two-dimensional closed-form solution methods 155
7.2.8 Three-dimensional closed-form solutions 161

7.3 Closed-form solutions for fluid flow through porous rocks 162
7.3.1 Some considerations on the Darcy law for rocks and discontinuities 162
7.3.2 Permeability tests based on a steady-state flow 167
7.3.3 Permeability tests based on a non-steady-state flow (transient flow tests) 169
7.4 Temperature distribution in the vicinity of geological active faults 177
7.5 Closed-form solutions for diffusion problems 180
7.5.1 Drying testing procedure 180
7.5.2 Saturation testing technique 185
7.6 Evaluation of creep-like deformation of semi-infinite soft rock layer 186
References 188

8 Methods for approximate solutions 192
8.1 Comparison of exact and approximate solutions 192
8.1.1 Exact (closed-form) solution 193
8.1.2 Finite difference method 193
8.1.3 Finite element method 194
8.1.4 Comparisons 197
8.2 1D hyperbolic problem: equation of motion 198
8.2.1 Weak form formulation 199
8.2.2 Discretization 199
8.2.3 Specific example 202
8.2.4 1D parabolic problem: creep problem 202
8.2.5 1D elliptic problem: static problem 202
8.2.6 Computational examples 203
8.3 Parabolic problems: heat flow, seepage and diffusion 203
8.3.1 Introduction 203
8.3.2 Governing equation 204
8.3.3 Weak form formulation 205
8.3.4 Discretization 205
8.3.5 Steady-state problem 207
8.3.6 Specific example 208
8.3.7 Example 1: simulation of a solid body with heat generation 208
8.3.8 Example 2: simulation of a diffusion problem 208
8.4 FEM for 1D pseudo-coupled parabolic problems: heat flow and thermal stress; swelling and swelling pressure 209
8.4.1 Introduction 209
8.4.2 Governing equations 210
8.4.3 Coupling of heat and stress fields 211
8.4.4 Weak form formulation 212
8.4.5 Discretization 213
8.4.6 Specific example 216
8.4.7 Example: simulation of heat generation and associated thermal stress 216
8.5 Hydro-mechanical coupling: seepage and effective stress problem 218

8.5.1 Introduction 218
8.5.2 Governing equations 218
8.5.3 Weak form formulation 219
8.5.4 Discretization 220
8.5.5 Specific example 224
8.5.6 Example: simulation of settlement under sudden loading 225
8.6 Biot problem: coupled dynamic response of porous media 226
8.6.1 Introduction 226
8.6.2 Governing equations 226
8.6.3 Weak form formulation 227
8.6.4 Discretization 228
8.6.5 Specific example 232
8.6.6 Example: simulation of dynamic response of saturated porous media 233
8.7 Introduction of boundary conditions in a simultaneous equation system 233
8.7.1 Formulation 233
8.7.2 Actual implementation and solution of Eq. (8.216b) 235
8.8 Rayleigh damping and its implementation 236
8.9 Non-linear problems 236
8.10 Multi-dimensional situations 236
8.10.1 Shape functions 237
8.10.2 Numerical integration 241
8.11 Special numerical methods for media having discontinuities 242
8.11.1 No-tension finite element method 242
8.11.2 Pseudo discontinuum finite element method 243
8.11.3 Smeared crack element 243
8.11.4 Finite element method with joint or interface element (FEM-J) 244
8.11.5 Discrete finite element method (DFEM) 245
8.11.6 Displacement discontinuity method (DDM) 246
8.11.7 Discrete element method (DEM) 246
8.11.8 Discontinuous deformation analysis method (DDA) 248
References 248

9 Applications of approximate methods in geo-engineering problems 250
9.1 Applications in continuum 250
9.1.1 The stress state of earth and earth's crust 250
9.1.2 Evaluation of the tunnel face effect 252
9.1.3 Three-dimensional simulation of the excavation of a railway tunnel supported with forepoles, rockbolts, shotcrete and steel ribs 254
9.1.4 Effect of bolting pattern in underground excavations 257
9.1.5 Numerical studies on the indentation (impression) experiment 257
9.1.6 The evaluation of the long-term response of an underground cavern 261
9.1.7 Long-term stability of the Derinkuyu underground city, Cappadocia, Turkey 262

9.1.8 Stability analyses of Tomb of Pharaoh Amenophis III, Luxor, Egypt 265
9.1.9 Dynamic response of a large underground cavern 265
9.1.10 Response and stability of abandoned room and pillar mine under static and earthquake loading 266
9.1.11 Modal analyses of shafts at the Horonobe Underground Laboratory 271
9.1.12 Temperature and stress distributions around an underground opening 272
9.1.13 Water-head variations in rock mass around an underground cavern 273
9.1.14 Breakout formation in boreholes in sedimentary rocks due to moisture loss 274

9.2 Applications in discontinuum 279

9.2.1 Earthquake fault rupture simulation 279
9.2.2 Pseudo-dynamic analyses on the interaction of structures and earthquake faults 282
9.2.3 Dynamic stability conditions of a single rock block 282
9.2.4 Stability of a slope against planar sliding 284
9.2.5 Stability of rock slope against columnar toppling 285
9.2.6 Stability of rock slope against flexural toppling and its stabilization 286
9.2.7 Retrofitting of unlined tunnels 288
9.2.8 Analysis of backfilling of abandoned mines 290
9.2.9 Simulation of creep-like deformation of the Babadağ landslide by DFEM 294
9.2.10 Simulation of creep-like deformation of a rock block at the Nakagusuku Castle 296

References 299

Appendix 1: Gauss divergence theorem 303
Appendix 2: Geometrical interpretation of the Taylor expansion 305
Appendix 3: Reynolds transport theorem 306
Appendix 4: The Gauss elimination method and its implementation 307
Appendix 5: Constitutive modeling of discontinuities and interfaces 310
Appendix 6: Thin band element for modeling discontinuities and interfaces in numerical analyses 314
Index 321

About the author

Born in 1955, Professor Aydan studied Mining Engineering at the Technical University of Istanbul, Turkey (B.Sc., 1979), Rock Mechanics and Excavation Engineering at the University of Newcastle upon Tyne, UK (M.Sc., 1982), and finally received his Ph.D. in Geotechnical Engineering from Nagoya University, Japan, in 1989. Prof. Aydan worked at Nagoya University as a research associate (1987–1991), and then at the Department of Marine Civil Engineering at Tokai University, first as Assistant Professor (1991–1993), then as Associate Professor (1993–2001), and finally as Professor (2001–2010). He then became Professor of the Institute of Oceanic Research and Development at Tokai University and is currently Emeritus Professor at the University of Ryukyus, Department of Civil Engineering, Nishihara, Okinawa, Japan. He has furthermore played an active role in numerous ISRM, JSCE, JGS, SRI, and Rock Mech. National Group of Japan committees and has organized several national and international symposia and conferences. Professor Aydan has received the 1998 Matsumae Scientific Contribution Award, the 2007 Erguvanlı Engineering Geology Best Paper Award, the 2011 Excellent Contributions Award from the International Association for Computer Methods in Geomechanics and Advances, and the 2011 Best Paper Award from the Indian Society for Rock Mechanics and Tunneling Technology, and he was awarded the 2013 Best Paper Award at the thirteenth Japan Symposium on Rock Mechanics and the sixth Japan-Korea Joint Symposium on Rock Engineering. He has been selected as ISRM Vice President for the term of 2019–2023. He was also made Honorary Professor in Earth Science by Pamukkale University in 2008 and received the 2005 Technology Award, the 2012 Frontier Award, and the 2015 Best Paper Award from the Japan National Group for Rock Mechanics.

Acknowledgments

The author sincerely acknowledges Prof. Reşat Ulusay, President of ISRM, and Prof. Xia-Ting Feng, Former President of ISRM, for inviting the author to contribute to the prestigious ISRM Book Series on "Continuum and Computational Mechanics for Geomechanical Engineers". The content of this book is an outcome of learning and teaching at the Istanbul Technical University in Turkey, Newcastle upon Tyne University in the United Kingdom, Nagoya University, Tokai University, and the University of the Ryukyus in Japan for more than four decades. The author would like to sincerely thank Emeritus Prof. Dr. Toshikazu Kawamoto of Nagoya University, Prof. Dr. Yasuaki Ichikawa of Okayama University, Prof. Dr. Takashi Kyoya of Tohoku University, Prof. Dr. Takashi Ito, Prof. Dr. Jun Tomiyama and Assoc. Prof. Dr. Yuya Suda and Prof. Dr. Naohiko Tokashiki of the University of the Ryukyus and Emeritus Prof. Dr. Masanori Hamada of Waseda University for the encouragement, collaboration and guidance for about four decades. Prof. Jun Tomiyama and Assoc. Prof. Dr. Yuya Suda helped and collaborated with the author during the classes at the University of the Ryukyus. The author would also like to thank his former students at Tokai University and the University of the Ryukyus for attending and shaping up the content of this book. In particular, special thanks are to his former students, Mr. Mitsuo Daido and Dr. Yoshimi Ohta of Tokai University, and Mr. Kouki Horicuhi of the University of the Ryukyus and a special guest scholar, Prof. Dr. Halil Kumsar of Pamukkale University, who attended the first classes in 1996 for many fruitful discussions about the content of this book.

Finally, I want to thank my wife, Reiko, my daughter, Ay, my son, Turan Miray, and my parents for their continuous help and understanding all the time, without whom this book could not have been completed.

Preface

Mankind constructed many structures in/on rocks in association with better living conditions and expectations of their societies. Without any doubt, these activities would continue in modern days and the future. Such activities may be extended to other planets in our solar system. The rock mechanics and rock engineering field utilizes some of the basic laws of continuum mechanics and the techniques developed in computational mechanics. This book is intended to describe the basic concepts behind these fundamental laws and their utilization in practice irrespective of whether rock and rock mass contain discontinuities.

The author studied the fundamentals of Continuum Mechanics and Computational Mechanics while he was a graduate student at Nagoya University (Japan) and he taught Continuum Mechanics and Computational Mechanics at Tokai University (Shizuoka, Japan) and the University of the Ryukyus (Okinawa, Japan) for undergraduate and graduate courses together with some emphases on Rock Mechanics and Rock Engineering. Therefore, this book has evolved over more than four decades of learning and teaching and it would address a wide spectra of readers ranging from undergraduate students to experts in Geomechanics.

It is expected to serve as a reference book in Geomechanics. This book consists of nine chapters and six appendices. Brief outlines of the chapters are as follows:

Chapter 1 describes the fundamental mathematical operations needed in continuum mechanics in a precise and simple way. Quantities in geomechanics can be expressed through tensors with different ranks up to the fourth order.

Chapter 2 explains the stress analyses. First, the definitions of stress vector and stress tensor are given. The derivation of the Cauchy relation, which relates the stress vector to the second-order stress tensor, is presented so that the readers would understand the fundamental logic behind the Cauchy relation, which is quite important in Continuum Mechanics. Next, the stress transformation law is described so that it becomes possible to determine principal stresses. Furthermore, the representation of principal stresses on the Mohr circle is presented so that it is possible to visualize the stress tensor geometrically.

Chapter 3 is concerned with deformation and strain. Finite strain tensors are derived using the Lagrangian and Eulerian descriptions and it is shown how they can be related to a small strain tensor. Some examples are given to illustrate how finite strain tensors are geometrically interpreted by considering some classical examples.

Chapter 4 describes the derivation of four fundamental conservation laws of continuum mechanics. Various actual applications of rock mechanics and rock engineering involve mass transportation phenomena such as seepage, diffusion, static and dynamic stability assessment of structures and heat flow. The fundamental laws for these phenomena are first developed for a one-dimensional case and then they are extended to multi-dimensional situations using integral form approaches.

The fundamental governing equations derived in Chapter 4 cannot be solved in their original forms, as the number of equations is less than the number of variables to be determined. If constitutive laws, which are fundamentally determined from experiments, are introduced, their solution becomes possible. In Chapter 5, the well-known constitutive laws for each phenomenon are introduced. The linear and non-linear constitutive laws involve time-dependent and time-independent behavior, fluid flow, heat conduction and diffusion of materials. The constitutive laws are first presented in the one-dimensional form and extended to multi-dimensional situations.

Mechanical, seepage, heat and diffusion properties related to constitutive laws described in Chapter 5 require tests on materials (such as rocks, concrete and metals) appropriate to physical and environmental conditions. In Chapter 6, the fundamental principles of available testing techniques used in the field of rock mechanics and rock engineering are explained. Laboratory tests on mechanical properties of rocks for time-dependent and time-independent behavior may be determined from uniaxial tensile and compression tests, triaxial compression experiments, three-/four-point bending experiments, direct shear and Brazilian experiments. The fundamentals of these testing techniques are described in the related subsections. Furthermore, the theory of tilting test and its application to determine the frictional properties of rock discontinuities are presented. Some techniques to measure thermal properties such as specific heat coefficient, thermal conductivity, thermal diffusion and thermal expansion coefficient are presented. The theory and applications of several techniques to measure the permeability of rocks and discontinuities such as falling-head test, transient pulse technique, and so on are described.

The solutions of governing equations of coupled or uncoupled motion, mass transportation and energy transport phenomena require some certain methods, which may be exact (closed-form solutions), approximate or both. When the resulting equations including initial and/or boundary conditions are simple to solve, the analytical methods are preferred. As the resulting equations and initial and/or boundary conditions are generally complex in many engineering problems, the use of numerical methods such as finite difference, finite element or boundary element becomes necessary. The analytical methods described in Chapter 7 basically solve the equations in their original form and they can be categorized as intuitive, complex variable or separation of variable techniques.

In Chapter 8, the solution of fundamental governing equations is based on the finite element method (FEM). Nevertheless, some illustrative examples are given to explain the similarities and dissimilarities of various numerical methods as well as exact solutions in the first part of this chapter. Although specific formulations for multi-dimensional situations are not explicitly presented, they fundamentally have the same forms except the shape functions used in one dimension and multi-dimensions. The main concepts of extension to the multi-dimensional problems are explained. In the last part, the main concepts of some numerical procedures developed for rock masses involving discontinuities are presented.

Chapter 9 describes some applications of the approximate methods to some typical structures and problems in rock mechanics and rock engineering together with some closed-form solutions. The applications involve both engineering issues associated with structures excavated in/on continuum and discontinuum-type rock masses. Some emphases are also given to numerical simulations of earthquake as well as the state of stress of the earth. Applications cover the static and dynamic responses of various structures as well as seepage, heat conduction and degradation of surrounding rocks.

Appendices 1–4 are concerned with the explanation of fundamental concepts used in continuum mechanics and numerical solution techniques in computational mechanics as simply as possible. The techniques include the Gauss divergence theorem, Taylor expansion, Reynolds transport theorem and Gauss elimination method. Appendices 5 and 6 are concerned with constitutive models and numerical models for discontinuities and interfaces. The multi-response theory is utilized for the constitutive modeling of interfaces/discontinuities while the thin-band element model is described for modeling discontinuities and interfaces in the finite element method, which provides unified numerical modeling.

Chapter 1

Fundamental operations

The fundamental mathematical operations needed in continuum mechanics (e.g. Mase 1970; Eringen 1980) are described in a precise but simplest way in this chapter.

1.1 Scalar

Scalar is a quantity having a magnitude and it remains the same irrespective of directions. It is defined as a rank-0 tensor. Examples of scalar quantities are volume, density, mass, temperature, energy and pressure. Scalar can be easily added, subtracted, multiplied and divided.

1.2 Vector

Vector is a quantity having both magnitude and direction as illustrated in Figure 1.1. It is defined as a rank-1 tensor. Examples of vectors are force, velocity, displacement, acceleration and moment.

1.3 Vector operations

1.3.1 Addition and subtraction

Addition and subtraction of vectors obey the geometrical parallelogram rule and they are illustrated as shown in Figure 1.2 and mathematically expressed using bold-faced symbols as follows:

$$\mathbf{c} = \mathbf{a} + \mathbf{b} \tag{1.1}$$

$$\mathbf{d} = \mathbf{a} - \mathbf{b} \tag{1.2}$$

Figure 1.1 Illustration of vectors.

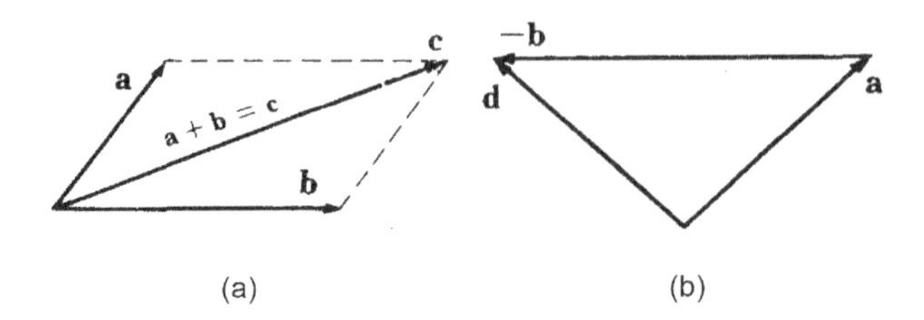

Figure 1.2 Geometrical illustration of addition and subtraction of vectors (a) and subtraction (b) of vectors.

1.3.2 Dot product

The dot product of two vectors is defined as follows:

$$c = \mathbf{a} \cdot \mathbf{b} = \|\mathbf{a}\|\|\mathbf{b}\| \cos\theta \tag{1.3}$$

The quantity above is geometrically interpreted as the projection of vector $(\mathbf{a})$ in the direction of vector $(\mathbf{b})$. θ is the angle between vectors **a** and **b**.

1.3.3 Cross product

The cross product of two vectors is defined as (Figure 1.3).

$$\mathbf{a} \times \mathbf{b} = \mathbf{c} \tag{1.4}$$

The magnitude of the quantity above corresponds to the area bounded by vectors $(\mathbf{a})$ and $(\mathbf{b})$ and the direction of the resulting vector $(\mathbf{c})$ is perpendicular to the plane constituted by vectors $(\mathbf{a})$ and $(\mathbf{b})$:

$$\|\mathbf{c}\| = \|\mathbf{a}\|\|\mathbf{b}\| \sin\theta \tag{1.5}$$

The amplitude of vector $(\mathbf{c})$ is equal to the area constituted by vectors **a** and **b**.

1.3.4 Unit vector

A unit vector is defined as a vector whose magnitude is 1. This vector is utilized to define or measure the vectorial quantities.

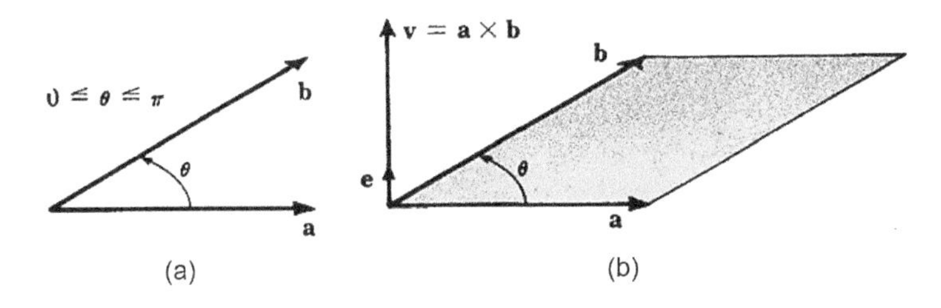

Figure 1.3 Illustration of a cross product of two vectors.

1.3.5 Coordinate systems and base vectors

Fundamentally, there are different coordinate systems such as Cartesian, cylindrical and spherical (Figure 1.4). Here, an orthogonal coordinate system (x_1, x_2, x_3) is considered. Three unit vectors $(\mathbf{e}_1, \mathbf{e}_2, \mathbf{e}_3)$ are assumed to be parallel to the coordinate axes and they are called base-vectors. For the chosen coordinate system, the following relations hold for dot product operations:

$$
\begin{aligned}
&\mathbf{e}_1 \cdot \mathbf{e}_1 = 1, \quad \mathbf{e}_2 \cdot \mathbf{e}_2 = 1, \quad \mathbf{e}_3 \cdot \mathbf{e}_3 = 1, \\
&\mathbf{e}_1 \cdot \mathbf{e}_2 = 0, \quad \mathbf{e}_2 \cdot \mathbf{e}_3 = 0, \quad \mathbf{e}_3 \cdot \mathbf{e}_1 = 0
\end{aligned}
\tag{1.6}
$$

As for the cross-products, the following relations hold among base vectors:

$$
\begin{aligned}
&\mathbf{e}_1 \times \mathbf{e}_1 = \mathbf{0}, \quad \mathbf{e}_2 \times \mathbf{e}_2 = \mathbf{0}, \quad \mathbf{e}_3 \times \varepsilon_3 = \mathbf{0}, \\
&\mathbf{e}_1 \times \mathbf{e}_2 = \mathbf{e}_3, \quad \mathbf{e}_2 \times \mathbf{e}_3 = \mathbf{e}_1, \quad \mathbf{e}_3 \times \mathbf{e}_1 = \mathbf{e}_2, \\
&\mathbf{e}_2 \times \mathbf{e}_1 = -\mathbf{e}_3, \quad \mathbf{e}_3 \times \mathbf{e}_2 = -\mathbf{e}_1, \quad \mathbf{e}_1 \times \mathbf{e}_3 = -\mathbf{e}_2,
\end{aligned}
\tag{1.7}
$$

In the chosen coordinate system, vector **a** can be expressed as follows:

$$
\mathbf{a} = a_1 \mathbf{e}_1 + a_2 \mathbf{e}_2 + a_3 \mathbf{e}_3 = \sum_{i=1}^{3} a_i \mathbf{e}_i = a_i \mathbf{e}_i, \quad (i = 1,2,3) \tag{1.8}
$$

The components of vector **a** can be given as follows:

$$
a_1 = \mathbf{a} \cdot \mathbf{e}_1, \quad a_2 = \mathbf{a} \cdot \mathbf{e}_2, \quad a_3 = \mathbf{a} \cdot \mathbf{e}_3, \quad a_i = \mathbf{a} \cdot \mathbf{e}_i \tag{1.9}
$$

1.3.6 Vector operations on a Cartesian coordinate system

1.3.6.1 Addition and subtraction

Addition and subtraction of two vectors are represented in the following form using base vectors and their components:

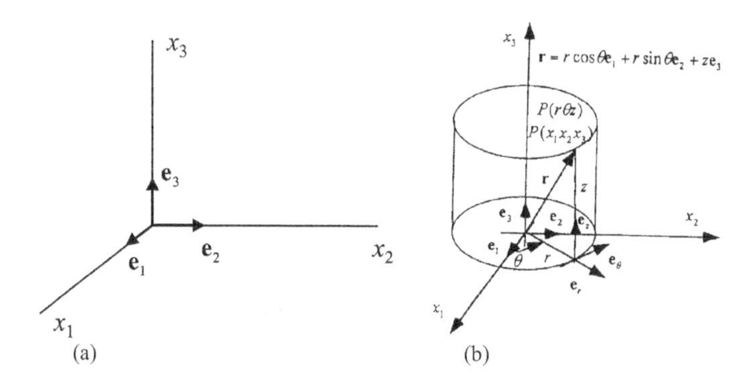

Figure 1.4 Illustration of some of the coordinate systems.

$$\mathbf{a} \pm \mathbf{b} = (a_1 \pm b_1)\mathbf{e}_1 + (a_2 \pm b_2)\mathbf{e}_2 + (a_3 \pm b_3)\mathbf{e}_3 = \sum_{i=1}^{3}(a_i \pm b_i)\mathbf{e}_i = (a_i \pm b_i)\mathbf{e}_i, \quad (i = 1,2,3) \tag{1.10}$$

1.3.6.2 Dot product

The dot product of two vectors is represented in the following form using base vectors and their components:

$$\mathbf{a} \cdot \mathbf{b} = a_1 b_1 + a_2 b_2 + a_3 b_3 = \sum_{i=1}^{3} a_i b_i = a_i b_i, \quad (i = 1,2,3) \tag{1.11}$$

1.3.6.3 Cross product

The cross product of two vectors is represented in the following form using base vectors and their components:

$$\mathbf{c} = \mathbf{a} \times \mathbf{b} = (a_2 b_3 - a_3 b_2)\mathbf{e}_1 + (a_3 b_1 - a_1 b_3)\mathbf{e}_2 + (a_1 b_2 - a_2 b_1)\mathbf{e}_3 \tag{1.12}$$

1.4 Tensors of rank *n*

Generally, the tensors of rank n have magnitude and directional components. They are difficult to visualize geometrically, and some of the examples include stress, strain, elasticity and viscosity relations.

1.4.1 Definition of tensors of rank *n*

The rank or order of a tensor is defined in terms of the number of independent (non-repeated) base vectors and magnitude. Examples are as follows:

a. Tensor of second order
   Tensors of rank 2 can be expressed in three different forms together with two independent base vectors as given below:

$$\begin{aligned}\mathbf{D} = {} & D_{11}\mathbf{e}_1\mathbf{e}_1 + D_{12}\mathbf{e}_1\mathbf{e}_2 + D_{13}\mathbf{e}_1\mathbf{e}_3 + D_{21}\mathbf{e}_2\mathbf{e}_1 + D_{22}\mathbf{e}_2\mathbf{e}_2 \\ & + D_{23}\mathbf{e}_2\mathbf{e}_3 + D_{31}\mathbf{e}_3\mathbf{e}_1 + D_{32}\mathbf{e}_3\mathbf{e}_2 + D_{33}\mathbf{e}_3\mathbf{e}_3\end{aligned} \tag{1.13a}$$

or

$$\begin{aligned}\mathbf{D} = {} & D_{11}\mathbf{e}_1\mathbf{e}_1 + D_{12}\mathbf{e}_1 \otimes \mathbf{e}_2 + D_{13}\mathbf{e}_1 \otimes \mathbf{e}_3 + D_{21}\mathbf{e}_2 \otimes \mathbf{e}_1 + D_{22}\mathbf{e}_2 \otimes \mathbf{e}_2 \\ & + D_{23}\mathbf{e}_2 \otimes \mathbf{e}_3 + D_{31}\mathbf{e}_3 \otimes \mathbf{e}_1 + D_{32}\mathbf{e}_3 \otimes \mathbf{e}_2 + D_{33}\mathbf{e}_3 \otimes \mathbf{e}_3\end{aligned} \tag{1.13b}$$

or

$$\mathbf{D} = \sum_{i=1}^{3}\sum_{j=1}^{3} D_{ij}\mathbf{e}_i\mathbf{e}_j = \sum_{i=1}^{3}\sum_{j=1}^{3} D_{ij}\mathbf{e}_i \otimes \mathbf{e}_j = D_{ij}\mathbf{e}_i\mathbf{e}_j = D_{ij}\mathbf{e}_i \otimes \mathbf{e}_j \quad (i = 1,2,3), (j = 1,2,3) \tag{1.13c}$$

where $\otimes$ is the tensor product. $\mathbf{e}_i\mathbf{e}_j = \mathbf{e}_i \otimes \mathbf{e}_j$; they are fundamentally the same and are called dyads.

b. Kronecker delta tensor
The Krocker delta tensor is known as the identity tensor, whose normal components have the value of 1, and it is expressed in the following form:

$$\mathbf{I} = \delta_{ij}\ \mathbf{e}_i\mathbf{e}_j;\ \delta_{ij} = 1 \cdot i = j;\ \delta_{ij} = 0 \cdot i \neq j \tag{1.14}$$

c. Tensor of rank 3
As the explicit representation of tensors of rank 3 would involve 3^3 expressions, they are expressed in two different forms together with three independent base vectors and summation symbols as follows:

$$\mathbf{C} = \sum_{i=1}^{3}\sum_{j=1}^{3}\sum_{k=1}^{3} C_{ijk}\ \mathbf{e}_i\mathbf{e}_j\mathbf{e}_k = \sum_{i=1}^{3}\sum_{j=1}^{3}\sum_{k=1}^{3} C_{ijk}\ \mathbf{e}_i \otimes \mathbf{e}_j \otimes \mathbf{e}_k \tag{1.15}$$

$$\mathbf{C} = C_{ijk}\ \mathbf{e}_i\mathbf{e}_j\mathbf{e}_k = C_{ijk}\ \mathbf{e}_i \otimes \mathbf{e}_j \otimes \mathbf{e}_k \quad (i = 1,2,3), (j = 1,2,3), (k = 1,2,3) \tag{1.16}$$

d. Tensor of rank 4
As the explicit representation of tensors of rank 4 would involve 81 (3^4) expressions in a three-dimensional space, they are expressed in two different forms together with four independent base vectors and summation symbols as follows:

$$\mathbf{E} = \sum_{i=1}^{3}\sum_{j=1}^{3}\sum_{k=1}^{3}\sum_{l=1}^{3} E_{ijkl}\ \mathbf{e}_i\mathbf{e}_j\mathbf{e}_k\mathbf{e}_l = \sum_{i=1}^{3}\sum_{j=1}^{3}\sum_{k=1}^{3}\sum_{l=1}^{3} E_{ijkl}\ \mathbf{e}_i \otimes \mathbf{e}_j \otimes \mathbf{e}_k \otimes \mathbf{e}_l \tag{1.17}$$

or

$$\mathbf{E} = E_{ijkl}\ \mathbf{e}_i\mathbf{e}_j\mathbf{e}_k\mathbf{e}_l = E_{ijkl}\ \mathbf{e}_i \otimes \mathbf{e}_j \otimes \mathbf{e}_k \otimes \mathbf{e}_l \quad (i = 1,2,3), (j = 1,2,3), (k = 1,2,3), (l = 1,2,3) \tag{1.18}$$

e. Permutation symbol
To express the cross product **(c = a × b)** in the indicial notation, a third-order tensor ε_{ijk} known as the permutation tensor or alternating tensor is used. This useful tensor is defined as

$$\varepsilon_{ijk} = \begin{cases} i \neq j \neq k;;1,2,3,1,2(\varepsilon_{ijk} = 1) \\ i \neq j \neq k;;3,2,1,3,2(\varepsilon_{ijk} = -1) \\ i = j \neq k, i = k \neq j, i \neq j = k; \varepsilon_{ijk} = 0 \end{cases}$$

For example, components of vector **c** are expressed using the permutation symbol as

$$c_i = \varepsilon_{ijk} a_j b_k$$

The permutation symbol is also used to express the box product ($\lambda = \mathbf{a} \times \mathbf{b} \cdot \mathbf{c}$) as

$$\lambda = \varepsilon_{ijk} a_i b_j c_k$$

The value of the box is equivalent to the volume constituted by vectors **c, a** and **b**.

1.4.2 Tensor operations

1.4.2.1 Multiplication of a tensor with a scalar

Multiplication of a tensor with a scalar results in a tensor with the same order, whose components are magnified by the scalar value as given below:

$$\mathbf{E} = \lambda \mathbf{D} \tag{1.19}$$

1.4.2.2 Operation of a tensor with vectors

a. The dot product of a vector and a tensor of rank 2 can be given in the following form:

$$\mathbf{c} = \mathbf{a} \cdot \mathbf{D} = \left(a_i \mathbf{e}_i\right) \cdot \left(D_{ij} \mathbf{e}_i \mathbf{e}_j\right) = a_i D_{ij} \left(\mathbf{e}_i \cdot \mathbf{e}_i\right) \mathbf{e}_j = a_i D_{ij} \mathbf{e}_j \tag{1.20}$$

$$\mathbf{d} = \mathbf{D} \cdot \mathbf{a} = \left(D_{ij} \mathbf{e}_i \mathbf{e}_j\right) \cdot \left(a_j \mathbf{e}_j\right) = D_{ij} a_j \left(\mathbf{e}_i\right) \mathbf{e}_j \cdot \mathbf{e}_j = D_{ij} a_j \mathbf{e}_i \tag{1.21}$$

Fundamentally, the resulting products are vectors, having different quantities unless the tensor of rank 2 is symmetric.

b. The dot product of two tensors of rank 2 is written in the following form:

$$\mathbf{C} = \mathbf{D} \cdot \mathbf{E} = (D_{ij} \mathbf{e}_i \mathbf{e}_j) \cdot (E_{jk} \mathbf{e}_j \mathbf{e}_k) = D_{ij} E_{jk} \mathbf{e}_i (\mathbf{e}_j \cdot \mathbf{e}_j) \mathbf{e}_k = D_{ij} E_{jk} \mathbf{e}_i \mathbf{e}_k \tag{1.22}$$

The rank of the resulting tensorial quantity is 2.

c. The double dot product of two tensors of rank 2 is written in the following form:

$$W = \mathbf{D} : \mathbf{E} = \left(D_{ij} \mathbf{e}_i \mathbf{e}_j\right) : \left(E_{ij} \mathbf{e}_i \mathbf{e}_j\right) = D_{ij} E_{ij} \left(\mathbf{e}_i \cdot \mathbf{e}_i\right)\left(\mathbf{e}_j \cdot \mathbf{e}_j\right) = D_{ij} E_{ij} \tag{1.23}$$

The resulting quantity is a scalar.

d. Tensor product of two tensors of rank 2:
The resulting tensorial quantity of the tensor product of two tensors of rank 2 is written in the following form and its rank is 4.

$$\mathbf{F} = \mathbf{D} \otimes \mathbf{E} = \left(D_{ij} \mathbf{e}_i \mathbf{e}_j\right) \otimes \left(E_{kl} \mathbf{e}_k \mathbf{e}_l\right) = D_{ij} E_{kl} \mathbf{e}_i \mathbf{e}_j \mathbf{e}_k \mathbf{e}_l = F_{ijkl} \mathbf{e}_i \mathbf{e}_j \mathbf{e}_k \mathbf{e}_l \tag{1.24}$$

1.5 Matrix representation of tensors

Base vectors may not be utilized when vectors or tensors are expressed. Herein, matrix operations are introduced to represent some vectorial or tensorial operations.

1.5.1 Matrix representation of vectors

A vector can be represented using either horizontal (1 × 3) or vertical matrix (3 × 1) as given below:

$$a = [a_1, a_2, a_3]; \quad a = \begin{bmatrix} a_1 \\ a_2 \\ a_3 \end{bmatrix} \tag{1.25}$$

The dot product of two vectors may be represented using the matrix operations as given below:

$$\mathbf{a} \cdot \mathbf{b} = [a_1, a_2, a_3] \begin{bmatrix} b_1 \\ b_2 \\ b_3 \end{bmatrix} \tag{1.26}$$

1.5.2 Matrix representation of tensors

Tensor of rank 2 can be represented in a matrix form as

$$\mathbf{A} = \begin{bmatrix} A_{11} & A_{12} & A_{13} \\ A_{21} & A_{22} & A_{23} \\ A_{31} & A_{32} & A_{33} \end{bmatrix} \tag{1.27}$$

Similarly, the Kronecker delta tensor can be written as

$$\mathbf{I} = \begin{bmatrix} 1 & 0 & 0 \\ 0 & 1 & 0 \\ 0 & 0 & 1 \end{bmatrix} \tag{1.28}$$

The dot product of a vector and a tensor of rank 2 is expressed using the matrix operations as follows:

$$\mathbf{a} \cdot \mathbf{A} = [a_1, a_2, a_3] \begin{bmatrix} A_{11} & A_{12} & A_{13} \\ A_{21} & A_{22} & A_{23} \\ A_{31} & A_{32} & A_{33} \end{bmatrix} \tag{1.29}$$

$$\mathbf{A} \cdot \mathbf{a} = \begin{bmatrix} A_{11} & A_{12} & A_{13} \\ A_{21} & A_{22} & A_{23} \\ A_{31} & A_{32} & A_{33} \end{bmatrix} \begin{bmatrix} a_1 \\ a_2 \\ a_3 \end{bmatrix} \tag{1.30}$$

1.6 Coordinate transformation

Let us consider the position of point P in two coordinate systems (oxy and $ox'y'$) (Figure 1.5). From the geometry, the following relations can be written:

$$\begin{bmatrix} x_1' \\ x_2' \end{bmatrix} = \begin{bmatrix} \cos\theta & \sin\theta \\ -\sin\theta & \cos\theta \end{bmatrix} \begin{bmatrix} x_1 \\ x_2 \end{bmatrix} \quad (1.31)$$

The inverse of the above relation can be shown to be

$$\begin{bmatrix} x_1 \\ x_2 \end{bmatrix} = \begin{bmatrix} \cos\theta & -\sin\theta \\ \sin\theta & \cos\theta \end{bmatrix} \begin{bmatrix} x_1' \\ x_2' \end{bmatrix} \quad (1.32)$$

Let us replace the components of the matrix in the following manner:

$$\beta_{11} = \cos\theta, \quad \beta_{12} = \sin\theta, \quad \beta_{21} = -\sin\theta, \quad \beta_{22} = \cos\theta \quad (1.33)$$

Equations (1.31) and (1.32) may be rewritten as follows:

$$x_i' = \beta_{ij} x_j \quad (1.34)$$

$$x_j = \beta_{jk} x_k' \quad (1.35)$$

If Eq. (1.35) is inserted into Eq. (1.34), the following relation is obtained:

$$x_i' = \beta_{ij}\beta_{jk} x_{k'} = \delta_{ik} x_k' \quad (1.36)$$

The matrix operation $\beta_{ij}\beta_{jk} = \delta_{ik}$ fundamentally corresponds to the identity tensor and is specifically given as

$$\begin{bmatrix} \cos\theta & \sin\theta \\ -\sin\theta & \cos\theta \end{bmatrix} \begin{bmatrix} \cos\theta & -\sin\theta \\ \sin\theta & \cos\theta \end{bmatrix} = \begin{bmatrix} 1 & 0 \\ 0 & 1 \end{bmatrix} \quad (1.37)$$

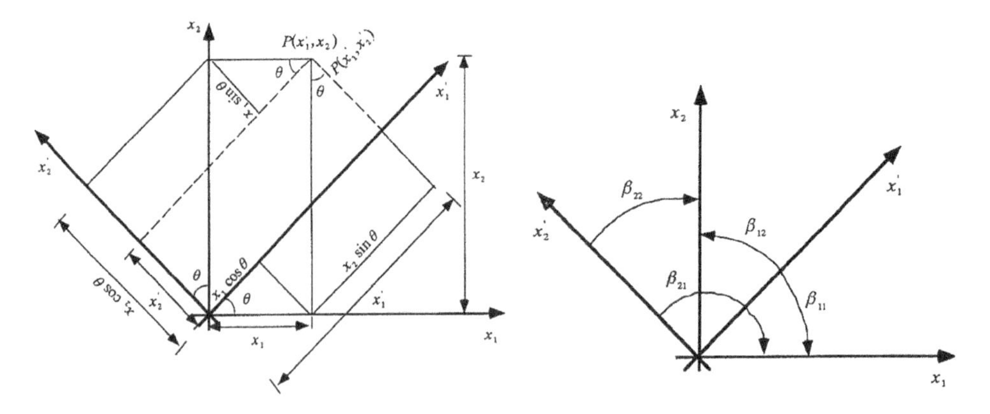

Figure 1.5 Coordinate systems.

1.7 Derivation of tensorial quantities

1.7.1 Derivative of a scalar function

The derivative of a scalar function is called the directional derivation and is represented as

$$df = \frac{\partial f}{\partial x_1} dx_1 + \frac{\partial f}{\partial x_2} dx_2 + \frac{\partial f}{\partial x_3} dx_3 \tag{1.38}$$

The expression above can also be rewritten as

$$df = \left(\frac{\partial f}{\partial x_1} \mathbf{e}_1 + \frac{\partial f}{\partial x_2} \mathbf{e}_2 + \frac{\partial f}{\partial x_3} \mathbf{e}_3 \right) \cdot \left(dx_1 \mathbf{e}_1 + dx_2 \mathbf{e}_1 + dx_3 \mathbf{e}_3 \right) \tag{1.39}$$

with the following definitions:

$$\nabla = \frac{\partial}{\partial x_1} \mathbf{e}_1 + \frac{\partial}{\partial x_2} \mathbf{e}_2 + \frac{\partial}{\partial x_3} \mathbf{e}_3 \tag{1.40}$$

$$d\mathbf{x} = dx_1 \mathbf{e}_1 + dx_2 \mathbf{e}_1 + dx_3 \mathbf{e}_3 \tag{1.41}$$

Equation (1.39) may also be written in the following form:

$$df = (\nabla f) \cdot d\mathbf{x}$$

1.7.2 Divergence

The dot product between the directional derivation operator (∇) and a vector ($\mathbf{a}$) results in the following form and it is interpreted as the divergence of vector ($\mathbf{a}$):

$$div\mathbf{a} = \nabla \cdot \mathbf{a} = \left(\frac{\partial}{\partial x_1} \mathbf{e}_1 + \frac{\partial}{\partial x_2} \mathbf{e}_2 + \frac{\partial}{\partial x_3} \mathbf{e}_3 \right) \cdot \left(a_1 \mathbf{e}_1 + a_2 \mathbf{e}_1 + a_3 \mathbf{e}_3 \right) = \frac{\partial a_1}{\partial x_1} + \frac{\partial a_2}{\partial x_2} + \frac{\partial a_3}{\partial x_3} \tag{1.42}$$

1.7.3 Rotation

The cross product between the directional derivation operator (∇) and a vector ($\mathbf{a}$) results in the following form and it is interpreted as the rotation of vector ($\mathbf{a}$):

$$curl\,\mathbf{a} = \nabla \times \mathbf{a} = \left[\frac{\partial a_3}{\partial x_2} - \frac{\partial a_2}{\partial x_3} \right] \mathbf{e}_1 + \left[\frac{\partial a_1}{\partial x_3} - \frac{\partial a_3}{\partial x_1} \right] \mathbf{e}_2 + \left[\frac{\partial a_2}{\partial x_1} - \frac{\partial a_1}{\partial x_2} \right] \mathbf{e}_3 \tag{1.43}$$

1.7.4 Gradient of a vector: second-order tensor

Derivation of a tensor product between the directional derivation operator (∇) and a vector ($\mathbf{a}$) can be carried out in two ways and results in the following forms and they are interpreted as the gradient of vector ($\mathbf{a}$)

$$\mathbf{D} = \nabla\mathbf{a} = \left(\frac{\partial}{\partial x_1}\mathbf{e}_1 + \frac{\partial}{\partial x_2}e_2 + \frac{\partial}{\partial x_3}e_3\right)(a_1\mathbf{e}_1 + a_2\mathbf{e}_1 + a_3\mathbf{e}_3) = \frac{\partial a_1}{\partial x_1}\mathbf{e}_1\mathbf{e}_1 + \frac{\partial a_2}{\partial x_1}\mathbf{e}_1\mathbf{e}_2$$

$$+\frac{\partial a_3}{\partial x_1}\mathbf{e}_1\mathbf{e}_3 + \frac{\partial a_1}{\partial x_2}\mathbf{e}_2\mathbf{e}_1 + \frac{\partial a_2}{\partial x_2}\mathbf{e}_2\mathbf{e}_2 + \frac{\partial a_3}{\partial x_2}\mathbf{e}_2\mathbf{e}_3 + \frac{\partial a_1}{\partial x_3}\mathbf{e}_3\mathbf{e}_1 + \frac{\partial a_2}{\partial x_3}\mathbf{e}_3\mathbf{e}_2 + \frac{\partial a_3}{\partial x_3}\mathbf{e}_3\mathbf{e}_3$$

$$= \frac{\partial a_j}{\partial x_i}\mathbf{e}_i\mathbf{e}_j = a_{j,i}\mathbf{e}_i\mathbf{e}_j \tag{1.44a}$$

$$E = \mathbf{a}\nabla = (a_1\mathbf{e}_1 + a_2\mathbf{e}_1 + a_3\mathbf{e}_3)\left(\frac{\partial}{\partial x_1}\mathbf{e}_1 + \frac{\partial}{\partial x_2}\mathbf{e}_2\frac{\partial}{\partial x_3}\mathbf{e}_3\right)\frac{\partial a_1}{\partial x_1}\mathbf{e}_1\mathbf{e}_1 + \frac{\partial a_1}{\partial x_2}\mathbf{e}_1\mathbf{e}_2 + \frac{\partial a_1}{\partial x_3}\mathbf{e}_1\mathbf{e}_3$$

$$+\frac{\partial a_2}{\partial x_1}\mathbf{e}_2\mathbf{e}_1 + \frac{\partial a_2}{\partial x_2}\mathbf{e}_2\mathbf{e}_2 + \frac{\partial a_2}{\partial x_3}\mathbf{e}_2\mathbf{e}_3 + \frac{\partial a_3}{\partial x_1}\mathbf{e}_3\mathbf{e}_1 + \frac{\partial a_3}{\partial x_2}\mathbf{e}_3\mathbf{e}_2 + \frac{\partial a_3}{\partial x_3}\mathbf{e}_3\mathbf{e}_3$$

$$= \frac{\partial a_i}{\partial x_j}\mathbf{e}_i\mathbf{e}_j = a_{i,j}\mathbf{e}_i\mathbf{e}_j \tag{1.44b}$$

1.7.5 Divergence of a tensor (second-order tensor)

The dot product between the directional derivation operator (∇) and a tensor (**D**) of rank 2 results in the following form and it is interpreted as the divergence of a tensor (**D**) of rank 2:

$$\mathbf{f} = \nabla\cdot\mathbf{D} = \left(\frac{\partial}{\partial x_1}\mathbf{e}_1 + \frac{\partial}{\partial x_2}\mathbf{e}_2 + \frac{\partial}{\partial x_3}\mathbf{e}_3\right)$$
$$\cdot(D_{11}\mathbf{e}_1\mathbf{e}_1 + D_{12}\mathbf{e}_1\mathbf{e}_2 + D_{13}\mathbf{e}_1\mathbf{e}_3 + D_{21}\mathbf{e}_2\mathbf{e}_1$$
$$+ D_{22}\mathbf{e}_2\mathbf{e}_2 + D_{23}\mathbf{e}_2e_3 + D_{31}\mathbf{e}_3\mathbf{e}_1 + D_{32}\mathbf{e}_3\mathbf{e}_2 + D_{33}\mathbf{e}_3\mathbf{e}_3) \tag{1.45a}$$

or

$$\left(\frac{\partial}{\partial x_j}\mathbf{e}_j\right)\cdot(D_{ji}\mathbf{e}_j\mathbf{e}_i) = \frac{\partial D_{ji}}{\partial x_j}\mathbf{e}_i = D_{ji,j}\mathbf{e}_i \tag{1.45b}$$

References

Eringen, A.C. 1980. *Mechanics of Continua*, Huntington, NY, Robert E. Krieger Publishing Co., 606 p.

Mase, G.E. 1970. *Theory and Problems of Continuum Mechanics*, Schaum's Outline Series, NY, McGraw-Hill Co., 221 p.

Chapter 2

Stress analysis

In this chapter, the definitions of stress vector and stress tensor are provided. The derivation of the Cauchy relation, which relates the stress vector to the second-order stress tensor, is presented so that the readers would understand the fundamental logic behind the Cauchy relation, which is quite important in continuum mechanics (e.g. Mase 1970; Eringen 1980).

2.1 Definition of a stress vector

A stress vector is defined as the limit of infinitesimal force acting over an infinitesimal area with a unit normal vector ($\mathbf{n}$) as given below (Figure 2.1a):

$$\mathbf{t}^{(\mathbf{n})} = \lim_{\Delta S \to 0} \frac{\Delta \mathbf{f}}{\Delta S} = \frac{d\mathbf{f}}{dS} \tag{2.1}$$

The action and reaction law of Newton on a given surface within a body of equilibrium requires that the following condition is satisfied (Figure 2.1b):

$$\mathbf{t}^{(\mathbf{n})} - \mathbf{t}^{(-\mathbf{n})} = \mathbf{0} \tag{2.2}$$

The stress vector is also known as the traction vector.

2.2 Stress tensor

A stress vector acting on a cubic body results in nine components of the second-order stress tensor (Figure 2.2). In a two-dimensional space, a stress tensor has four components. It is represented using tensorial notation as

$$\boldsymbol{\sigma} = \sigma_{ij} \mathbf{e}_i \mathbf{e}_j \tag{2.3}$$

The first and second subscripts of a component of stress tensor correspond to the surface and the axis, respectively.

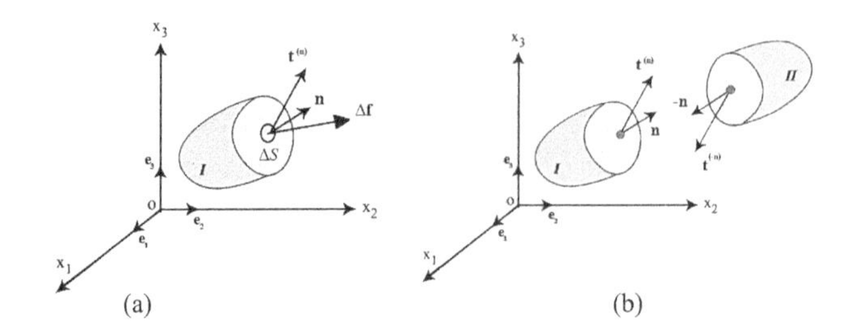

Figure 2.1 Illustration of a stress (traction) vector (a) and unit normal vectors (b) .

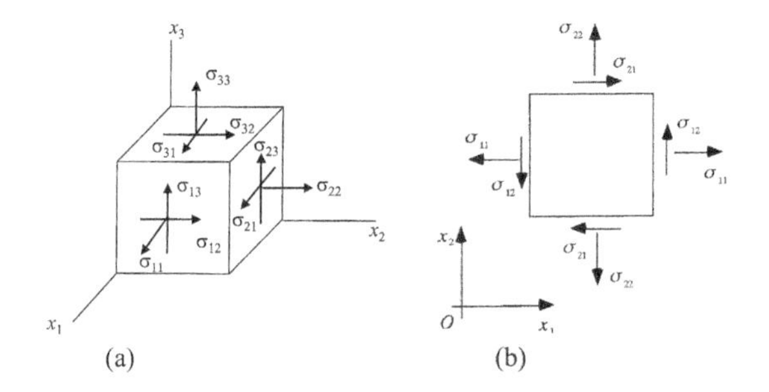

Figure 2.2 Illustration of stress tensor components: (a) 3D illustration and (b) 2D illustration.

2.3 Relationship between a stress vector and a stress tensor: Cauchy's law

Cauchy's law states that a stress vector and a stress tensor can be related to each other in the following form:

$$\mathbf{t} = \mathbf{n} \cdot \boldsymbol{\sigma} = \sigma_{ji} n_j \mathbf{e}_i \tag{2.4}$$

To derive this relation, let us consider a two-dimensional body in equilibrium (Figure 2.3), and the effect of the top part is taken into account as the traction acts on the plane with the surface area Δs.

The unit normal and traction vectors and stress tensor can be given as

$$\mathbf{n} = n_1 \mathbf{e}_1 + n_2 \mathbf{e}_2 \tag{2.5}$$
$$\mathbf{t} = t_1 \mathbf{e}_1 + t_2 \mathbf{e}_2 \tag{2.6}$$
$$\boldsymbol{\sigma} = \sigma_{11} \mathbf{e}_1 \mathbf{e}_1 + \sigma_{12} \mathbf{e}_1 \mathbf{e}_2 + \sigma_{21} \mathbf{e}_2 \mathbf{e}_1 + \sigma_{22} \mathbf{e}_2 \mathbf{e}_2 \tag{2.7}$$

where

$$n_1 = \cos\theta;\ n_2 = \cos(90 - \theta) = \sin\theta$$

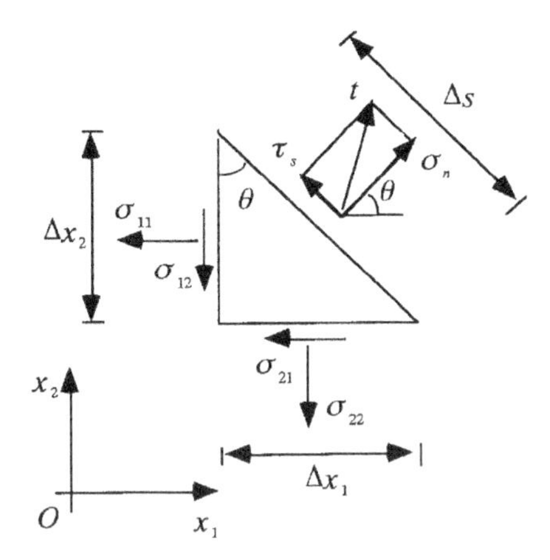

Figure 2.3 Stress components and traction vectors.

The force equilibrium in directions x_1 and x_2 can be written as:
x_1 direction

$$\sum F_{x1} = t_1 \Delta s - \sigma_{11} n_1 \Delta s + \sigma_{21} n_2 \Delta s = 0;\ t_1 = \sigma_{11} n_1 + \sigma_{21} n_2 \tag{2.8}$$

x_2 direction

$$\sum F_{x2} = t_2 \Delta s - \sigma_{12} n_1 \Delta s + \sigma_{22} n_2 \Delta s = 0;\ t_2 = \sigma_{12} n_1 + \sigma_{22} n_2 \tag{2.9}$$

By substituting Eqs. (2.5)–(2.7) into Eqs. (2.8) and (2.9), the following relation can be obtained:

$$\mathbf{t} = \left(n_1 \mathbf{e}_1 + n_2 \mathbf{e}_2\right) \cdot \left(\sigma_{11} \mathbf{e}_1 \mathbf{e}_1 + \sigma_{12} \mathbf{e}_1 \mathbf{e}_2 + \sigma_{21} \mathbf{e}_2 \mathbf{e}_1 + \sigma_{22} \mathbf{e}_2 \mathbf{e}_2\right) \tag{2.10a}$$

or

$$\mathbf{t} = \mathbf{n} \cdot \boldsymbol{\sigma} = \sigma_{ji} n_j \mathbf{e}_i \quad (j = 1,2; i = 1,2) \tag{2.10b}$$

Thus, Eq. (2.10) corresponds to Cauchy's law given in Eq. (2.4).

Normal and shear components of the traction vector on the plane can be given through the following relations:

$$\sigma_N = \mathbf{t} \cdot \mathbf{n} \tag{2.11}$$
$$\sigma_S = \mathbf{t} \cdot \mathbf{s} \tag{2.12}$$

where

$$\mathbf{s} = s_1 \mathbf{e}_1 + s_2 \mathbf{e}_2;\ s_1 = \sin\theta;\ s_2 = \cos\theta \tag{2.13}$$

2.4 Stress transformation

Stress transformation becomes necessary when stress tensors are to be related to each other in different coordinate systems. This transformation law is derived by requiring the stress vector to be the same, independent of the coordinate system. Let us introduce coordinate systems $ox_1x_2x_3$ and $ox_1'x_2'x_3'$ together with stress tensors (σ_{ij} and σ'_{km}) as shown in Figure 2.4.

The following relation is obtained:

$$\mathbf{t} = \sigma_{ji} n_j \mathbf{e}_i = \sigma'_{km} n'_k \mathbf{e}'_m \tag{2.14}$$

Taking dot products of the both sides of Eq. (2.14) with base vector $\mathbf{e}'_m$ yields

$$\sigma_{ji} n_j (\mathbf{e}'_m \cdot \mathbf{e}_i) = \sigma'_{km} n'_k \text{ or } \sigma_{ji} n_j \beta_{mi} = \sigma'_{km} n'_k \tag{2.15}$$

A unit normal vector on an un-dashed coordinate system can be related to a dashed coordinate system through the transformation law as

$$n_j = \beta_{kj} n'_k \tag{2.16}$$

By substituting Eq. (2.16) into Eq. (2.15), the following relation is obtained:

$$\left(\sigma_{ji} \beta_{kj} \beta_{mi} - \sigma'_{km}\right) n'_k = 0 \tag{2.17}$$

As n'_k is arbitrarily chosen, Eq. (2.17) requires the following identity:

$$\sigma'_{km} = \sigma_{ji} \beta_{kj} \beta_{mi} \tag{2.18}$$

Equation (2.18) may also be represented in the matrix form as follows:

$$[\sigma'] = [\beta][\sigma][\beta]^T \tag{2.19}$$

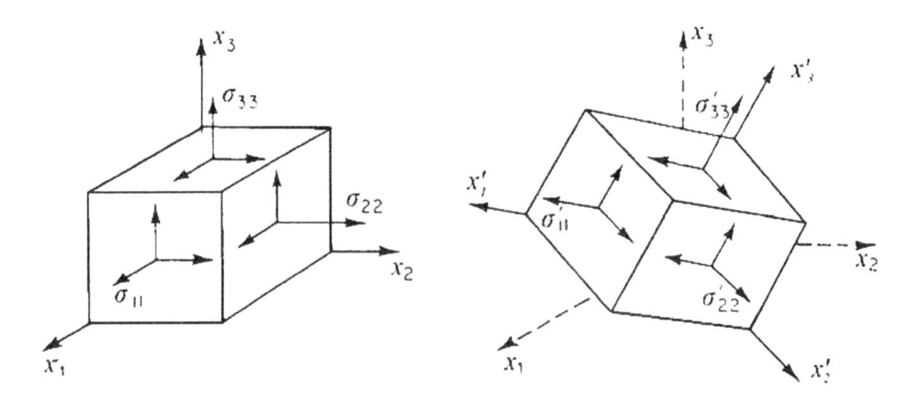

Figure 2.4 Stress components in two coordinate systems.

Using a similar procedure, stress tensor in an un-dashed system can be related to that in a dashed system as follows:

$$\sigma_{ij} = \sigma'_{km} \beta_{jk} \beta_{im} \tag{2.20}$$

In the matrix form, it is written as

$$[\sigma] = [\beta]^T [\sigma'][\beta] \tag{2.21}$$

2.5 Principal stresses and stress invariants

Stress tensor can be represented by three orthogonal components, as shear stress components disappear at the planes on which principal stresses act. Thus, principal stresses can be related to stress tensor on a given coordinate system by requiring that the stress vectors for each situation are equivalent

$$\mathbf{t} = \sigma_{ji} n_j \mathbf{e}_i = \sigma n_i \mathbf{e}_i \text{ or } \left(\sigma_{ji} n_j - \sigma n_i\right) \mathbf{e}_i = \mathbf{0} \tag{2.22}$$

with the following relation

$$n_i = \delta_{ij} n_j \tag{2.23}$$

Equation (2.22) can be rewritten as

$$(\sigma_{ji} - \sigma \delta_{ij}) n_j \mathbf{e}_i = \mathbf{0} \tag{2.24}$$

Equation (2.24) requires the following condition to be satisfied:

$$\left|\sigma_{ji} - \sigma \delta_{ij}\right| = 0 \tag{2.25}$$

Taking the determinant given in Eq. (2.25) results in the following relation, from which three roots are obtained and they correspond to the values of principal stresses.

$$\sigma^3 - I_1 \sigma^2 + I_2 \sigma - I_3 = 0 \tag{2.26}$$

where

$$I_1 = \sigma_{ii} = tr(\boldsymbol{\sigma}) = \sigma_{11} + \sigma_{22} + \sigma_{33};\ I_2 = \frac{1}{2}\left(\sigma_{ii}\sigma_{jj} - \sigma_{ij}\sigma_{ij}\right);$$
$$I_3 = \left|\sigma_{ij} = \det(\boldsymbol{\sigma})\right|;\quad \left(i = 1,2,3;\ j = 1,2,3\right)$$

2.6 Geometrical representation of stress tensor on the Mohr circle for the 2D condition

Mohr (1882) devised a method to represent graphically the stress components using the Mohr circle method (Figure 2.5).

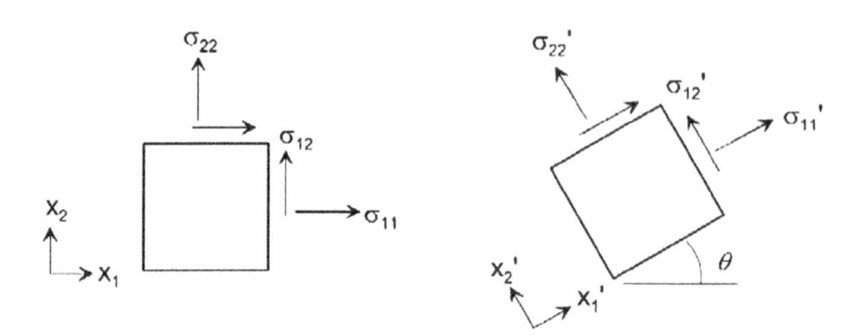

Figure 2.5 Stress components in two coordinate systems.

$$\begin{bmatrix} \sigma'_{11} & \sigma'_{12} \\ \sigma'_{21} & \sigma'_{22} \end{bmatrix} = \begin{bmatrix} \beta_{11} & \beta_{12} \\ \beta_{21} & \beta_{22} \end{bmatrix} \begin{bmatrix} \sigma'_{11} & \sigma'_{12} \\ \sigma'_{21} & \sigma'_{22} \end{bmatrix} \begin{bmatrix} \beta_{11} & \beta_{21} \\ \beta_{12} & \beta_{22} \end{bmatrix} \tag{2.27}$$

where

$$\begin{bmatrix} \beta_{11} & \beta_{12} \\ \beta_{21} & \beta_{22} \end{bmatrix} = \begin{bmatrix} c & s \\ -s & c \end{bmatrix}; c = \cos\theta; s = \sin\theta; \cos^2\theta = \frac{1}{2}(1+\cos 2\theta) \tag{2.28a}$$

$$\sin^2\theta = \frac{1}{2}(1-\cos 2\theta); \sin 2\theta = 2\cos\theta\sin\theta \tag{2.28b}$$

Carrying out the matrix operation given by Eq. (2.27) together with Eq. (2.28) and the symmetry property of the stress tensor, one can easily get the following relations:

$$\sigma'_{11} = \sigma_{11}c^2 + \sigma_{22}s^2 + 2\sigma_{12}cs \text{ or } 2\sigma'_{11} = \frac{\sigma_{11}+\sigma_{22}}{2} + \frac{\sigma_{11}-\sigma_{22}}{2}\cos 2\theta + \sigma_{12}\sin 2\theta \tag{2.29a}$$

$$\sigma'_{22} = \sigma_{11}s^2 + \sigma_{22}c^2 - 2\sigma_{12}cs \text{ or } \sigma'_{22} = \frac{\sigma_{11}+\sigma_{22}}{2} - \frac{\sigma_{11}-\sigma_{22}}{2}\cos 2\theta - \sigma_{12}\sin 2\theta \tag{2.29b}$$

$$\sigma'_{12} = (\sigma_{22}-\sigma_{11})cs + \sigma_{12}\left(c^2 - s^2\right) \text{ or } \sigma'_{12} = -\frac{\sigma_{11}-\sigma_{22}}{2}\sin 2\theta + \sigma_{12}\cos 2\theta \tag{2.29c}$$

As shear stress on the plane should disappear to obtain principal stresses, the angle of rotation should take the following form for Eq. (2.29c):

$$\theta = \frac{1}{2}\tan^{-1}\left(\frac{2\sigma_{12}}{\sigma_{11}-\sigma_{22}}\right) \tag{2.30}$$

Furthermore, one can also derive the following trigonometric relations in terms of stress tensor components:

$$\cos 2\theta = \frac{1}{\sqrt{\left(\frac{\sigma_{11}-\sigma_{22}}{2}\right)^2 + \sigma_{12}^2}} \frac{\sigma_{11}-\sigma_{22}}{2}; \sin 2\theta = \frac{\sigma_{12}}{\sqrt{\left(\frac{\sigma_{11}-\sigma_{22}}{2}\right)^2 + \sigma_{12}^2}} \tag{2.31}$$

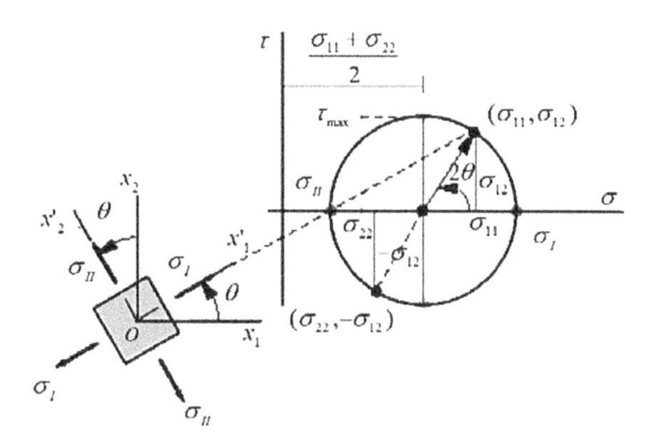

Figure 2.6 Graphical representation of stress tensor components on Mohr's circle.

By substituting Eq. (2.31) in Eqs. (2.29a) and (2.29b), one can easily obtain the following relations for principal stresses

$$\sigma'_{11} = \sigma_I = \frac{\sigma_{11} + \sigma_{22}}{2} + \sqrt{\left(\frac{\sigma_{11} - \sigma_{22}}{2}\right)^2 + \sigma_{12}^2};$$
$$\sigma'_{22} = \sigma_{II} = \frac{\sigma_{11} + \sigma_{22}}{2} - \sqrt{\left(\frac{\sigma_{11} - \sigma_{22}}{2}\right)^2 + \sigma_{12}^2} \qquad (2.32)$$

Accordingly, the maximum shear stress is obtained using Eq. (2.32) as

$$\tau_{max} = \frac{\sigma_I - \sigma_{II}}{2} = \sqrt{\left(\frac{\sigma_{11} - \sigma_{22}}{2}\right)^2 + \sigma_{12}^2} \qquad (2.33)$$

Figure 2.6 shows the graphical presentation of components of stress tensor and principal stresses on Mohr's circle for two-dimensional (2D) conditions.

References

Eringen, A.C. 1980. *Mechanics of Continua*, Huntington, NY, Robert E. Krieger Publishing Co., 606 p.

Mase, G. 1970. *Theory and Problems of Continuum Mechanics*, Schaum Outline Series, New York, McGraw-Hill Co., 230 p.

Mohr, O. 1882. Uber die Darstellung des Spannungszudstandes und des Deformationszustandes eines Korperelementes und über die Anwendung derselben in der Festiakeitslehve. *Zivilingenieur*, 28, 112–155.

Chapter 3

Deformation and strain

This chapter deals with deformation and strain (e.g. Mase 1970; Eringen 1980). Finite strain tensors are derived using Lagrangian and Eulerian descriptions and it has been shown how they can be related to a small strain tensor. Some examples are given to illustrate how finite strain tensors are geometrically interpreted by considering some classical examples.

3.1 Preliminaries

Let us consider a body in the two-dimensional space of two coordinate systems represented by OX_1X_2 and ox_1x_2 as shown in Figure 3.1. The coordinate system OX_1X_2 is introduced at a time step ($t = A$) before the deformation of the body. On the other hand, the coordinate system ox_1x_2 is introduced at a time step ($t = B$) after the deformation of the body. Let us also introduce base vectors $\mathbf{E}_i, \mathbf{e}_i$ associated with each coordinate system.

If we describe the deformation of the body using the coordinate system OX_1X_2, it is called the Lagrangian description. On the other hand, if we describe the deformation of the body using the coordinate system ox_1x_2, it is called the Eulerian description. Herein, strain tensors are derived in two coordinate systems.

Let us introduce the following relations as preliminaries:

a. Lagrangian description:
   Position vectors

$$\mathbf{X} = X_1\mathbf{E}_1 + X_2\mathbf{E}_2 \tag{3.1}$$

$$\mathbf{x} = x_1\mathbf{E}_1 + x_2\mathbf{E}_2 \tag{3.2}$$

Displacement vector

$$\mathbf{u} = u_1\mathbf{E}_1 + u_2\mathbf{E}_2 \tag{3.3}$$

b. Eulerian description:
   Position vectors

$$\mathbf{X} = X_1\mathbf{e}_1 + X_2\mathbf{e}_2 \tag{3.4}$$

$$\mathbf{x} = x_1\mathbf{e}_1 + x_2\mathbf{e}_2 \tag{3.5}$$

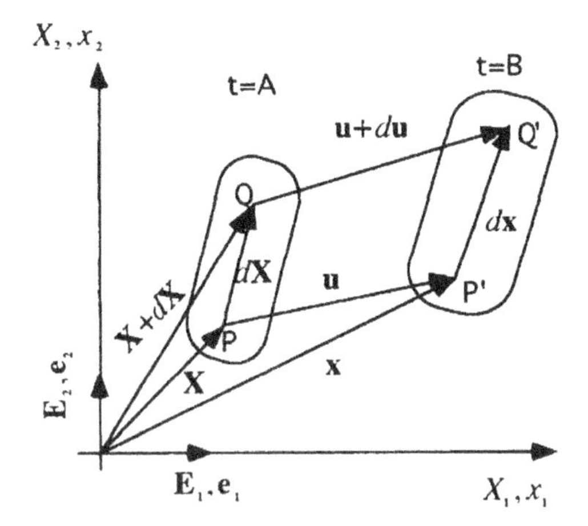

Figure 3.1 Coordinate system and notations.

Displacement vector

$$\mathbf{u} = u_1\mathbf{e}_1 + u_2\mathbf{e}_2 \tag{3.6}$$

Relations between position vectors before and after deformation may be given as

$$\mathbf{x} = \mathbf{X} + \mathbf{u} \tag{3.7}$$

$$d\mathbf{x} = d\mathbf{X} + d\mathbf{u} \tag{3.8}$$

The following tensorial operation should be noted.

$$(\mathbf{ab}) \cdot \mathbf{c} = \mathbf{a}(\mathbf{b} \cdot \mathbf{c}) \tag{3.9}$$

3.2 Derivation of a strain tensor using the Lagrangian description

In this sub-section, a strain tensor is derived according to the Lagrangian description. The position vector of an arbitrary point in the body after deformation using the coordinate system OX_1X_2 may be given as follows:

$$x_1 = x_1(X_1, X_2); \quad x_2 = x_2(X_1, X_2) \tag{3.10}$$

Points P and Q in the body before deformation move to new positions denoted by P' and Q' after deformation. The power of length before and after deformation may be written as

$$dS^2 = d\mathbf{X} \cdot d\mathbf{X} = dX_1dX_1 + dX_2dX_2 = dX_idX_i; \quad i = 1,2 \tag{3.11}$$

$$ds^2 = d\mathbf{x} \cdot d\mathbf{x} = dx_1dx_1 + dx_2dx_2 = dx_idx_i; \quad i = 1,2 \tag{3.12}$$

From these equations, we write the following relation for length change:

$$ds^2 - dS^2 = d\mathbf{x} \cdot d\mathbf{x} - d\mathbf{X} \cdot d\mathbf{X} \tag{3.13a}$$

or

$$ds^2 - dS^2 = (dx_1 dx_1 + dx_2 dx_2) - (dX_1 dX_1 + dX_2 dX_2) = dx_i dx_i - dX_i dX_i; \quad i = 1,2 \tag{3.13b}$$

The length vector after deformation may be given using the Lagrangian description as follows:

$$d\mathbf{x} = \nabla_X \mathbf{x} \cdot d\mathbf{X} = \left(\frac{\partial x_1}{\partial X_1} \mathbf{E}_1\mathbf{E}_1 + \frac{\partial x_1}{\partial X_2} \mathbf{E}_1\mathbf{E}_2 + \frac{\partial x_2}{\partial X_1} \mathbf{E}_2\mathbf{E}_1 + \frac{\partial x_2}{\partial X_2} \mathbf{E}_2\mathbf{E}_2 \right) \cdot \left(dX_1 \mathbf{E}_1 + dX_2 \mathbf{E}_2 \right) \tag{3.14}$$

or

$$d\mathbf{x} = \mathbf{F} \cdot d\mathbf{X} = \left(\frac{\partial x_1}{\partial X_1} dX_1 + \frac{\partial x_1}{\partial X_2} dX_2 \right) \mathbf{E}_1 + \left(\frac{\partial x_2}{\partial X_1} dX_1 + \frac{\partial x_2}{\partial X_2} dX_2 \right) \mathbf{E}_2 \tag{3.15}$$

where

$$\mathbf{F} = \frac{\partial x_1}{\partial X_1} \mathbf{E}_1\mathbf{E}_1 + \frac{\partial x_1}{\partial X_2} \mathbf{E}_1\mathbf{E}_2 + \frac{\partial x_2}{\partial X_1} \mathbf{E}_2\mathbf{E}_1 + \frac{\partial x_2}{\partial X_2} \mathbf{E}_2\mathbf{E}_2 \tag{3.16}$$

F is called the deformation gradient tensor. Variation of the displacement vector is similarly given as

$$d\mathbf{u} = \nabla_X \mathbf{u} \cdot d\mathbf{X} = \left(\frac{\partial u_1}{\partial X_1} \mathbf{E}_1\mathbf{E}_1 + \frac{\partial u_1}{\partial X_2} \mathbf{E}_1\mathbf{E}_2 + \frac{\partial u_2}{\partial X_1} \mathbf{E}_2\mathbf{E}_1 + \frac{\partial u_2}{\partial X_2} \mathbf{E}_2\mathbf{E}_2 \right) \cdot \left(dX_1 \mathbf{E}_1 + dX_2 \mathbf{E}_2 \right) \tag{3.17}$$

or

$$d\mathbf{u} = \mathbf{H} \cdot d\mathbf{X} = \left(\frac{\partial u_1}{\partial X_1} dX_1 + \frac{\partial u_1}{\partial X_2} dX_2 \right) \mathbf{E}_1 + \left(\frac{\partial u_2}{\partial X_1} dX_1 + \frac{\partial u_2}{\partial X_2} dX_2 \right) \mathbf{E}_2 \tag{3.18}$$

where

$$\mathbf{H} = \frac{\partial u_1}{\partial X_1} \mathbf{E}_1\mathbf{E}_1 + \frac{\partial u_1}{\partial X_2} \mathbf{E}_1\mathbf{E}_2 + \frac{\partial u_2}{\partial X_1} \mathbf{E}_2\mathbf{E}_1 + \frac{\partial u_2}{\partial X_2} \mathbf{E}_2\mathbf{E}_2 \tag{3.19}$$

H is called the displacement gradient tensor. If the relation above is inserted into Eq. (3.13), the following relation may be obtained:

$$ds^2 - dS^2 = (\mathbf{F} \cdot d\mathbf{X}) \cdot (\mathbf{F} \cdot d\mathbf{X}) - d\mathbf{X} \cdot d\mathbf{X} = (d\mathbf{X} \cdot \mathbf{F}_c) \cdot (\mathbf{F} \cdot d\mathbf{X}) - d\mathbf{X} \cdot d\mathbf{X} \tag{3.20}$$

provided that

$$\mathbf{F}_c = \mathbf{F}^T; \mathbf{I}_c = \mathbf{I}; \quad d\mathbf{X} = \mathbf{I} \cdot d\mathbf{X} \tag{3.21}$$

Equation (3.20) can be rewritten as

$$ds^2 - dS^2 = d\mathbf{X} \cdot (\mathbf{F}_c \cdot \mathbf{F}) \cdot d\mathbf{X} - d\mathbf{X} \cdot (\mathbf{I}_c \cdot \mathbf{I}) \cdot d\mathbf{X} = d\mathbf{X} \cdot (\mathbf{F}_c \cdot \mathbf{F} - \mathbf{I}_c \cdot \mathbf{I}) \cdot d\mathbf{X} \tag{3.22}$$

If we use the following relations

$$d\mathbf{x} = d\mathbf{X} + d\mathbf{u} = \mathbf{I} \cdot d\mathbf{X} + \mathbf{H} \cdot d\mathbf{X} = (\mathbf{H} + \mathbf{I}) \cdot d\mathbf{X} \tag{3.23}$$

$$\mathbf{F}_c \cdot \mathbf{F} = \mathbf{F}^T \cdot \mathbf{F} = (\mathbf{H} + \mathbf{I})^T \cdot (\mathbf{H} + \mathbf{I}) \tag{3.24}$$

$$(\mathbf{H} + \mathbf{I})^T = \mathbf{H}^T + \mathbf{I}^T = \mathbf{H}^T + \mathbf{I} \tag{3.25}$$

$$\mathbf{F}^T \cdot \mathbf{F} = \mathbf{H}^T \cdot \mathbf{H} + \mathbf{H}^T \cdot \mathbf{I} + \mathbf{I} \cdot \mathbf{H} + \mathbf{I} \cdot \mathbf{I} = \mathbf{H}^T \cdot \mathbf{H} + \mathbf{H}^T + \mathbf{H} + \mathbf{I} \tag{3.26}$$

The Lagrangian strain tensor is defined as

$$\mathbf{L} = \frac{1}{2}\left[\mathbf{F}^T \cdot \mathbf{F} - \mathbf{I}\right] = \frac{1}{2}\left[\mathbf{H} + \mathbf{H}^T + \mathbf{H}^T \cdot \mathbf{H}\right] \tag{3.27a}$$

or

$$\mathbf{L} = \frac{1}{2}\left[(\nabla_x \mathbf{u}) + (\nabla_x \mathbf{u})^T + (\nabla_x \mathbf{u})^T \cdot (\nabla_x \mathbf{u})\right] \tag{3.27b}$$

The Lagrangian strain tensor is also known as the Green strain tensor. It is interpreted as a finite strain tensor. In index notation, it is expressed as follows:

$$L_{jk} = \frac{1}{2}\left[\frac{\partial u_j}{\partial X_k} + \frac{\partial u_k}{\partial X_j} + \frac{\partial u_i}{\partial X_j}\frac{\partial u_i}{\partial X_k}\right] = \frac{1}{2}\left[u_{j,k} + u_{k,j} + u_{i,j}u_{i,k}\right] \tag{3.28}$$

3.3 Derivation of a train tensor using the Eulerian description

In this sub-section, a strain tensor is derived according to the Eulerian description. The position vector of an arbitrary point in the body after deformation using the coordinate system ox_1x_2 may be given as follows:

$$X_1 = X_1(x_1, x_2); \quad X_2 = X_2(x_1, x_2) \tag{3.29}$$

The length vector before deformation can be given using the Eulerian description as follows:

$$d\mathbf{X} = \nabla_x \mathbf{X} \cdot d\mathbf{x} = \left(\frac{\partial X_1}{\partial x_1} \mathbf{e_1 e_1} + \frac{\partial X_1}{\partial x_2} \mathbf{e_1 e_2} + \frac{\partial X_2}{\partial x_1} \mathbf{e_2 e_1} + \frac{\partial X_2}{\partial x_2} \mathbf{e_2 e_2} \right) \cdot \left(dx_1 \mathbf{e}_1 + dx_2 \mathbf{e}_2 \right) \tag{3.30}$$

or

$$d\mathbf{X} = \mathbf{J} \cdot d\mathbf{x} = \left(\frac{\partial X_1}{\partial x_1} dx_1 + \frac{\partial X_1}{\partial x_2} dx_2 \right) \mathbf{e}_1 + \left(\frac{\partial X_2}{\partial x_1} dx_1 + \frac{\partial X_2}{\partial x_2} dx_2 \right) \mathbf{e}_2 \tag{3.31}$$

where

$$\mathbf{J} = \frac{\partial X_1}{\partial x_1} \mathbf{e_1 e_1} + \frac{\partial X_1}{\partial x_2} \mathbf{e_1 e_2} + \frac{\partial X_2}{\partial x_1} \mathbf{e_2 e_1} + \frac{\partial X_2}{\partial x_2} \mathbf{e_2 e_2} \tag{3.32}$$

$\mathbf{J}$ is denoted as the deformation gradient tensor. The variation of displacement is given according to the Eulerian description as follows:

$$d\mathbf{u} = \nabla_x \mathbf{u} \cdot d\mathbf{x} = \left(\frac{\partial u_1}{\partial x_1} \mathbf{e_1 e_1} + \frac{\partial u_1}{\partial x_2} \mathbf{e_1 e_2} + \frac{\partial u_2}{\partial x_1} \mathbf{e_2 e_1} + \frac{\partial u_2}{\partial x_2} \mathbf{e_2 e_2} \right) \cdot \left(dx_1 \mathbf{e_1} + dx_2 \mathbf{e_2} \right) \tag{3.33}$$

or

$$d\mathbf{u} = \mathbf{K} \cdot d\mathbf{x} = \left(\frac{\partial u_1}{\partial x_1} dx_1 + \frac{\partial u_1}{\partial x_2} dx_2 \right) \mathbf{e}_1 + \left(\frac{\partial u_2}{\partial x_1} dx_1 + \frac{\partial u_2}{\partial x_2} dx_2 \right) \mathbf{e}_2 \tag{3.34}$$

where

$$\mathbf{K} = \frac{\partial u_1}{\partial x_1} \mathbf{e_1 e_1} + \frac{\partial u_1}{\partial x_2} \mathbf{e_1 e_2} + \frac{\partial u_2}{\partial x_1} \mathbf{e_2 e_1} + \frac{\partial u_2}{\partial x_2} \mathbf{e_2 e_2} \tag{3.35}$$

$\mathbf{K}$ is the displacement gradient tensor in the Eulerian description. If it is inserted into Eq. (3.13), the following relation could be written as:

$$ds^2 - dS^2 = d\mathbf{x} \cdot d\mathbf{x} - (\mathbf{J} \cdot d\mathbf{x}) \cdot (\mathbf{J} \cdot d\mathbf{x}) = d\mathbf{x} \cdot d\mathbf{x} - (d\mathbf{X} \cdot \mathbf{J}_c) \cdot (\mathbf{J} \cdot d\mathbf{X}) \tag{3.36}$$

provided that

$$\mathbf{J}_c = \mathbf{J}^T; \mathbf{I}_c = \mathbf{I}; d\mathbf{x} = \mathbf{I} \cdot d\mathbf{x} \tag{3.37}$$

Equation (3.36) can be rewritten as

$$ds^2 - dS^2 = d\mathbf{x} \cdot (\mathbf{I}_c \cdot \mathbf{I}) \cdot d\mathbf{x} - d\mathbf{x} \cdot (\mathbf{J}_c \cdot \mathbf{J}) \cdot d\mathbf{x} = d\mathbf{x} \cdot (\mathbf{I}_c \cdot \mathbf{I} - \mathbf{J}_c \cdot \mathbf{J}) \cdot d\mathbf{x} \tag{3.38}$$

Introducing the following relations

$$d\mathbf{X} = d\mathbf{x} - d\mathbf{u} = \mathbf{I} \cdot d\mathbf{x} - \mathbf{K} \cdot d\mathbf{x} = (\mathbf{I} - \mathbf{K}) \cdot d\mathbf{x} \tag{3.39}$$

$$\mathbf{J}_c \cdot \mathbf{J} = \mathbf{J}^T \cdot \mathbf{J} = (\mathbf{I} - \mathbf{K})^T \cdot (\mathbf{I} - \mathbf{K}) \tag{3.40}$$

$$(\mathbf{I}-\mathbf{K})^T = \mathbf{I}^T - \mathbf{K}^T = \mathbf{I} - \mathbf{K}^T \tag{3.41}$$

$$\mathbf{J}^T \cdot \mathbf{J} = \mathbf{K}^T \cdot \mathbf{K} - \mathbf{K}^T \cdot \mathbf{I} - \mathbf{I} \cdot \mathbf{K} + \mathbf{I} \cdot \mathbf{I} = \mathbf{K}^T \cdot \mathbf{K} - \mathbf{K}^T - \mathbf{K} + \mathbf{I} \tag{3.42}$$

the Eulerian strain tensor is defined as

$$\mathbf{E} = \frac{1}{2}\left[\mathbf{I} - \mathbf{J}^T \cdot \mathbf{J}\right] = \frac{1}{2}\left[\mathbf{K} + \mathbf{K}^T - \mathbf{K}^T \cdot \mathbf{K}\right] \tag{3.43a}$$

or

$$\mathbf{E} = \frac{1}{2}\left[(\nabla_x \mathbf{u}) + (\nabla_x \mathbf{u})^T - (\nabla_x \mathbf{u})^T \cdot (\nabla_x \mathbf{u})\right] \tag{3.43b}$$

The Eulerian strain tensor is also known as the Almansi strain tensor and it is a finite strain tensor used for large deformation of materials. In index notation, it is rewritten as

$$E_{jk} = \frac{1}{2}\left[\frac{\partial u_j}{\partial x_k} + \frac{\partial u_k}{\partial x_j} - \frac{\partial u_i}{\partial x_j}\frac{\partial u_i}{\partial x_k}\right] = \frac{1}{2}\left[u_{j,k} + u_{k,j} - u_{i,j}u_{i,k}\right] \tag{3.44}$$

3.4 Relationship between the small strain theory and the finite strain theory

From the strain definitions given by Eqs. (3.28) and (3.44), the terms corresponding to the power of strain components are noted. Therefore, the finite strain tensors are geometrically non-linear and their use in practice becomes troublesome. When the strain is small, say less than 10%, the non-linear components may be omitted. Furthermore, if coordinate systems are assumed to be the same, say $\mathbf{x} = \mathbf{X}$, strain tensors given by (3.28) and (3.44) are reduced to the following form:

$$E_{jk} = L_{jk} = \frac{1}{2}\left[\frac{\partial u_j}{\partial x_k} + \frac{\partial u_k}{\partial x_j}\right] = \frac{1}{2}\left[u_{j,k} + u_{k,j}\right] \tag{3.45}$$

The strain tensor given by Eq. (3.45) is the most commonly used form in practice.

3.5 Geometrical interpretations of a strain tensor

3.5.1 Uniaxial deformation

Let us consider a body deformed uniaxially as shown in Figure 3.2. The length change for this specific example may be written as

$$ds^2 - dS^2 = 2L_{11}dX_1^2 \tag{3.46}$$

Provided that

$$\begin{aligned} &ds^2 = (dX_1 + du_1)^2 + dX_2^2; \quad dS^2 = dX_1^2 + dX_2^2 \\ &dX_1 du_1 + du_1^2 = 2L_{11}dX_1^2 \end{aligned} \tag{3.47}$$

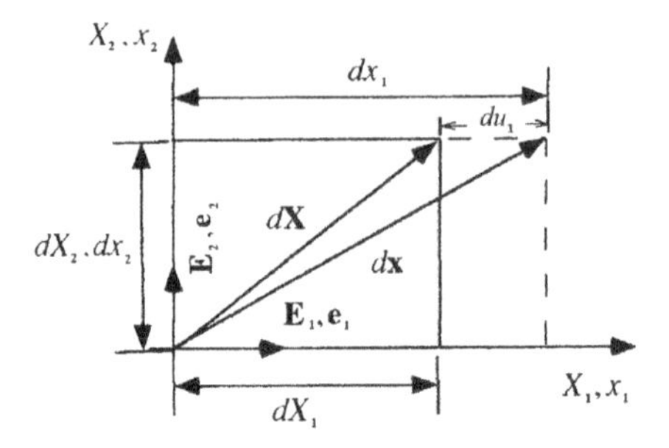

Figure 3.2 One-dimensional normal deformation.

and $du_1^2 \approx 0$, one can obtain the following:

$$\frac{du_1}{dX_1} = L_{11} \tag{3.48}$$

3.5.2 Simple shear deformation

Let us consider a body deformed in simple shear as shown in Figure 3.3. The length change for this example may be written as

$$ds^2 - dS^2 = 2L_{12}dX_1dX_2 \tag{3.49}$$

Provided that

$$\begin{aligned} &ds^2 = (dX_1 + du_1)^2 + dX_2^2; \quad dS^2 = dX_1^2 + dX_2^2 \\ &dX_1du_1 + du_1^2 = 2L_{12}dX_1dX_2 \end{aligned} \tag{3.50}$$

and $du_1^2 \approx 0$, the following relation may be written:

$$\frac{du_1}{dX_2} = L_{12} \tag{3.51}$$

The relation above can be easily interpreted as the angle variation

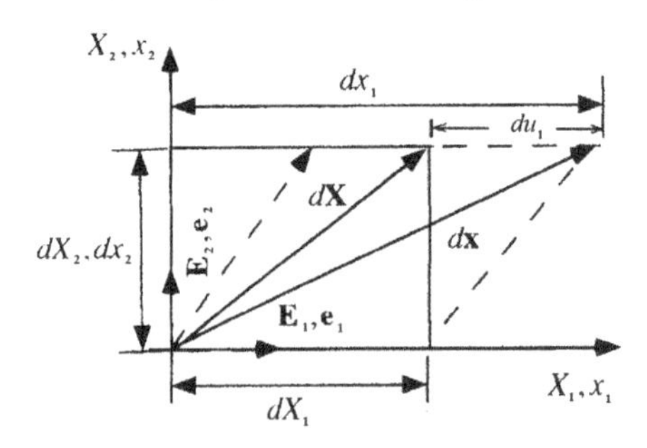

Figure 3.3 Simple shear deformation.

References

Eringen, A.C. 1980. *Mechanics of Continua*, Huntington, NY, Robert E. Krieger Publishing Co., 606 p.

Mase, G. 1970. *Theory and Problems of Continuum Mechanics*, Schaum's Outline Series, New York, McGraw-Hill Co., 221 p.

Chapter 4

Fundamental conservation laws

Various actual applications of rock mechanics and rock engineering involve mass transportation phenomena such as seepage, diffusion, static and dynamic stability assessment of structures, wave propagation, vibration and heat flow. Principles of fundamental laws presented herein follow the laws of continuum mechanics (e.g. Eringen 1980; Mase 1970). In this section, the fundamental governing equation of each phenomenon is presented. The governing equations are first developed for one-dimensional (1D) cases and then they are extended to multi-dimensional situations.

4.1 Fundamental conservation laws for one-dimensional cases

4.1.1 Mass conservation law

The mass conservation law is simply stated as

gained mass = input flux − output flux + mass generated

Let us consider an infinitely small cubic element as shown in Figure 4.1. The statement above can be written in the following form for the x-direction as:

$$\Delta m = q_x \Delta y \Delta z \Delta t - q_{x+\Delta x} \Delta y \Delta z \Delta t + g \Delta x \Delta y \Delta z \Delta t \tag{4.1}$$

where m is the mass, q is the flux and g is the mass generated per unit volume per unit time.

Terms Δm and q are explicitly written as

$$\Delta m = \Delta \rho \Delta x \Delta y \Delta z \tag{4.2}$$

$$q = \rho v \tag{4.3}$$

where ρ is the density and v is the velocity. The quantity $q_{x+\Delta x}$ could be written in the following form using the Taylor expansion

$$q_{x+\Delta x} = q_x + \frac{\partial q}{\partial x} \Delta x + 0^2 \tag{4.4}$$

Inserting this relation in Eq. (4.1) and dividing both sides by $\Delta x \Delta y \Delta z \Delta t$ yields the following:

$$\frac{\Delta \rho}{\Delta t} = -\frac{\Delta(\rho v)}{\Delta x} + g \tag{4.5}$$

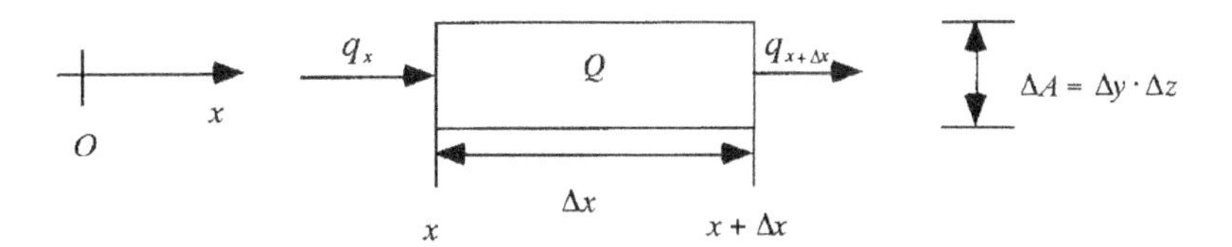

Figure 4.1 Illustration of the mass conservation law in one dimension.

Taking the limit results in the mass conservation law for 1D cases as

$$\frac{\partial \rho}{\partial t} = -\frac{\partial(\rho v)}{\partial x} + g \tag{4.6}$$

4.1.2 Momentum conservation law

The momentum balance for a 1D case can be written for a typical infinitely small control volume (Figure 4.2)[1]:

$$-\sigma_x \Delta y \Delta z \Delta t + \sigma_{x+\Delta x} \Delta y \Delta z \Delta t + b \Delta x \Delta y \Delta z \Delta t - \Delta(\rho v) \Delta x \Delta y \Delta z - \Delta(\rho v^2) \Delta y \Delta z \Delta t = 0 \tag{4.7}$$

where σ is the stress, b is the body force, ρ is the density and v is the velocity. Stress $\sigma_{x+\Delta x}$ can be expressed using the Taylor expansion as

$$\sigma_{x+\Delta x} = \sigma_x + \frac{\partial \sigma}{\partial x} \Delta x + 0^2 \tag{4.8}$$

Dividing by $\Delta x \Delta y \Delta z \Delta t$ and taking the limits yield the following:

$$\frac{\partial \sigma}{\partial x} + b = \frac{\partial(\rho v)}{\partial t} + \frac{\partial(\rho v^2)}{\partial x} \tag{4.9}$$

The above equation is rewritten as

$$\frac{\partial \sigma}{\partial x} + b = v\left(\frac{\partial(\rho)}{\partial t} + \frac{\partial(\rho v)}{\partial x}\right) + \rho\left(\frac{\partial v}{\partial t} + v\frac{\partial v}{\partial x}\right) \tag{4.10}$$

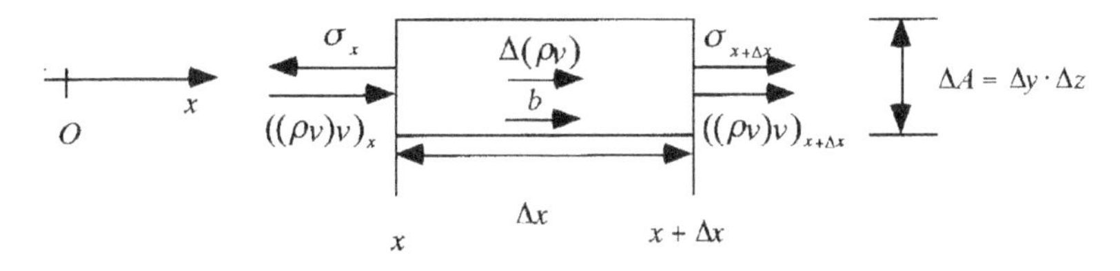

Figure 4.2 Illustration of the momentum conservation law.

1 Simple momentum concept: $p = m \cdot v$

Introducing the following operator, which is called the "material derivative operator"

$$\frac{d}{dt} = \frac{\partial}{\partial t} + v\frac{\partial}{\partial x} \tag{4.11}$$

and using the mass conservation law for no mass generation, Eq. (4.10) becomes

$$\frac{\partial \sigma}{\partial x} + b = \rho \frac{dv}{dt} \tag{4.12}$$

If we define acceleration (a) as

$$a = \frac{dv}{dt} \tag{4.13}$$

Equation (4.12) can be rewritten as

$$\frac{\partial \sigma}{\partial x} + b = \rho a \tag{4.14}$$

Acceleration (a) may also be expressed in terms of displacement (u) as

$$a = \frac{d^2 u}{dt^2} = \ddot{u} \tag{4.15}$$

Accordingly, Eq. (4.14) can be reexpressed in terms of u as

$$\frac{\partial \sigma}{\partial x} + b = \rho \ddot{u} \tag{4.16}$$

4.1.3 Energy conservation laws

The energy balance law for a 1D case can be written for a typical infinitely small control volume (Figure 4.3):

$$\begin{aligned}\Delta(U+K)\Delta x\Delta y\Delta z = {} & \left((U+K)v\right)_x \Delta y\Delta z\Delta t - \left((U+K)v\right)_{x+\Delta x} \Delta y\Delta z\Delta t \\ & + q_x \Delta y\Delta z\Delta t - q_{x+\Delta x}\Delta y\Delta z\Delta t - (\sigma v)_x \Delta y\Delta z\Delta t \\ & + (\sigma v)_{x+\Delta x} \Delta y\Delta z\Delta t + (bv)\Delta x\Delta y\Delta z\Delta t + Q\Delta x\Delta y\Delta z\Delta t \end{aligned} \tag{4.17}$$

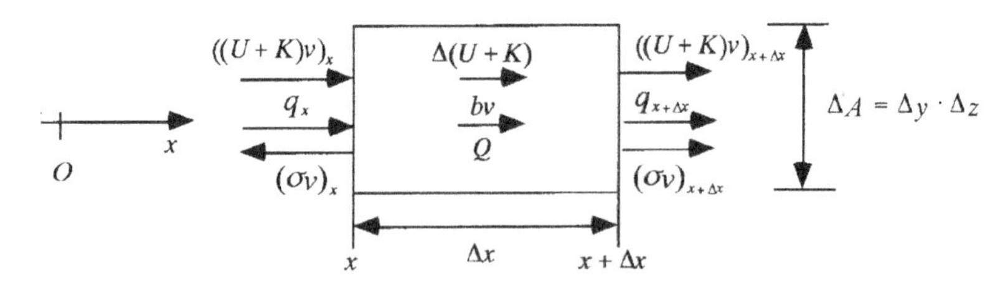

Figure 4.3 Illustration of the energy conservation law.

where U is the internal energy, K is the kinetic energy, v is the velocity, q is the flux, σ is the stress, b is the body force and Q is the energy generated per unit volume per unit time. Energy $(U+K)_{x+\Delta x}$ and momentum $(\sigma v)_{x+\Delta x}$ can be expressed using the Taylor expansion as

$$\big((U+K)v\big)_{x+\Delta x} = \big((U+K)v\big)_x + \frac{\partial\big((U+K)v\big)}{\partial x}\Delta x + 0^2 \tag{4.18}$$

$$(\sigma v)_{x+\Delta x} = (\sigma v)_x + \frac{\partial(\sigma v)}{\partial x}\Delta x + 0^2 \tag{4.19}$$

Dividing Eq. (4.17) by $\Delta x\Delta y\Delta z\Delta t$ and taking the limits yield the following:

$$\frac{\partial(U+K)}{\partial t} = -\frac{\partial\big((U+K)v\big)}{\partial x} + -\frac{\partial q}{\partial x} + \frac{\partial(\sigma v)}{\partial x} + (bv) + Q \tag{4.20}$$

Expressing the internal energy U and the kinetic energy K as

$$U = \rho e; \quad K = \frac{1}{2}\rho v^2 \tag{4.21}$$

and rearranging Eq. (4.20) yield

$$\left[\frac{\partial \rho}{\partial t} + \frac{\partial(\rho v)}{\partial x}\right]\left[e + \frac{1}{2}v^2\right] + \rho\frac{\partial(e+\frac{1}{2}v^2)}{\partial t} + \rho v\frac{\partial(e+\frac{1}{2}v^2)}{\partial x} = -\frac{\partial q}{\partial x} + \frac{\partial(\sigma v)}{\partial x} + (bv) + Q \tag{4.22}$$

where e is the specific internal energy per unit mass. Using the mass conservation law and momentum balance law, Eq. (4.22) becomes

$$\rho\left(\frac{\partial e}{\partial t} + v\frac{\partial e}{\partial x}\right) = -\frac{\partial q}{\partial x} + \sigma\frac{\partial v}{\partial x} + Q \tag{4.23}$$

Introducing the following operator

$$\frac{d}{dt} = \frac{\partial}{\partial t} + v\frac{\partial}{\partial x} \tag{4.24}$$

Equation (4.23) can be rewritten as

$$\rho\frac{de}{dt} = -\frac{\partial q}{\partial x} + \sigma\frac{\partial v}{\partial x} + Q \tag{4.25}$$

Denoting the gradient of velocity as the strain rate given below

$$\dot{\varepsilon} = \frac{\partial v}{\partial x} = \frac{\partial \dot{u}}{\partial x}; \quad \dot{u} = \frac{dv}{dt} \tag{4.26}$$

Equation (4.25) is rewritten in the following form:

$$\rho\frac{de}{dt}=-\frac{\partial q}{\partial x}+\sigma\dot{\varepsilon}+Q \tag{4.27}$$

If we express free energy e by cT, where c is the specific heat capacity and T is the temperature, we have the following form for the energy balance law

$$\rho\frac{d(cT)}{dt}=-\frac{\partial q}{\partial x}+\sigma\frac{\partial v}{\partial x}+Q \tag{4.28}$$

If we, further, introduce Fourier's law as a constitutive law between heat flux q and temperature T as

$$q=-k\frac{\partial T}{\partial x} \tag{4.29}$$

Equation (4.28) takes the following well-known 1D energy balance law in thermodynamics by assuming that the heat conductivity coefficient and specific heat capacity are constant

$$\rho c\frac{dT}{dt}=k\frac{\partial^2 T}{\partial x^2}+\sigma\dot{\varepsilon}+Q \tag{4.30}$$

4.1.4 Fundamental governing equations for coupled hydro-mechanical phenomena

The mathematics is kept to a minimum in this subsection. However, more detailed derivations may be found in Aydan (2017, 2001a, b).

4.1.4.1 1D Mass conservation law for a mixture of solid and fluid (two-phase body)

The mass conservation law for a mixture without mass generation is stated as:

$$\text{gained mass} = \text{input flux} - \text{output flux} \tag{4.31}$$

Let us consider an infinitely small cubic element as shown in Figure 4.4. The final equation for the mass conservation law of mixture for the 1D case is fundamentally the same as given below:

$$\frac{\partial \rho}{\partial t}=-\frac{\partial q}{\partial x} \tag{4.32}$$

However, the average density of the mixture is defined as given below:

$$\rho=(1-n)\rho_s+n\rho_f \tag{4.33}$$

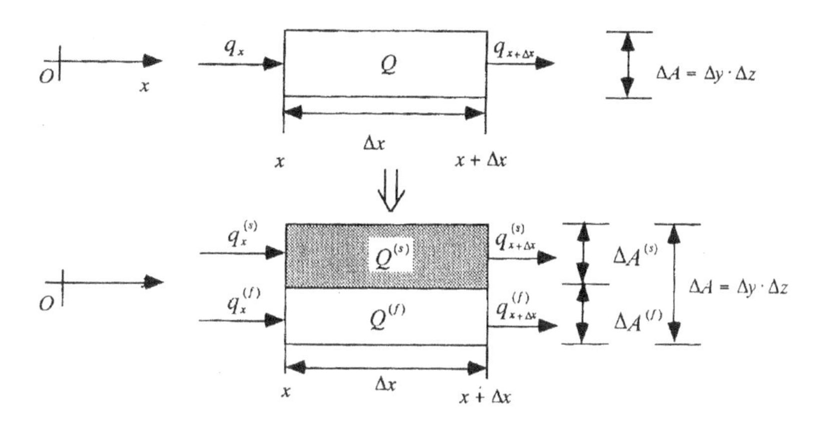

Figure 4.4 Illustration of the mass conservation law for two-phase mass in one dimension.

where n is the volume fraction of a fluid, ρ_f is the density of the fluid and ρ_s is the density of the solid.

Similarly, flux is written as

$$q = \rho v = (1-n)\rho_s v_s + n\rho_f v_f \tag{4.34}$$

Let us also define a relative velocity between the solid phase and fluid phase as

$$v_r = n(v_f - v_s) \quad \text{or} \quad nv_f = v_r + nv_s) \tag{4.35}$$

The mass conservation laws for each constituent can be written as:

Solid phase:

$$\frac{\partial}{\partial t}((1-n)\rho_s) = -\frac{\partial}{\partial x}((1-n)\rho_s v_s) \tag{4.36}$$

Let us introduce the following operator

$$\frac{d_s}{dt} = \frac{\partial}{\partial t} + v_s \frac{\partial}{\partial x} \tag{4.37}$$

The above equation then becomes

$$\frac{d_s n}{dt} \frac{(1-n)}{\rho_s} \frac{d_s \rho_s}{dt} + (1-n)\frac{\partial v_s}{\partial x} \tag{4.38}$$

Fluid phase:

$$\frac{\partial}{\partial t}\left(n\rho_f\right) = -\frac{\partial}{\partial x}\left(\rho_f v_r + n\rho_f v_s\right) \tag{4.39}$$

Thus, the above equation becomes

$$\frac{d_s n}{dt} = -\frac{n}{\rho_f}\frac{d_s \rho_f}{dt} - \frac{1}{\rho_f}\frac{\partial(\rho_f v_r)}{\partial x} - n\frac{\partial}{\partial x}(\rho_f v_f) \tag{4.40}$$

Equating Eqs. (4.38) and (4.40), we have

$$\frac{(1-n)}{\rho_s}\frac{d_s \rho_s}{dt} + \frac{\partial v_s}{\partial x} = -\frac{n}{\rho_f}\frac{d_s \rho_f}{dt} - \frac{1}{\rho_f}\frac{\partial(\rho_f v_r)}{\partial x} \tag{4.41}$$

Let us write constitutive laws for the volumetric response of each constituent in the following forms

$$\frac{1}{\rho_s}\frac{d_s \rho_s}{dt} = \frac{1}{K_s}\frac{d_s p}{dt} \tag{4.42}$$

$$\frac{1}{\rho_f}\frac{d_s \rho_f}{dt} = \frac{1}{K_f}\frac{d_s p}{dt} \tag{4.43}$$

By inserting these constitutive laws in Eq. (4.41), we obtain

$$\frac{\partial \varepsilon_v}{\partial t} = -\left(\frac{(1-n)}{K_s} + \frac{n}{K_f}\right)\frac{d_s p}{dt} - \frac{1}{\rho_f}\frac{\partial}{\partial x}(\rho_f v_r) \tag{4.44}$$

where

$$\frac{\partial v_s}{\partial x} = \frac{\partial}{\partial x}\left(\frac{\partial u_s}{\partial t}\right) = \frac{\partial}{\partial t}\left(\frac{\partial u_s}{\partial x}\right) = \frac{\partial \varepsilon_v}{\partial t} \tag{4.45}$$

Let us introduce Darcy's law given by

$$v_r = -\frac{k}{\mu}\frac{\partial p}{\partial x} = -K\frac{\partial p}{\partial x} \tag{4.46}$$

Thus, Eq. (4.44) becomes

$$\frac{\partial \varepsilon_v}{\partial t} = -\left(\frac{(1-n)}{K_s} + \frac{n}{K_f}\right)\frac{d_s p}{dt} + \frac{1}{\rho_f}\frac{\partial}{\partial x}\left(\rho_f K\frac{\partial p}{\partial x}\right) \tag{4.47}$$

Some particular cases of the above equation are as follows:

Undrained condition and slow deformation: $v_r = 0$ and $v_s = 0$

$$\frac{\partial \varepsilon_v}{\partial t} = -\left(\frac{(1-n)}{K_s} + \frac{n}{K_f}\right)\frac{\partial p}{\partial t} = -C_u\frac{\partial p}{\partial t} \tag{4.48}$$

Drained condition and slow deformation: $v_s = 0$

$$\frac{\partial \varepsilon_v}{\partial t} = -\left(\frac{(1-n)}{K_s}\right)\frac{\partial p}{\partial t} = -C_d \frac{\partial p}{\partial t} \tag{4.49}$$

If solid and fluid are incompressible materials, Eq. (4.47) becomes

$$\frac{\partial \varepsilon_v}{\partial t} = -\frac{\partial v_r}{\partial x} = \frac{\partial}{\partial x}\left(K \frac{\partial p}{\partial x}\right) = K \frac{\partial^2 p}{\partial x^2} \tag{4.50}$$

Equation (4.50) corresponds to the equations of Terzaghi or Biot used in coupled problems, in which densities are assumed to be constant.

4.1.4.2 One-dimensional momentum conservation law for a mixture of solid and fluid (two-phase body)

Total force equilibrium in terms of total stress without the inertia form yields the following relation (Figure 4.5):

$$\frac{\partial \sigma}{\partial x} + b = 0 \tag{4.51}$$

Let us introduce the concept of effective stress as

$$\sigma = \sigma' - \alpha p \tag{4.52}$$

where α is a physical non-dimensional quantity. When α is equal to 1, it corresponds to Terzaghi's effective stress law. On the other hand, if it is not equal to 1,

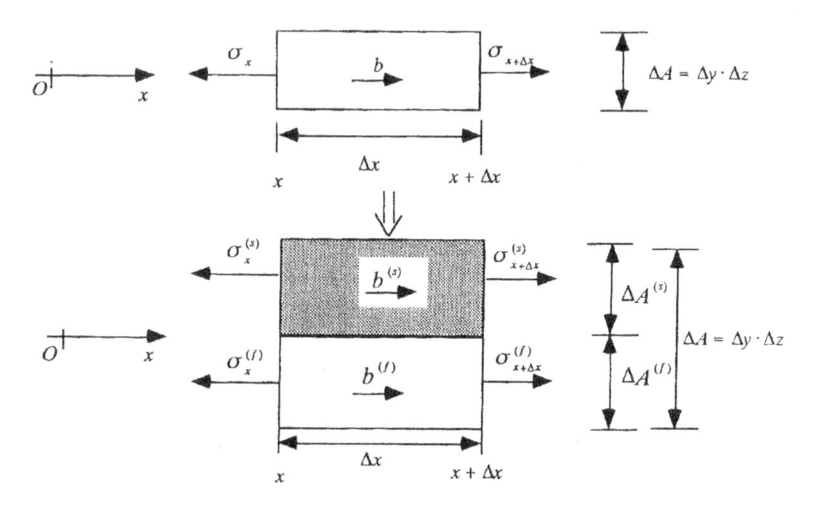

Figure 4.5 Illustration of the momentum conservation law for two-phase mass in one dimension.

it corresponds to Biot's effective stress law. Accordingly, the equilibrium equation becomes

$$\frac{\partial \sigma'}{\partial x} - \alpha \frac{\partial p}{\partial x} + b = 0 \tag{4.53}$$

Taking the time derivative of the above expression, we have

$$\frac{\partial \dot{\sigma}'}{\partial x} - \alpha \frac{\partial \dot{p}}{\partial x} + \dot{b} = 0 \tag{4.54}$$

Equation (4.54) is used together with Eq. (4.47) for the hydro-mechanical response of porous rock masses.

Under dynamic conditions, which require the consideration of inertia components, Eq. (4.51) is replaced by the following equation

$$\frac{\partial \sigma}{\partial x} + \rho g = (1-n)\rho_s \frac{d_s v_s}{dt} + n\rho_f \frac{d_f v_f}{dt} \tag{4.55}$$

$$\frac{\partial \sigma_f}{\partial x} + \rho_f g = \rho_f \ddot{u}_f + \frac{1}{n}\tau_{sf} \tag{4.56}$$

where

$$\sigma = (1-n)\sigma_s + n\sigma_f \tag{4.57}$$

$$\tau_{sf} = -\frac{n}{K} v_r \tag{4.58}$$

The coefficient K is called the hydraulic conductivity coefficient (wrongly called permeability in many publications) which appears in Darcy's law. Equations (4.55) and (4.56) can be re-written as

Total system

$$\frac{\partial \sigma}{\partial x} + \rho g = \rho \frac{\partial^2 u_s}{\partial t^2} + \rho_f \frac{\partial^2 w}{\partial t^2} \tag{4.59}$$

Fluid phase

$$\frac{\partial \sigma_f}{\partial x} + \rho_f g = \rho_f \frac{\partial^2 u_s}{\partial t^2} + \frac{\rho_f}{n}\frac{\partial^2 w}{\partial t^2} + \frac{1}{K}\frac{\partial w}{\partial t} \tag{4.60}$$

where $w = n(v_f - v_s)$.

4.1.4.3 One-dimensional (1D) seepage

If fluid density does not change with position and there is no gained or lost mass, the above equations may be rewritten as:

$$\frac{\partial \rho}{\partial t} = -\rho \frac{\partial}{\partial x}(\bar{v}) \tag{4.61}$$

Although it is known that the relationship between average velocity ($\bar{v}$) and pressure gradient ($-\partial p / \partial x$) becomes non-linear with increasing values of Reynolds numbers, Darcy's law is widely used for analyzing test results on soil and rock. Darcy's law can be given as

$$\bar{v} = -\frac{k}{\eta}\frac{dp}{dx} \tag{4.62}$$

where k is permeability, η is viscosity and p is pressure. Inserting Eq. (4.62) into Eq. (4.61) yields the governing equations for fluid flow as:

$$\frac{\partial \rho}{\partial t} = \frac{\rho k}{\eta}\frac{\partial}{\partial x}\left(\frac{\partial p}{\partial x}\right) \tag{4.63}$$

If the following relation exists between fluid density and pressure:

$$\rho = cp \tag{4.64}$$

Equation (4.63) takes the following form, which is the governing equation used in seepage of fluids through rock mass:

$$\frac{\partial p}{\partial t} = \frac{k}{c\eta}\frac{\partial^2 p}{\partial x^2} \quad \text{or} \quad \frac{\partial p}{\partial t} = \beta\frac{\partial^2 p}{\partial x^2} \tag{4.65}$$

where $\beta = \dfrac{k}{c\eta}$

4.2 Multi-dimensional conservation laws

4.2.1 Mass conservation laws for seepage and diffusion phenomena

Water is always present in rock mass and it strongly affects the stability of rock engineering structures. Furthermore, rock excavations, which are generally of large scale, may disturb the groundwater regime and it may have some environmental impacts. In any case, the governing equation for seepage flow in rock mass would be derived from the mass conservation law if rock mass is considered as a porous medium. The final form of the governing equation for seepage flow may be written as

$$S\frac{\partial h}{\partial t} = -\nabla \cdot \mathbf{q} \pm Q \tag{4.66}$$

where $S, h, \mathbf{q}$ and Q are storativity coefficient, water head, flux vector and source or sink, respectively (e.g. Aydan 1998; Zimmerman et al. 1986). ∇ is the directional derivative operator.

As for the diffusion phenomenon of a certain substance, the governing equation may be obtained as

$$\frac{d\varphi}{dt} = -\nabla \cdot \mathbf{f} \pm P \tag{4.67}$$

where φ, $\mathbf{f}$ and P are the concentration of the substance, diffusion flux vector and source or sink, respectively.

4.2.2 Momentum conservation law

The momentum conservation law for rock mass can be derived as in continuum mechanics and the final form takes the following form:

$$\rho \frac{\partial \mathbf{v}}{\partial t} = -\nabla \cdot \boldsymbol{\sigma} + \mathbf{b} \tag{4.68}$$

where $\rho, \mathbf{v}, \boldsymbol{\sigma}$ and $\mathbf{b}$ are the density, velocity, stress tensor and body force, respectively.

4.2.3 Angular momentum conservation law

The final form of the angular momentum equation implies that the stress tensor is a symmetric tensor, that is,

$$\sigma_{ij} = \sigma_{ji} \tag{4.69}$$

4.2.4 Energy conservation law

The energy conservation law of rock mass may be written in the following form:

$$\rho c \frac{\partial T}{\partial t} = -\nabla \cdot \mathbf{q}_h + \boldsymbol{\sigma} \cdot \dot{\boldsymbol{\varepsilon}} + Q_h \tag{4.70}$$

where $c, T, \mathbf{q}_h, \dot{\boldsymbol{\varepsilon}}$ and Q_h are specific heat, temperature, strain rate tensor and heat source or sink, respectively.

4.3 Derivation of governing equations in the integral form

4.3.1 Mass conservation law

Mass is defined as

$$m = \int_{\Omega} \rho(\mathbf{x}, t) d\Omega \tag{4.71}$$

The mass conservation law requires

$$\frac{dm}{dt} = \frac{d}{dt}\int_{\Omega} \rho \, d\Omega + \int_{\Omega} \rho \frac{d(d\Omega)}{dt} \tag{4.72}$$

The control volume at any time is related to the original control volume as

$$d\Omega = Jd\Omega \quad \text{and} \quad \frac{d(d\Omega)}{dt} = (\nabla \cdot \mathbf{x})d\Omega \tag{4.73}$$

With the use of the Reynolds transport theorem together with Eq. (4.73), Eq. (4.72) becomes

$$\frac{dm}{dt} = \int_{\Omega} \left(\frac{d\rho}{dt} + \rho \nabla \cdot \mathbf{v} \right) d\Omega \tag{4.74}$$

To satisfy this condition, the integrand should be zero, so that we have the following:

$$\frac{d\rho}{dt} + \rho \nabla \cdot \mathbf{v} = 0 \tag{4.75}$$

Time derivatives in the Lagrangian and Eulerian descriptions are given in the following forms:

Lagrangian description

$$\frac{d()}{dt} = \frac{\partial()}{\partial t} \tag{4.76a}$$

Eulerian description

$$\frac{d()}{dt} = \frac{\partial()}{\partial t} + v \cdot \nabla() \tag{4.76b}$$

With the use of Eq. (4.76a) in Eq. (4.75), we obtain the following relation

$$\frac{\partial \rho}{\partial t} + \mathbf{v} \cdot \nabla \rho + \rho \nabla \cdot \mathbf{v} = 0 \quad \text{o} \quad \frac{\partial \rho}{\partial t} + \nabla \cdot (\rho v) = 0 \tag{4.77}$$

Using index notation, we have

$$\mathbf{v} = v_k \mathbf{e}_k \text{ and } \nabla = \frac{\partial}{\partial x_k} \mathbf{e}_k \tag{4.78}$$

Equation (4.77b) can be rewritten as

$$\frac{\partial \rho}{\partial t} + \frac{\partial (\rho v_k)}{\partial x_k} = \frac{\partial \rho}{\partial t} + (\rho v_k)_{,k} = 0 \tag{4.79}$$

If $\frac{d\rho}{dt} = 0$, the media is incompressible so that

$$\rho(\nabla \cdot \mathbf{v}) = 0 \text{ or } \nabla \cdot \mathbf{v} = 0 \tag{4.80}$$

This equation is often called the continuity equation.

4.3.2 Momentum conservation law

Momentum can be derived using the following form:

$$\mathbf{p} = \int_{\Omega} \rho \mathbf{v} d\Omega \tag{4.81}$$

Preliminary relations are

$$\int_{\Omega} \nabla \cdot \boldsymbol{\sigma} d\Omega = \int_{\Gamma} \boldsymbol{\sigma} \cdot \mathbf{n} d\Gamma; \frac{d(d\Omega)}{dt} = (\nabla \cdot \mathbf{x}) d\Omega; \mathbf{t} = \boldsymbol{\sigma} \cdot \mathbf{n} \tag{4.82}$$

Conservation of momentum is written in the following form with the use of preliminary relations given in Eq. (4.82), which is also known as the Reynolds transport theorem

$$\frac{d}{dt} \int_{\Omega} \rho \mathbf{v} d\Omega = \int_{\Gamma} \mathbf{t} d\Gamma + \int_{\Omega} \mathbf{b} d\Omega \tag{4.83}$$

Equation (4.83) could be rewritten as

$$\int_{\Omega} \left(\frac{d(\rho \mathbf{v})}{dt} + (\rho \mathbf{v}) \nabla \cdot \mathbf{v} \right) d\Omega = \int_{\Omega} \nabla \cdot \boldsymbol{\sigma} d\Omega + + \int_{\Omega} \mathbf{b} d\Omega \tag{4.84}$$

Carrying out the derivation in Eq. (4.84), we have the following

$$\int_{\Omega} \left(\left(\frac{d\rho}{dt} + \rho(\nabla \cdot \mathbf{v}) \right) \mathbf{v} \right) d\Omega + \int_{\Omega} \rho \frac{d\mathbf{v}}{dt} d\Omega = \int_{\Omega} (\nabla \cdot \boldsymbol{\sigma} + \mathbf{b}) d\Omega \tag{4.85}$$

The first term on the left-hand side disappears by virtue of the mass conservation law and takes the following form:

$$\int_{\Omega} \rho \frac{d\mathbf{v}}{dt} d\Omega = \int_{\Omega} (\nabla \cdot \boldsymbol{\sigma} + \mathbf{b}) d\Omega \tag{4.86}$$

Equation (4.86) could be rewritten as

$$\int_{\Omega} \left[\rho \frac{d\mathbf{v}}{dt} - (\nabla \cdot \boldsymbol{\sigma} + \mathbf{b}) \right] d\Omega = 0 \tag{4.87}$$

To satisfy Eq. (4.87), the integrand should be zero so that we have the following relation:

$$\rho \frac{d\mathbf{v}}{dt} = \nabla \cdot \boldsymbol{\sigma} + \mathbf{b} \tag{4.88}$$

Furthermore, the derivation on the left-hand side could be related to the acceleration or displacement vectors as follows:

$$\frac{d\mathbf{v}}{dt} = \mathbf{a} \text{ or } \frac{d\mathbf{v}}{dt} = \frac{d^2\mathbf{u}}{dt^2} \tag{4.89}$$

4.3.3 Angular momentum conservation law

Before the derivation of the angular momentum conservation law, some preliminary relations are necessary and they are given as follows:

$$\mathbf{v}\times\mathbf{v} = \mathbf{0};\ \int_{\Gamma} \mathbf{r}\times\mathbf{n}\,d\Gamma = \int_{\Omega} \nabla\times\mathbf{r}\,d\Omega = \int_{\Omega} \mathbf{I}\,d\Omega;\ \mathbf{r}\times\boldsymbol{\sigma}\cdot\mathbf{n} = (\mathbf{r}\times\boldsymbol{\sigma})\cdot\mathbf{n};\ \frac{d\mathbf{r}}{dt} = \mathbf{v} \tag{4.90}$$

$$\begin{aligned}\int_{\Gamma} \mathbf{r}\times\mathbf{t}\,d\Gamma &= \int_{\Gamma} \mathbf{r}\times(\boldsymbol{\sigma}\cdot\mathbf{n})\,d\Gamma = \int_{\Omega} \nabla\cdot(\mathbf{r}\times\boldsymbol{\sigma})\,d\Omega \\ &= \int_{\Omega} (\nabla\times\mathbf{r})\cdot\boldsymbol{\sigma}\,d\Omega + \int_{\Omega} \mathbf{r}\times(\nabla\cdot\boldsymbol{\sigma})\,d\Omega\end{aligned} \tag{4.91}$$

The angular momentum law requires the following condition to be valid

$$\frac{d}{dt}\int_{\Omega} \mathbf{r}\times\rho\mathbf{v}\,d\Omega = \int_{\Gamma} \mathbf{r}\times\mathbf{t}\,d\Gamma + \int_{\Omega} \mathbf{r}\times\mathbf{b}\,d\Omega \tag{4.92}$$

$$\begin{aligned}\int_{\Omega} \frac{d\mathbf{r}}{dt}\times\rho\mathbf{v}\,d\Omega + \int_{\Omega} \mathbf{r}\times\frac{d(\rho\mathbf{v})}{dt}\,d\Omega + \int_{\Omega} \mathbf{r}\times(\rho\mathbf{v})\frac{d(d\Omega)}{dt} &= \int_{\Omega} (\nabla\times\mathbf{r})\cdot\boldsymbol{\sigma}\,d\Omega \\ &+ \int_{\Omega} \mathbf{r}\times(\nabla\cdot\boldsymbol{\sigma}+\mathbf{b})\,d\Omega\end{aligned} \tag{4.93}$$

or

$$\begin{aligned}\int_{\Omega} \mathbf{v}\times\rho\mathbf{v}\,d\Omega + \int_{\Omega} \mathbf{r}\times\left(\frac{d\rho}{dt}\mathbf{v} + \rho\frac{d\mathbf{v}}{dt}\right)d\Omega + \int_{\Omega} (\mathbf{r}\times\rho\mathbf{v})\nabla\cdot\mathbf{v}\,d\Omega &= \int_{\Omega} (\nabla\times\mathbf{r})\cdot\boldsymbol{\sigma}\,d\Omega \\ &+ \int_{\Omega} \mathbf{r}\times(\nabla\cdot\boldsymbol{\sigma}+\mathbf{b})\,d\Omega\end{aligned} \tag{4.94}$$

Thus, the equation above takes the following form:

$$\begin{aligned}&\int_{\Omega} \mathbf{v}\times\rho\mathbf{v}\,d\Omega + \int_{\Omega} \mathbf{r}\times\left(\frac{d\rho}{dt} + \rho\nabla\cdot\mathbf{v}\right)\mathbf{v}\,d\Omega + \int_{\Omega}\left(\mathbf{r}\times\left(\rho\frac{d\mathbf{v}}{dt} - \nabla\cdot\boldsymbol{\sigma} - \mathbf{b}\right)\right)d\Omega \\ &-\int_{\Omega} (\nabla\times\mathbf{r})\cdot\boldsymbol{\sigma}\,d\Omega = \mathbf{0}\end{aligned} \tag{4.95}$$

Under the mass conservation law, momentum conservation law and preliminary relations, Eq. (4.95) reduce to the following form:

$$\int_{\Omega} (\nabla \times \mathbf{r}) \cdot \boldsymbol{\sigma} d\Omega = \int_{\Omega} \boldsymbol{\varepsilon} \cdot \boldsymbol{\sigma} d\Omega = \mathbf{0} \tag{4.96}$$

where ε is known as the permutation tensor (symbol) and it is a rank-3 tensor. It is given in index notation as

$$\varepsilon_{ijk} \sigma_{jk} = 0 \tag{4.97}$$

As explained in Chapter 1, the rank-3 permutation tensor has the following properties:

$$\varepsilon_{ijk} = \begin{cases} i \neq j \neq k;;1,2,3,1,2\left(\varepsilon_{ijk} = 1\right) \\ i \neq j \neq k;;3,2,1,3,2\left(\varepsilon_{ijk} = -1\right) \\ i = j \neq k, i = k \neq j, i \neq j = k;\ \varepsilon_{ijk} = 0 \end{cases} \tag{4.98}$$

With this property, it is shown that the stress tensor is symmetric as given below:

$$\sigma_{jk} = \sigma_{kj} \tag{4.99}$$

4.3.4 Energy conservation law

The energy conservation law requires that the time variation of the sum of internal and kinetic energies should be equal to the work done by external tractions and body force and energy production due to some causes within the body as given below:

$$\frac{d}{dt}(U + K)) = W + H \tag{4.100}$$

where

$$U = \int_{\Omega} \rho e d\Omega;\ K = \frac{1}{2}\int_{\Omega} \rho \mathbf{v} \cdot \mathbf{v}\, d\Omega;\ W = \int_{\Gamma} \mathbf{t} \cdot \mathbf{v}\, d\Gamma + \int_{\Omega} \mathbf{b} \cdot \mathbf{v}\, dQ;\ H = -\int_{\Gamma} \mathbf{q} \cdot \mathbf{n}\, d\Gamma \pm \int_{\Omega} Q d\Omega \tag{4.101}$$

With the use of Eq. (4.101), one may write the following relations:

$$\frac{dU}{dt} = \int_{\Omega} \left(\frac{d(\rho e)}{dt} + \rho e \nabla \cdot \mathbf{v} \right) d\Omega \tag{4.102}$$

$$\frac{dK}{dt} = \frac{1}{2}\int_{\Omega}\left(\frac{d(\rho\mathbf{v})}{dt}\cdot\mathbf{v} + \rho\mathbf{v}\cdot\frac{d\mathbf{v}}{dt} + (\rho\mathbf{v}\cdot\mathbf{v})\nabla\cdot\mathbf{v}\right)d\Omega \tag{4.103}$$

$$W = \int_{\Omega}(\nabla\cdot\boldsymbol{\sigma} + \mathbf{b})\cdot\mathbf{v}\,d\Omega + \int_{\Omega}\boldsymbol{\sigma} : \nabla\mathbf{v}\,d\Omega \tag{4.104}$$

$$H = -\int_{\Omega}(\nabla\cdot\mathbf{q} \pm Q)\,d\Omega \tag{4.105}$$

Inserting Eq. (4.102) into Eq. (4.103) into the left-hand side of Eq. (4.100) results in

$$\begin{aligned}\frac{d}{dt}(U + K) = &\frac{1}{2}\int_{\Omega}\left(\frac{d\rho}{dt} + \rho\nabla\cdot\mathbf{v}\right)\mathbf{v}\cdot\mathbf{v}\,d\Omega + \int_{\Omega}\left(\frac{d\rho}{dt} + \rho\nabla\cdot\mathbf{v}\right)e\,d\Omega\\ &+ \int_{\Omega}\rho\frac{de}{dt}d\Omega + \int_{\Omega}\rho\mathbf{v}\cdot\frac{d\mathbf{v}}{dt}\end{aligned} \tag{4.106}$$

Furthermore, inserting Eqs. (4.104) and (4.105) into the right-hand side of Eq. (4.100) and requiring the mass conservation and momentum conservation laws to be satisfied, the final form for Eq. (4.100) could be derived as follows:

$$\int_{\Omega}\left(\rho\frac{de}{dt} - \boldsymbol{\sigma} : \nabla\mathbf{v} + \nabla\cdot\mathbf{q} \pm Q\right)d\Omega = 0 \tag{4.107}$$

To satisfy Eq. (4.107), the integrand should be zero, so that the following relation is obtained:

$$\rho\frac{de}{dt} = -\nabla\cdot\mathbf{q} + \boldsymbol{\sigma} : \nabla\mathbf{v} \pm Q \tag{4.108}$$

As $\boldsymbol{\varepsilon} = \nabla\mathbf{v}$, Eq. (4.108) is rewritten as

$$\rho\frac{de}{dt} = -\nabla\cdot\mathbf{q} + \boldsymbol{\sigma} : \dot{\boldsymbol{\varepsilon}} \pm Q \tag{4.109}$$

If the internal energy e is related to temperature (T) together with the specific heat coefficient c, Eq. (4.109) becomes

$$\rho c\frac{dT}{dt} = -\nabla\cdot\mathbf{q} + \boldsymbol{\sigma} : \dot{\boldsymbol{\varepsilon}} \pm Q \tag{4.110}$$

Equation (4.110) is known as the first law of thermodynamics.

References

Aydan, Ö. 1998. Finite element analysis of transient pulse method tests for permeability measurements. *The 4th European Conference on Numerical Methods in Geotechnical Engineering-NUMGE98*, Udine, 719–727.

Aydan, Ö. 2001a. Modelling and analysis of fully coupled hydro-thermo-diffusion phenomena. *International Symposium on Clay Science for Engineering*, Balkema, Is-Shizuoka, 353–360.

Aydan, Ö. 2001b. A finite element method for fully coupled hydro-thermo-diffusion problems and its applications to geoscience and geoengineering. *10th IACMAG Conference*, Austin, 781–786.

Aydan, Ö. (2017). *Time Dependency in Rock Mechanics and Rock Engineering*, London, CRC Press, Taylor and Francis Group, 241 p.

Eringen, A.C. 1980. *Mechanics of Continua*, Huntington, NY, Robert E. Krieger Publishing Co., 606 p.

Mase, G. 1970. *Theory and Problems of Continuum Mechanics*, Schaum's Outline Series, New York, McGraw-Hill Co., 221 p.

Zimmerman, R.W., W.H. Somerton and M.S. King 1986. Compressibility of porous rocks. *Journal of Geophysical Research Atmospheres*, 91, 12765–12777.

Chapter 5

Constitutive laws

The fundamental governing equations cannot be solved in their original forms as the number of equations is less than the number of variables to be determined. If constitutive laws, which are fundamentally determined from experiments, are introduced, their solutions become possible (e.g. Aydan 2017; Mase 1970; Eringen 1980; Jaeger and Cook 1979). In this chapter, the well-known constitutive laws are introduced.

5.1 One dimensional (1D) constitutive laws

5.1.1 1D linear constitutive laws

In this sub-section, linear constitutive laws related to heat, seepage, diffusion and mechanical behavior of rocks and rock masses are given.

5.1.1.1 Fourier's law

Fourier's law states that heat flux q is linearly proportional to the gradient of temperature T, that is (Figure 5.1):

$$q = -k\frac{\partial T}{\partial x} \tag{5.1}$$

where k is the thermal heat conductivity.

5.1.1.2 Fick's law

Fick's law is essentially the same as that of Fourier's and it is used in diffusion problems. Fick's law states that the mass flux q is linearly proportional to the gradient of mass concentration C, that is (Figure 5.2):

$$q = -D\frac{\partial C}{\partial x} \tag{5.2}$$

where D is the diffusion coefficient

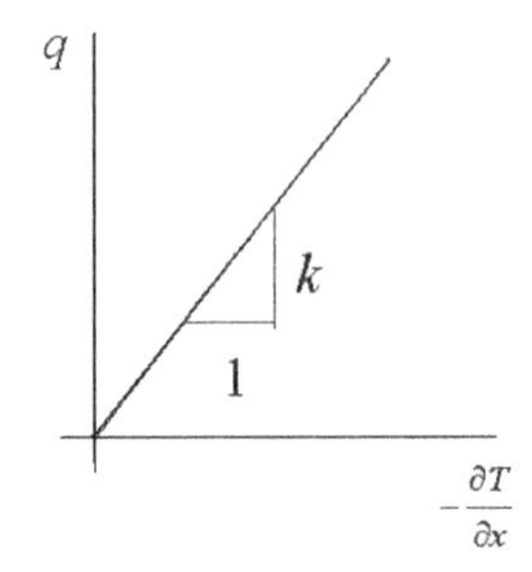

Figure 5.1 One-dimensional illustration of Fourier's law.

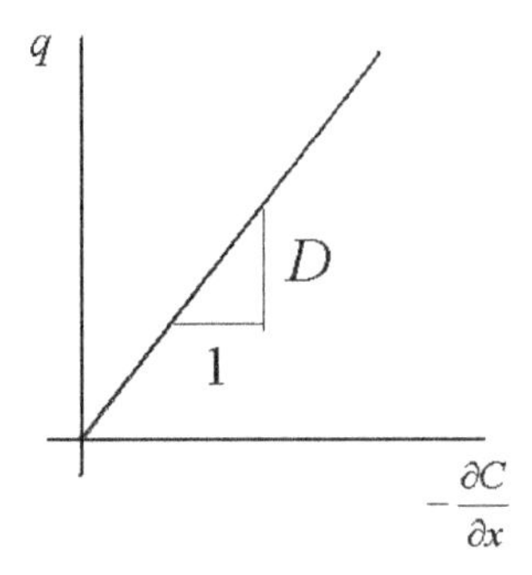

Figure 5.2 One-dimensional illustration of Fick's law.

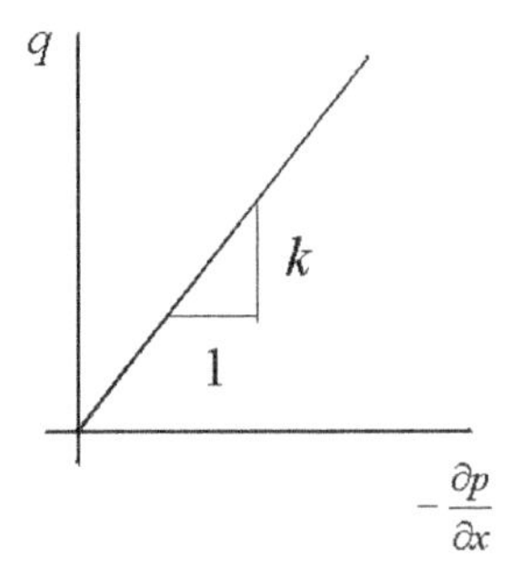

Figure 5.3 One-dimensional illustration of Darcy's law.

5.1.1.3 Darcy's law

Darcy's law is also essentially the same as that of Fourier's and it is used in groundwater seepage problems. Darcy's law states that the seepage flux q is linearly proportional to the gradient of groundwater pressure p, that is (Figure 5.3):

$$q = -k\frac{\partial p}{\partial x} \tag{5.3}$$

where k is the hydraulic conductivity and it is sometimes wrongly used as the permeability coefficient.

5.1.1.4 Hooke's law

Hooke's law is used in the theory of elasticity of solids. Hooke's law states that stress σ is linearly proportional to the strain ε, that is (Figures 5.4 and 5.6a):

$$\sigma = E\varepsilon \tag{5.4}$$

where E is the elasticity modulus or Young's modulus.

5.1.1.5 Newton's law

Newton's law (Figures 5.5 and 5.6b) is linear and is given as

$$\sigma = \eta\dot{\varepsilon} \tag{5.5}$$

If this law is integrated over time, it takes the following form with a condition, that is, $\varepsilon = 0$ at $t = 0$:

$$\varepsilon = \frac{\sigma}{\eta}t \tag{5.6}$$

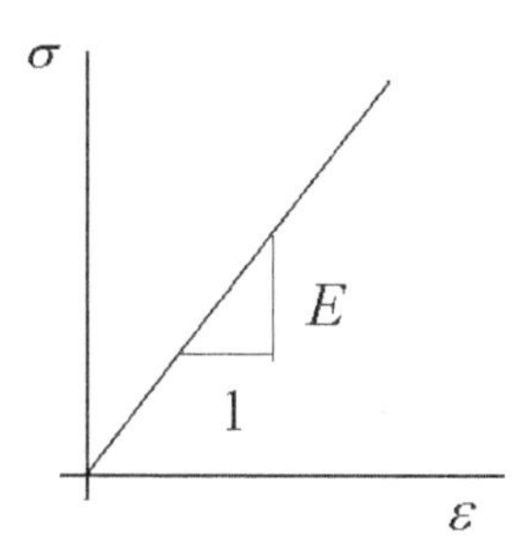

Figure 5.4 One-dimensional illustration of Hooke's law.

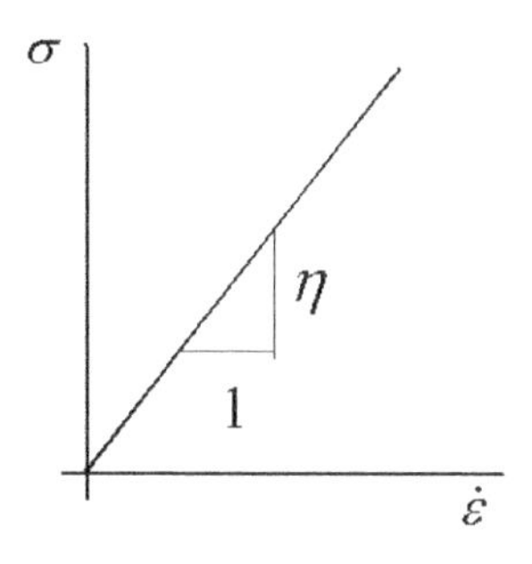

Figure 5.5 One-dimensional illustration of Newton's law.

If we assume that the strain rate is given in the following form:

$$\dot{\varepsilon} = \frac{\sigma}{\eta} \tag{5.7}$$

The equation above can be written as

$$\varepsilon = \dot{\varepsilon} \cdot t \tag{5.8}$$

This has a similarity to the steady-state creep response.

5.1.1.6 Maxwell law

A substance according to the Maxwell law (Figure 5.6c) is assumed to consist of elastic and viscous components connected in a series fashion. Therefore, the total strain and its derivatives are given as

$$\varepsilon = \varepsilon_e + \varepsilon_v \text{ and } \dot{\varepsilon} = \dot{\varepsilon}_e + \dot{\varepsilon}_v \tag{5.9}$$

The constitutive relations for elastic and viscous responses are

$$\varepsilon_e = \frac{\sigma}{E} \text{ and } \dot{\varepsilon}_v = \frac{\sigma}{\eta} \tag{5.10}$$

If $\sigma = \sigma_o$ for $t \geq 0$ and $\varepsilon = \varepsilon_o$ with $\varepsilon_o = \sigma_o / E$, the above function becomes

$$\varepsilon = \frac{\sigma_o}{E} + \frac{\sigma_o}{\eta} t \tag{5.11}$$

This equation also has a similarity to the steady-state creep response.

5.1.1.7 Kelvin–Voigt law

A substance according to the Kelvin–Voigt law (Figure 5.6d) is assumed to consist of elastic and viscous components connected in a parallel fashion. Therefore, the total stress is given as

$$\sigma = E\varepsilon + \eta\dot{\varepsilon} \tag{5.12}$$

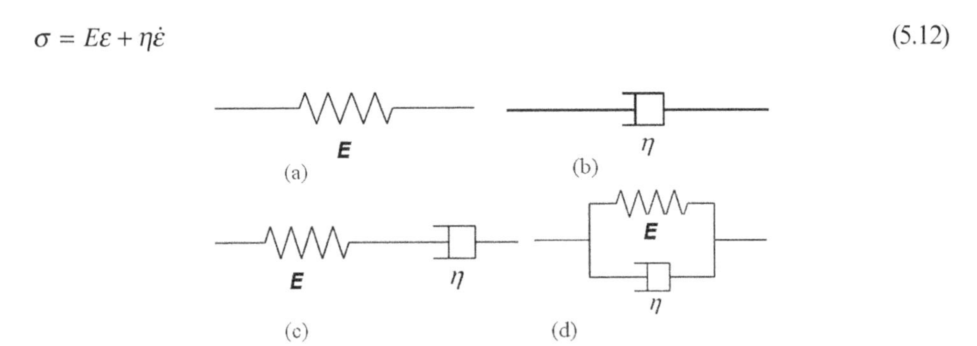

Figure 5.6 Simple rheological models. (a) Hooke model; (b) Newton model; (c) Maxwell model and (d) Kelvin–Voigt model.

If stress is applied, σ_o at $t = 0$ with $\varepsilon = 0$, and then sustained thereafter, the following relation is obtained

$$\varepsilon = \frac{\sigma_o}{E}\left(1 - e^{-t/t_r}\right) \text{ with } t_r = \frac{E}{\eta} \tag{5.13}$$

It is interesting to note that the above response would be similar to the transient creep stage. Figure 5.7 shows the creep strain responses for different simple rheological models.

5.1.1.8 Generalized Kelvin model

The model (Figure 5.8a) has a Hookean element and Kelvin element connected in a series fashion. The total strain of the model is

$$\varepsilon = \varepsilon_h + \varepsilon_k \tag{5.14}$$

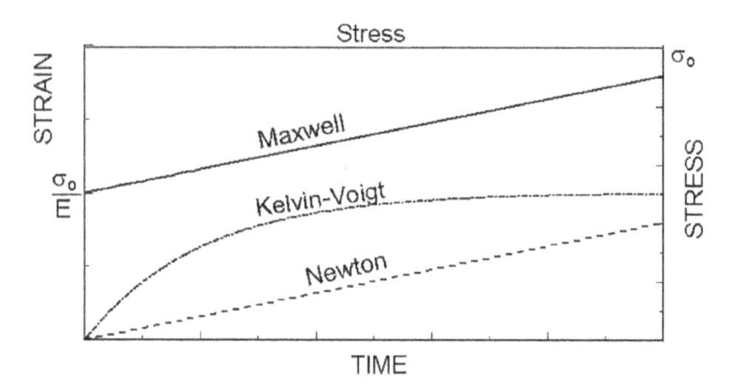

Figure 5.7 Creep strain response of simple rheological models.

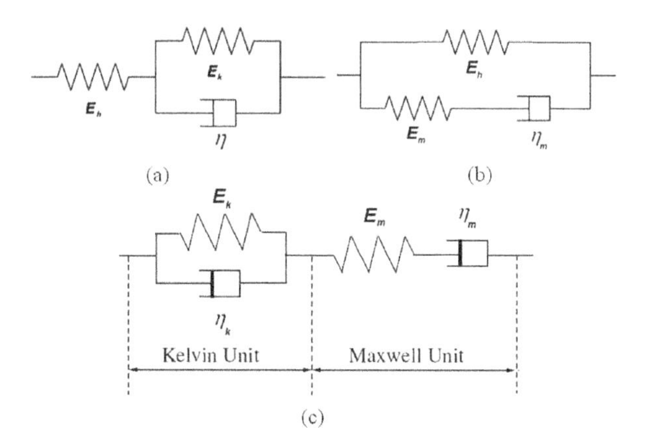

Figure 5.8 More complex rheological models. (a) Generalized Kelvin model; (b) Zener model and (c) Burgers model.

The stress relations of each element are given as

$$\varepsilon_h = \frac{\sigma}{E_h} \text{ and } \sigma = E_k \varepsilon_k + \eta \dot{\varepsilon}_k \tag{5.15}$$

Thus, one gets the following equation:

$$\sigma = \eta\left(\dot{\varepsilon} - \frac{\dot{\sigma}}{E_h}\right) + E_k\left(\varepsilon - \frac{\sigma}{E_h}\right) \tag{5.16}$$

If stress is applied, σ_o at $t=0$ with $\varepsilon = 0$ and $\varepsilon_e = \sigma_o / E_h$, and it is sustained thereafter, the following relation is obtained:

$$\varepsilon = \frac{\sigma_o}{E_h} + \frac{\sigma_o}{E_k}\left(1 - e^{-t/t_r}\right) \text{ with } t_r = \frac{\eta}{E_k} \tag{5.17}$$

As noted from this relation, the instantaneous strain is due to the elastic response and the transient creep stage can be modeled.

5.1.1.9 Zener model

Zener model (see Figure 5.8b) is also known as the standard linear solid model and it consists of a Hookean element and a Maxwell element connected in a parallel fashion.

The total stress may be given in the following form:

$$\sigma = \sigma_h + \sigma_m \tag{5.18}$$

The constitutive laws of Hookean and Maxwell elements are

$$\sigma_h = E_h \varepsilon;\ \varepsilon = \varepsilon_s + \varepsilon_d;\ \dot{\varepsilon} = \dot{\varepsilon}_s + \dot{\varepsilon}_d;\ \dot{\varepsilon}_s = \frac{\dot{\sigma}_m}{E_m};\ \dot{\varepsilon}_d = \frac{\sigma_m}{\eta_m} \tag{5.19}$$

Thus, one can easily get the following differential equation:

$$\frac{d\varepsilon}{dt} + \frac{1}{\eta_m} \cdot \frac{E_h E_m}{E_h + E_m} \varepsilon = \frac{1}{E_h + E_m}\left(\frac{d\sigma}{dt} + \frac{E_m}{\eta_m}\sigma\right) \tag{5.20}$$

If stress is applied, σ_o at $t=0$ with $\varepsilon_o = \sigma_o / (E_h + E_m)$, and it is sustained thereafter, the following relation is obtained:

$$\varepsilon = \frac{\sigma_o}{E_h}\left[1 - \frac{E_m}{E_h + E_m} e^{-t/t_r}\right] \text{ with } t_r = \eta_m \frac{E_m + E_h}{E_m E_h} \tag{5.21}$$

The creep response to be determined from this model involves the instantaneous strain and transient creep.

5.1.1.10 Burgers model

The Burgers model (Figure 5.8c) consists of the Maxwell and Kelvin elements connected in a series fashion. The constitutive relations for each element can be given as

$$\dot{\varepsilon}_m = \frac{\dot{\sigma}}{E_m} + \frac{\sigma}{\eta_m} \text{ and } \sigma = E_k \varepsilon_k + \eta \dot{\varepsilon}_k \tag{5.22}$$

The total strain is given by

$$\varepsilon = \varepsilon_m + \varepsilon_k \tag{5.23}$$

If stress is applied, σ_o at $t = 0$ with $\varepsilon = 0$ and $\varepsilon_m = \sigma_o / E_m$, and it is sustained thereafter, the following relation is obtained:

$$\varepsilon = \frac{\sigma_o}{E_m} + \frac{\sigma_o}{E_k}\left(1 - e^{-t/t_k}\right) + \frac{\sigma_o}{\eta_m} t \text{ with } t_k = \frac{\eta_k}{E_k} \tag{5.24}$$

As noted, this model can simulate the instantaneous strain due to elastic response and transient and steady-state creep stages. Figure 5.9 shows and compares the creep strain responses for different complex rheological models.

Figure 5.10 compares the experimental responses with those from intuitive and rheological models described above. As noted from these figures, each model has its own merits and demerits (Aydan 2017).

5.1.2 1D non-linear constitutive laws for solids

5.1.2.1 Elasto-plastic law

In this sub-section, a constitutive law based on the concepts of the classical plasticity theory is described. The elasto-plastic behavior of materials is illustrated in Figure 5.11. The classical plasticity theory is based upon the following assumptions:

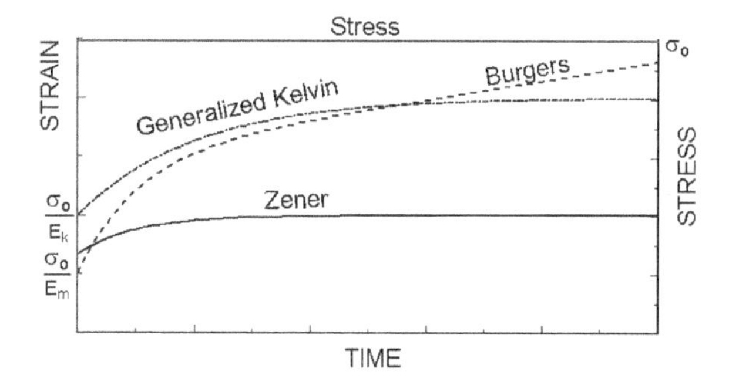

Figure 5.9 Creep responses from complex rheological models.

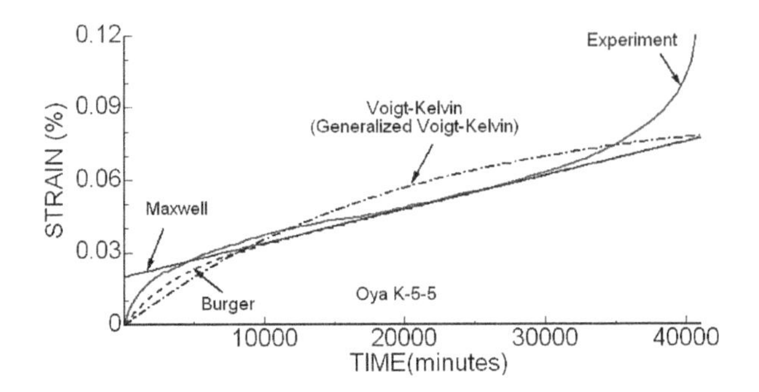

Figure 5.10 Comparison of rheological models with experimental responses.

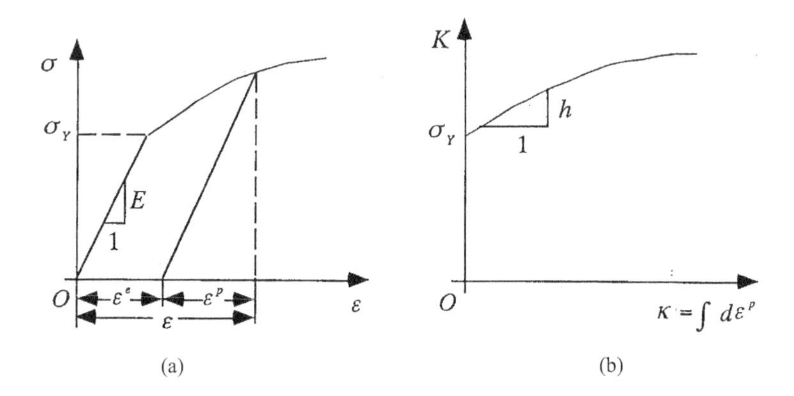

Figure 5.11 Illustration of the elasto-plastic behavior and some fundamental parameters. (a) Elasto-plastic behavior and (b) hardening function.

- Yield function is of the following form

$$F(\sigma,\kappa) = f(\sigma) - K(\kappa) = 0 \tag{5.25}$$

- Flow rule, which is given below, holds

$$d\varepsilon^p = \lambda \frac{\partial G}{\partial \sigma} \tag{5.26}$$

- Prager's consistency condition, which is given below, holds

$$dF = \frac{\partial F}{\partial \sigma} d\sigma + \frac{\partial F}{\partial \kappa} \frac{\partial \kappa}{\partial \varepsilon^p} d\varepsilon^p = 0 \tag{5.27}$$

- Strain increment is a linear sum of elastic and plastic increments as given below:

$$d\varepsilon = d\varepsilon^e + d\varepsilon^p \tag{5.28}$$

- For an elastic component, Hooke's law holds

$$d\sigma = D^e d\varepsilon^e \tag{5.29}$$

Inserting Eq. (5.26) into Eq. (5.27) and denoting h as

$$h = -\frac{\partial F}{\partial \kappa}\frac{\partial \kappa}{\partial \varepsilon^p}\frac{\partial G}{\partial \sigma} \tag{5.30}$$

yield the following relation between plastic strain increment $d\varepsilon^p$ and stress increment $d\sigma$:

$$d\varepsilon = \frac{1}{h}\frac{\partial G}{\partial \sigma}\left(\frac{\partial F}{\partial \sigma}d\sigma\right) = C^p d\sigma \tag{5.31}$$

The above relation is called Melan's formula. The following relation could be written as

$$d\sigma = D^e d\varepsilon - \frac{1}{h}D^e\frac{\partial G}{\partial \sigma}\left(\frac{\partial F}{\partial \sigma}d\sigma\right) \tag{5.32}$$

Multiplying the expression above by $\frac{\partial F}{\partial \sigma}$ yields

$$\frac{\partial F}{\partial \sigma}d\sigma = \frac{\frac{\partial F}{\partial \sigma}D^e d\varepsilon}{1 + \frac{1}{h}\frac{\partial F}{\partial \sigma}\left(D^e\frac{\partial G}{\partial \sigma}\right)} \tag{5.33}$$

Utilizing the above relation in Eq. (5.32) yields the incremental elasto-plastic law as

$$d\sigma = \left(D^e - \frac{D^e\frac{\partial G}{\partial \sigma}\frac{\partial F}{\partial \sigma}D^e}{h + \frac{\partial F}{\partial \sigma}\left(D^e\frac{\partial G}{\partial \sigma}\right)}\right)d\varepsilon \tag{5.34}$$

5.1.2.2 Visco-plastic models

a. Bingham model—elastic-perfectly visco-plastic model
   The visco-plastic model of the Bingham type assumes that the material behaves elastically below the yield stress level and visco-plastic above the yield stress level as given below:

$$\varepsilon = \frac{\sigma}{E} \text{ if } \sigma < \sigma_o \tag{5.35}$$

$$\varepsilon = \frac{\sigma - \sigma_o}{\eta} t + \frac{\sigma}{E} \text{ if } \sigma \geq \sigma_o \tag{5.36}$$

The equation above corresponds to the perfectly visco-plastic material if σ_o corresponds to the yield threshold value of stress. Furthermore, the fluidity coefficient is defined as

$$\gamma = \frac{1}{\eta} \tag{5.37}$$

b. Elastic-visco-plastic model of hardening type (Perzyna type)
The elastic-visco-plastic model of hardening type (Perzyna type) (Figure 5.12) assumes that the material behaves elastically below the yield stress level and visco-plastic above the yield stress level σ_Y (Perzyna 1966). The yield strength of the visco-plastic material in relation to the visco-plastic strain of hardening type is written as

$$Y = \sigma_Y + H\varepsilon_{vp} \tag{5.38}$$

Furthermore, the total strain is assumed to be a sum of elastic strain and visco-plastic strain as

$$\varepsilon = \varepsilon_e + \varepsilon_{vp} \tag{5.39}$$

Thus, the stress–strain relations are written in the following form:

$$\sigma_p = \sigma = E\varepsilon \text{ if } \sigma_p < Y \tag{5.40}$$
$$\sigma_p = \sigma_Y + H\varepsilon_{vp} \text{ if } \sigma_p \geq Y \tag{5.41}$$

Total stress at any time can be written as

$$\sigma = \sigma_p + \sigma_d \tag{5.42}$$

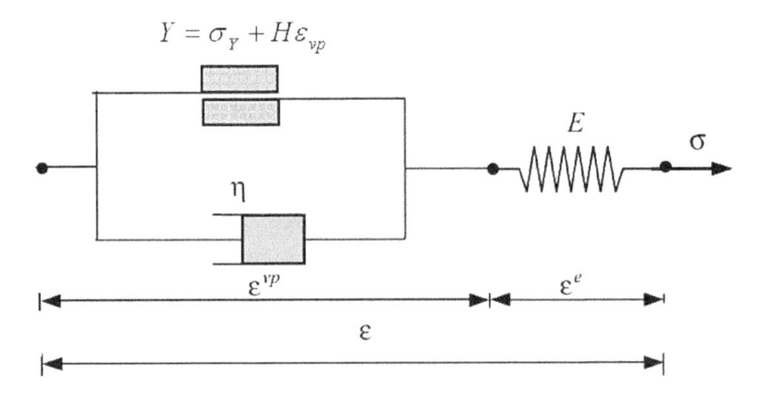

Figure 5.12 Elastic-visco-plastic model (Perzyna type).

The viscous component of stress is related to the visco-plastic strain rate as follows:

$$\sigma_d = C_p \frac{d\varepsilon_{vp}}{dt} \tag{5.43}$$

Thus, one can obtain the following differential equation for visco-plastic response

$$\sigma = \sigma_Y + H\varepsilon_{vp} + C_p \frac{d\varepsilon_{vp}}{dt} \tag{5.44}$$

Replacing the visco-plastic strain with the use of total strain and elastic strain in the above equation, one can easily obtain the following relation:

$$HE\varepsilon + \frac{1}{C_p} E \frac{d\varepsilon}{dt} = H\sigma + E(\sigma - \sigma_Y) + \frac{1}{C_p} \frac{d\sigma}{dt} \tag{5.45}$$

Let us assume that a constant stress $\sigma = \sigma_A$ is applied and kept constant (creep test). The differential equation above is reduced to the following form:

$$\frac{d\varepsilon}{dt} + \frac{H}{E}\varepsilon = \frac{H}{C_p E}\sigma_A + \frac{1}{C_p}(\sigma_A - \sigma_Y) \tag{5.46}$$

The solution of the differential equation above is obtained as follows:

$$\varepsilon = Ce^{\frac{H}{C_p}t} + \frac{1}{E}\sigma_A + \frac{1}{H}(\sigma_A - \sigma_Y) \tag{5.47}$$

When $\varepsilon = \varepsilon_e = \sigma_A / E$ at $t = 0$, the final form of the equation above becomes

$$\varepsilon = \frac{\sigma_A}{E} + \frac{(\sigma_A - \sigma_Y)}{H}\left(1 - e^{\frac{H}{C_p}t}\right) \tag{5.48}$$

Figure 5.13 illustrates the elastic-visco-plastic strain response for visco-plastic hardening and Bingham type visco-plastic behaviors.

c. Elasto-visco-plastic model of the hardening type
Instead of using the elasticity model for linear (recoverable) response, some rheological models described in the previous section can be adopted. For a non-linear (permanent) response, the models described can be utilized. For example, if the linear response is modeled using the Kelvin–Voigt type model, the following relation would hold for the linear part ($\sigma < \sigma_y$)

$$\varepsilon_r = \frac{\sigma}{E} \text{ and } \sigma = E\varepsilon_r + \eta\dot{\varepsilon}_r \text{ with } \varepsilon = \varepsilon_r \tag{5.49}$$

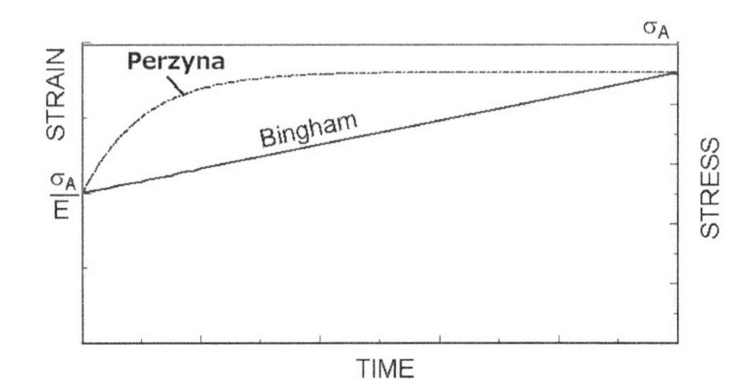

Figure 5.13 Responses obtained from elastic-visco-plastic models.

As for the non-linear (permanent) part $\sigma \geq \sigma_y$, the following can be written

$$\sigma = \sigma_Y + H\varepsilon_p + C_p \frac{d\varepsilon_p}{dt} \tag{5.50}$$

The total strain is assumed to consist of linear (recoverable) and non-linear (permanent) components as given below:

$$\varepsilon = \varepsilon_r + \varepsilon_p \tag{5.51}$$

d. Aydan–Nawrocki type elasto-visco-plastic constitutive law
In analogy to the derivation of an incremental elasto-plastic constitutive law, the following relations are set (Aydan and Nawrocki 1998):
- Yield function

$$F(\sigma, \kappa_p, \kappa_v) = f(\sigma) - K(\kappa_p, \kappa_v) = 0 \tag{5.52}$$

It should be noted here that the yield function is a function of permanent plastic and viscous hardening parameters (Figure 5.14).
- Flow rule

$$d\varepsilon^p = \lambda \frac{\partial G}{\partial \sigma}, \quad d\varepsilon^p = \dot{\lambda} \frac{\partial G}{\partial \sigma} + \lambda \frac{\partial \dot{G}}{\partial \sigma} \tag{5.53}$$

- Prager's consistency condition

$$dF = \frac{\partial F}{\partial \sigma} \cdot d\sigma + \frac{\partial F}{\partial \kappa_p} \frac{\partial \kappa_p}{\partial \varepsilon} \cdot d\varepsilon^p + \frac{\partial F}{\partial \kappa_v} \frac{\partial \kappa_v}{\partial \varepsilon^p} \cdot d\varepsilon^p = 0 \tag{5.54}$$

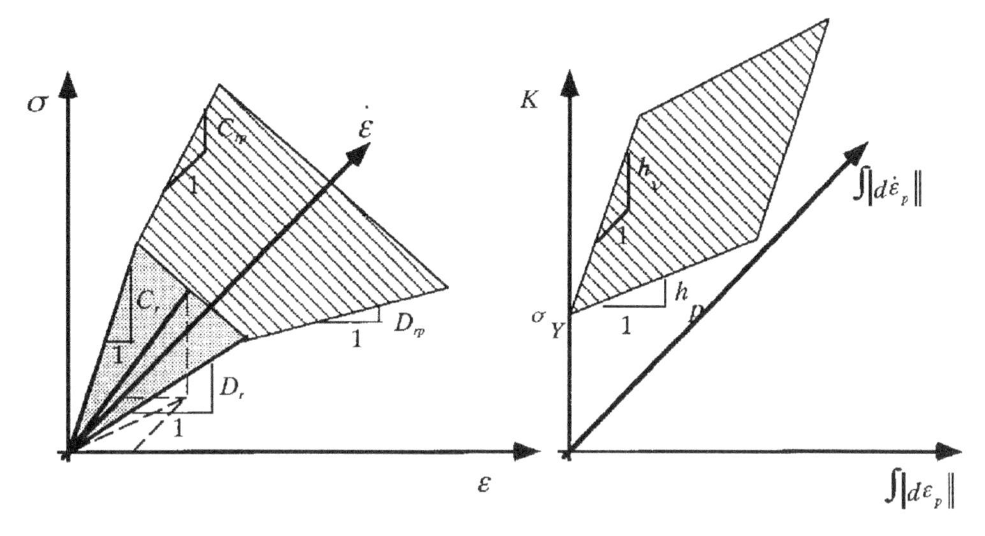

Figure 5.14 The elasto-visco-plastic model for a one-dimensional response.

- Linear decomposition of the strain increment $d\varepsilon$ and strain rate increment $d\dot{\varepsilon}$ into their reversible and permanent components $d\varepsilon^r$ and $d\varepsilon^p$.

$$d\varepsilon = d\varepsilon^r + d\varepsilon^p, \quad d\dot{\varepsilon} = d\dot{\varepsilon}^r + d\dot{\varepsilon}^p, \tag{5.55}$$

- Voigt–Kelvin's law

$$d\sigma = D^r d\varepsilon^r + C^r d\dot{\varepsilon}^r \tag{5.56}$$

where σ is the stress tensor, $K(\kappa_p, \kappa_v)$ is the hardening function, G is the plastic potential, λ is the proportionality coefficient, κ_p is the plastic hardening parameter, κ_v is the viscous hardening parameter, ε is the strain tensor, ε^r is the reversible strain tensor, ε^p is the permanent strain tensor, $\dot{\varepsilon}^p$ is the permanent strain rate tensor, D^r is the elasticity tensor, C^r is the viscosity tensor and $(\cdot)$ denotes the dot product.

In elastic-visco-plastic formulations of Perzyna type, the *flow rule* is always assumed to be of the following form:

$$d\varepsilon^p = \dot{\lambda} \frac{\partial G}{\partial \sigma} \tag{5.57}$$

The above equation implies that any plastic straining is always time dependent.

Here, the permanent strain rate increment given by Eq. (5.53) is simplified to the following form by assuming that $\dot{\lambda} = 0$:

$$d\varepsilon^p = \lambda \frac{\partial \dot{G}}{\partial \sigma} \tag{5.58}$$

The above equation implies that the plastic potential function shrinks (or expands) in the time domain while keeping its original form in stress space.[1]

Substituting Eqs. (5.53) and (5.58) into Eq. (5.54) and rearranging the resulting expression together with the denotion of its denominator by h_{rp} (hardening modulus)

$$h_{rp} = -\left[\frac{\partial F}{\partial \kappa_p}\frac{\partial \kappa_p}{\partial \varepsilon^p}\cdot\frac{\partial G}{\partial \sigma} + \frac{\partial F}{\partial \kappa_v}\frac{\partial \kappa_v}{\partial \varepsilon^p}\cdot\frac{\partial \dot{G}}{\partial \sigma}\right] \tag{5.59}$$

we have λ as:

$$\lambda = \frac{1}{h_{rp}}\frac{\partial F}{\partial \sigma}\cdot d\sigma \tag{5.60}$$

Now, let us insert the above relation into Eqs. (5.53) and (5.58), we have the constitutive relations between the permanent strain rate increment $d\dot{\varepsilon}^p$ and the stress increment $d\sigma$ as:

$$d\varepsilon^p = \frac{1}{h_{rp}}\frac{\partial G}{\partial \sigma}\left(\frac{\partial F}{\partial \sigma}\cdot d\sigma\right) = \frac{1}{h_{rp}}\left(\frac{\partial G}{\partial \sigma}\otimes\frac{\partial F}{\partial \sigma}\right)d\sigma \tag{5.61}$$

$$d\dot{\varepsilon}^p = \frac{1}{h_{rp}}\frac{\partial \dot{G}}{\partial \sigma}\left(\frac{\partial F}{\partial \sigma}\cdot d\sigma\right) = \frac{1}{h_{rp}}\left(\frac{\partial \dot{G}}{\partial \sigma}\otimes\frac{\partial F}{\partial \sigma}\right)d\sigma \tag{5.62}$$

where ($\otimes$) denotes the tensor product. The inverse of the above relations could not be obtained whether the plastic potential G is of the associated or non-associated type. Therefore, the following technique is used to establish the relation between $d\sigma$ and $d\varepsilon$, $d\dot{\varepsilon}$. Using the relations (5.55), (5.56), (5.61) and (5.62), one can write the following:

$$d\sigma = D^r d\varepsilon - D^r\frac{1}{h_{rp}}\frac{\partial G}{\partial \sigma}\left(\frac{\partial F}{\partial \sigma}\cdot d\sigma\right) + C^r d\varepsilon - C^r\frac{1}{h_{rp}}\frac{\partial \dot{G}}{\partial \sigma}\left(\frac{\partial F}{\partial \sigma}\cdot d\sigma\right) \tag{5.63}$$

Taking the dot products of both sides of the above expression by $\partial F/\partial\sigma$ yields

$$\frac{\partial F}{\partial \sigma}\cdot d\sigma = \frac{\frac{\partial F}{\partial \sigma}\cdot\left(D^r d\varepsilon\right) + \frac{\partial F}{\partial \sigma}\cdot\left(C^r d\varepsilon\right)}{1 + \frac{1}{h_{rp}}\frac{\partial F}{\partial \sigma}\cdot\left(D^r\frac{\partial G}{\partial \sigma}\right) + \frac{1}{h_{rp}}\frac{\partial F}{\partial \sigma}\cdot\left(C^r\frac{\partial \dot{G}}{\partial \sigma}\right)} \tag{5.64}$$

Substituting the above relation in (5.63) gives the incremental elasto-visco-plastic constitutive law as:

$$d\sigma = D^{rp} d\varepsilon + C^{rp} d\varepsilon \tag{5.65}$$

[1] It should be noted that the permanent strain rate increment consists of time-independent and time-dependent parts.

where

$$D^{rp} = D^{r} - \frac{D^{r}\dfrac{\partial G}{\partial \sigma} \otimes \dfrac{\partial F}{\partial \sigma} D^{r}}{h_{rp} + \dfrac{\partial F}{\partial \sigma} \cdot \left(D^{r} \dfrac{\partial G}{\partial \sigma} \right) + \dfrac{\partial F}{\partial \sigma} \cdot \left(C^{r} \dfrac{\partial \dot{G}}{\partial \sigma} \right)}$$

$$C^{rp} = C^{r} - \frac{C^{e}\dfrac{\partial \dot{G}}{\partial \sigma} \otimes \dfrac{\partial F}{\partial \sigma} C^{r}}{h_{rp} + \dfrac{\partial F}{\partial \sigma} \cdot \left(D^{r} \dfrac{\partial G}{\partial \sigma} \right) + \dfrac{\partial F}{\partial \sigma} \cdot \left(C^{r} \dfrac{\partial \dot{G}}{\partial \sigma} \right)}$$

5.2 Multi-dimensional constitutive laws

5.2.1 Fourier's law

Fourier's law states that heat flux q_i is linearly proportional to the gradient of temperature T, that is:

$$q_i = -K_{ij} \frac{\partial T}{\partial x_i} \tag{5.66}$$

where K_{ij} is called thermal heat conductivity tensor.

5.2.2 Fick's law

Fick's law is essentially the same as that of Fourier's and it is used in diffusion problems. Fick's law states that mass flux q_i is linearly proportional to the gradient of mass concentration C, that is:

$$q_i = -D_{ij} \frac{\partial C}{\partial x_i} \tag{5.67}$$

where D_{ij} is called diffusion coefficient tensor

5.2.3 Darcy's law

Darcy's law is essentially the same as that of Fourier's and it is used in ground-water seepage problems. Darcy's law states that seepage flux q_i is linearly proportional to the gradient of groundwater pressure p, that is:

$$q_i = -K_{ij} \frac{\partial p}{\partial x_i} \tag{5.68}$$

where K_{ij} is called hydraulic conductivity tensor.

5.2.4 Hooke's law

When rock or rock mass behaves linearly without any rate dependency, the simplest constitutive law is Hooke's law. This law is written in the following form:

$$\sigma_{ij} = D_{ijkl}\varepsilon_{kl} \tag{5.69}$$

Where $\sigma_{ij}, \varepsilon_{kl}$ and D_{ijkl} are stress, strain, and elasticity tensors, respectively.

If the material is homogenous and isotropic, Eq. (5.69) may be written as

$$\sigma_{ij} = 2\mu\varepsilon_{ij} + \lambda\delta_{ij}\varepsilon_{kk} \tag{5.70}$$

Where δ_{ij} is the Kronecker delta tensor. λ and μ are Lame coefficients, which are given in terms of elasticity (Young's) modulus (E) and Poisson's ratio (υ) as

$$\lambda = \frac{E\upsilon}{(1+\upsilon)(1-2\upsilon)}; \mu = \frac{E}{2(1+\upsilon)} \tag{5.71}$$

5.2.5 Newton's law

Newton's law is used in fluid mechanics. Newton's law states that stress σ_{ij} is linearly proportional to strain rate $\dot{\varepsilon}_{kl}$, that is:

$$\sigma_{ij} = C_{ijkl}\dot{\varepsilon}_{kl} \tag{5.72}$$

where C_{ijkl} is the viscosity tensor.

5.2.6 Kelvin–Voigt's law

Kelvin–Voigt's law is used in the field of visco-elasticity. Voigt's law states that stress σ_{ij} is linearly proportional to strain ε_{ij} and strain rate $\dot{\varepsilon}_{ij}$, that is:

$$\sigma_{ij} = D_{ijkl}\varepsilon_{kl} + C_{ijkl}\dot{\varepsilon}_{kl} \tag{5.73}$$

Where $\dot{\varepsilon}_{kl}$ and C_{ijkl} are strain rate and viscosity tensors, respectively.

If the material is homogenous and isotropic, Eq. (5.73) may be written in analogy to Eq. (5.70) as

$$\sigma_{ij} = 2\mu\varepsilon_{ij} + \lambda\delta_{ij}\varepsilon_{kk} + 2\mu^*\dot{\varepsilon}_{ij} + \lambda^*\delta_{ij}\dot{\varepsilon}_{kk} \tag{5.74}$$

Coefficients λ^* and μ^* may be called viscous Lame coefficients. This constitutive law was first proposed by Aydan (1995a).

5.2.7 Navier-Stokes law

Navier-Stokes constitutive law can be visualized as a simple case of Kelvin–Voigt's law. If the material behavior is associated with the volumetric strain (ε_v) and pressure

(p) under the steady-state condition without any shear resistance like fluids, Eq. (5.74) can be re-written as

$$\sigma_{ij} = -p\delta_{ij} + 2\mu^* \dot{\varepsilon}_{ij} + \lambda^* \delta_{ij}\dot{\varepsilon}_{kk} \tag{5.75}$$

If coefficient $\lambda^* = 0$, the above relation reduces to the following form

$$\sigma_{ij} = -p\delta_{ij} + 2\mu^* \dot{\varepsilon}_{ij} \text{ or } \sigma_{ij} = -p\delta_{ij} + \mu^* \dot{\gamma}_{ij} \tag{5.76}$$

The above constitutive law corresponds to the constitutive law commonly used in fluid mechanics known as Navier-Stokes Law.

There are different visco-elasticity models in the literature. As it is difficult to cover, all models here, the reader is advised to consult the available literature on the topic of visco-elasticity (e.g. Cristescu and Hunsche 1998; Jaeger and Cook 1979; Perzyna 1966).

5.3 Non-linear behavior (elasto-plasticity and elasto-visco-plasticity) for solids

Every material in nature starts to yield after a certain stress or strain state and rock or rock mass is no exception. The terms used to describe the material behavior such as elasticity and visco-elasticity are replaced by the terms of elasto-plasticity or elasto-visco-plasticity as soon as material behavior deviates from linearity. The relation between total stress and strain or strain rate can no longer be used and every relationship must be written in the incremental form. For example, if the conventional plasticity models are used, the elasto-plastic constitutive law between incremental stress and strain tensors would take the following form:

5.3.1 *Elasto-plastic law*

The derivation of an incremental elasto-plastic constitutive law, based on the conventional elasto-plastic theory, starts with the following:

- Yield function

$$F(\sigma,\kappa) = f(\sigma) - K(\kappa) = 0 \tag{5.77}$$

- Flow rule

$$d\varepsilon^p = \lambda \frac{\partial G}{\partial \sigma} \tag{5.78}$$

- Prager's consistency condition

$$dF = \frac{\partial F}{\partial \sigma} \cdot d\sigma + \frac{\partial F}{\partial \kappa}\frac{\partial \kappa}{\partial \varepsilon^p} \cdot d\varepsilon^p = 0 \tag{5.79}$$

- Linear decomposition of the strain increment $d\varepsilon$ into its elastic and plastic components $d\varepsilon^e$ and $d\varepsilon^p$.

$$d\varepsilon = d\varepsilon^e + d\varepsilon^p \tag{5.80}$$

- Hooke's law

$$d\sigma = \mathbf{D}^e d\varepsilon^e \tag{5.81}$$

where
σ: stress tensor
$K(\kappa)$: hardening function
G: plastic potential
λ: proportionality coefficient
κ: hardening parameter
$d\varepsilon$: strain tensor
$d\varepsilon^e$: elastic strain tensor
$d\varepsilon^p$: plastic strain tensor
$\mathbf{D}^e$: elasticity tensor
and $(\cdot)$ denotes the dot product.

Substituting Eq. (5.78) into Eq. (5.79), and rearranging the resulting expression together with the denotion of its denominator by h (hardening modulus)

$$h = -\frac{\partial F}{\partial \kappa}\frac{\partial \kappa}{\partial \varepsilon^p} \cdot \frac{\partial G}{\partial \sigma} \tag{5.82}$$

we have λ as:

$$\lambda = \frac{1}{h}\frac{\partial F}{\partial \sigma} \cdot d\sigma \tag{5.83}$$

Now, let us insert the above relation into Eq. (5.79), we have the constitutive relation between the plastic strain increment $d\varepsilon^p$ and the stress increment $d\sigma$, which is also known as the Melan's formula, as:

$$d\varepsilon^p = \frac{1}{h}\frac{\partial G}{\partial \sigma}\left(\frac{\partial F}{\partial \sigma} \cdot d\sigma\right) = \frac{1}{h}\left(\frac{\partial G}{\partial \sigma} \otimes \frac{\partial F}{\partial \sigma}\right) d\sigma = \mathbf{C}^p d\sigma \tag{5.84}$$

where $(\otimes)$ denotes the tensor product. The inverse of the above relation could not be done as the determinant of the plasticity matrix $|\mathbf{C}^p| = 0|$ irrespective of whether the plastic potential G is of the associated or non-associated type. Therefore, the following technique is used to establish the relation between $d\varepsilon$ and $d\sigma$. Using the relations (5.82), (5.83) and (5.84), one can write the following:

$$d\sigma = \mathbf{D}^e d\varepsilon - \mathbf{D}^e \frac{1}{h}\frac{\partial G}{\partial \sigma}\left(\frac{\partial F}{\partial \sigma} \cdot d\sigma\right) \tag{5.85}$$

Taking the dot products of both sides of the above expression by $\partial F/\partial \sigma$ yields

$$\frac{\partial F}{\partial \sigma}\cdot d\sigma = \frac{\frac{\partial F}{\partial \sigma}\cdot\left(\mathbf{D}^e d\varepsilon\right)}{1+\frac{1}{h}\frac{\partial F}{\partial \sigma}\cdot\left(\mathbf{D}^e\frac{\partial G}{\partial \sigma}\right)} \tag{5.86}$$

Substituting the above relation in (5.85) gives the incremental elasto-plastic constitutive law as:

$$d\sigma = \left(\mathbf{D}^e - \frac{d\mathbf{D}^e\frac{\partial G}{\partial \sigma}\otimes\frac{\partial F}{\partial \sigma}\mathbf{D}^e}{h+\frac{\partial F}{\partial \sigma}\cdot\left(\mathbf{D}^e\frac{\partial G}{\partial \sigma}\right)}\right)d\varepsilon \tag{5.87}$$

The hardening modulus h is generally determined as a function of a hardening parameter κ by employing either a work-hardening model or strain-hardening model. The hardening parameter κ is defined for both cases as follows:

$$\kappa = W^p = \int \sigma\cdot d\varepsilon^p \quad \text{work-hardening} \tag{5.88}$$

$$\kappa = \int \| d\varepsilon^p \| \quad \text{strain hardening} \tag{5.89}$$

where W^p is plastic work.

The materials (i.e. steel, glass-fibers) exhibit a non-dilatant plastic behavior and isotropically harden. Therefore, a work hardening model is generally used together with the effective stress-strain concept, which is defined as:

$$\sigma_e = \sqrt{\frac{3}{2}(\mathbf{s}\cdot\mathbf{s})} \quad d\varepsilon_e^p = \sqrt{\frac{2}{3}\left(d\varepsilon^p\cdot d\varepsilon^p\right)} \tag{5.90}$$

where $\mathbf{s}$ is deviatoric stress tensor.

As the volumetric plastic strain increment $d\bar{\varepsilon}_v^p = 0$ together with the co-axiality of the stress and plastic strain, the hardening parameter κ of work-hardening type can be re-written in the following form:

$$d\kappa = dW^p = \sigma\cdot d\varepsilon^p = \mathbf{s}\cdot de^p = \sigma_e d\varepsilon_e^p \tag{5.91}$$

where de^p is the deviatoric strain increment. The hardening modulus h for this case take the following form with the use of Euler's theorem[2] by taking a homogeneous plastic potential G of order m:

$$h = -\frac{\partial F}{\partial \kappa}\frac{\partial \kappa}{\partial \varepsilon^p}\cdot\frac{\partial G}{\partial \sigma} = \frac{\partial K}{\partial W^p}\sigma\cdot\frac{\partial G}{\partial \sigma} = m\frac{\partial K}{\partial W^p}G \tag{5.92}$$

2 $\mathbf{x}\cdot\partial f/\partial \mathbf{x} = mf$.

If $F = G$ and $f(\sigma) = \sigma_e$, then the hardening modulus h becomes:

$$h = m\frac{\partial K}{\partial W^p} f(\sigma) = m\frac{\partial \sigma_e}{\partial \varepsilon_e} \tag{5.93}$$

The hardening modulus can, then, be easily obtained from a gradient of the plot of a uniaxial test in σ_1 and ε_1^p space, since the effective stress and strain in the uniaxial state become:

$$\sigma_e = \sigma_1 \quad \varepsilon_e^p = \varepsilon_1^p \tag{5.94}$$

5.3.2 Elasto-visco-plasticity

These approaches assume that the materials are assumed to be elastic before yielding and behave in a visco-plastic manner following yielding. In visco-plastic evaluations, $\mathbf{e}_p$ is replaced by $\mathbf{e}_{vp}$.

5.3.2.1 Power-type models

When the Norton type constitutive law is used for creep response, the visco-plastic strain rate ($\mathbf{e}_{vp} = \boldsymbol{\varepsilon}_{vp}$) is expressed as follows:

$$\frac{d\boldsymbol{\varepsilon}_{vp}}{dt} = \left(\frac{\sigma_{eq}}{\sigma_o}\right)^n \frac{\partial \sigma_{eq}}{\partial \boldsymbol{\sigma}} \tag{5.95}$$

Perzyna-type elastic-visco-plastic laws are used for representing non-linear rate dependency involving plasticity.

$$\frac{d\boldsymbol{\varepsilon}_{vp}}{dt} = \lambda \mathbf{s} \tag{5.96}$$

where λ is the proportionality coefficient and it is interpreted as the fluidity coefficient. This parameter is obtained from uniaxial creep experiments as

$$\lambda = \frac{\dot{\varepsilon}_c}{\sigma} \tag{5.97}$$

5.3.2.2 Elasto-visco-plasticity

Another approach was proposed by Aydan and Nawrocki (1998), in which the material behavior is visco-elastic before yielding and becomes visco-plastic after yielding. The derivation of this constitutive law involves the following factors:

Yield function

$$F(\boldsymbol{\sigma}, \kappa_p, \kappa_v) = f(\boldsymbol{\sigma}) - K(\kappa_p, \kappa_v) = 0 \tag{5.98}$$

It should be noted that the yield function is a function of permanent plastic and viscous hardening parameters (Figure 5.14).

Flow rule

$$d\varepsilon^p = \lambda \frac{\partial G}{\partial \sigma}, \; d\dot{\varepsilon}^p = \dot{\lambda} \frac{\partial G}{\partial \sigma} + \lambda \frac{\partial \dot{G}}{\partial \sigma} \tag{5.99}$$

Prager's consistency condition

$$dF = \frac{\partial F}{\partial \sigma} \cdot d\sigma + \frac{\partial F}{\partial \kappa_p} \frac{\partial \kappa_p}{\partial \varepsilon_p} \cdot d\varepsilon_p + \frac{\partial F}{\partial \kappa_v} \frac{\partial \kappa_v}{\partial \dot{\varepsilon}_p} \cdot d\dot{\varepsilon}_p = 0 \tag{5.100}$$

Linear decomposition of the strain increment ($d\varepsilon$) and strain rate increment ($d\dot{\varepsilon}$) into their reversible ($d\varepsilon^r$) and permanent components ($d\varepsilon^p$)

$$d\varepsilon = d\varepsilon^r + d\varepsilon^p; \; d\dot{\varepsilon} = d\dot{\varepsilon}^r + d\dot{\varepsilon}^p \tag{5.101}$$

The incremental Kelvin–Voigt law is written as

$$d\sigma = \mathbf{D}^r d\varepsilon^r + \mathbf{C}^r d\dot{\varepsilon}^r \tag{5.102}$$

where σ is the stress tensor, ε is the strain tensor, $K(\kappa_p, \kappa_v)$ is the hardening function, G is the plastic potential, λ is the proportionality coefficient, κ_p is the plastic hardening parameter, κ_v is the viscous hardening parameter, $d\varepsilon^r$ is the reversible incremental strain tensor, $d\dot{\varepsilon}^r$ is the reversible incremental strain rate tensor, $d\varepsilon^p$ is the permanent incremental strain tensor, $d\dot{\varepsilon}^p$ is the permanent incremental strain rate tensor, $\mathbf{D}^r$ is the elasticity tensor, $\mathbf{C}^r$ is the viscosity tensor and$(\cdot)$ denotes the dot product.

In elastic-visco-plastic formulations of the Perzyna type, the flow rule is assumed to be of the following form:

$$d\dot{\varepsilon}^p = \dot{\lambda} \frac{\partial G}{\partial \sigma} \tag{5.103}$$

The flow rule above implies that any plastic straining is time dependent. Aydan and Nawrocki (1998) suggested the following form:

$$d\dot{\varepsilon}^p = \lambda \frac{\partial \dot{G}}{\partial \sigma} \tag{5.104}$$

This flow rule of Aydan and Nawrocki (1998) implies that the plastic potential function shrinks (or expands) in a time domain while keeping its original form in stress space and the permanent strain increment consists of time-dependent and time-independent parts.

Substituting Eq. (5.104) into Eq. (5.100) and rearranging the resulting equations yield the following:

$$dF = \frac{1}{h_{rp}} \frac{\partial F}{\partial \sigma} \cdot d\sigma \tag{5.105}$$

where h_{rp} is called the hardening modulus and is given specifically as follows:

$$h_{rp} = -\left[\frac{\partial F}{\partial \kappa_p} \frac{\partial \kappa_p}{\partial \varepsilon_p} \cdot \frac{\partial G}{\partial \sigma} + \frac{\partial F}{\partial \kappa_v} \frac{\partial \kappa_v}{\partial \dot{\varepsilon}_p} \cdot \frac{\partial \dot{G}}{\partial \sigma} \right] \tag{5.106}$$

Inserting the relations above into Eq. (5.105) yields the constitutive relations between permanent strain increment and permanent strain rate increment and stress increment as

$$d\dot{\varepsilon}_p = \frac{1}{h_{rp}} \frac{\partial G}{\partial \sigma} \left(\frac{\partial F}{\partial \sigma} \cdot d\sigma \right) = \frac{1}{h_{rp}} \left(\frac{\partial G}{\partial \sigma} \otimes \frac{\partial F}{\partial \sigma} \right) \cdot d\sigma \tag{5.107}$$

$$d\dot{\varepsilon}_p = \frac{1}{h_{rp}} \frac{\partial \dot{G}}{\partial \sigma} \left(\frac{\partial F}{\partial \sigma} \cdot d\sigma \right) = \frac{1}{h_{rp}} \left(\frac{\partial \dot{G}}{\partial \sigma} \otimes \frac{\partial F}{\partial \sigma} \right) \cdot d\sigma \tag{5.108}$$

where (⊗) denotes the tensor product. The inverse of the relations above cannot be determined whether the plastic potential function is of associated or non-associated type. Therefore, the following technique is used to establish the relationship between stress increment and strain and strain rate increments. Using the relations given above, one can derive the following relations:

$$d\sigma = \mathbf{D}^r d\varepsilon - \mathbf{D}^r \frac{1}{h_{rp}} \frac{\partial G}{\partial \sigma} \left(\frac{\partial F}{\partial \sigma} \cdot d\sigma \right) + \mathbf{C}^r d\dot{\varepsilon} - \mathbf{C}^r \frac{1}{h_{rp}} \frac{\partial \dot{G}}{\partial \sigma} \left(\frac{\partial F}{\partial \sigma} \cdot d\sigma \right) \tag{5.109}$$

Taking the dot products of both sides of the expression above by $\partial F / \partial \sigma$ yields

$$\frac{\partial F}{\partial \sigma} \cdot d\sigma = \frac{\frac{\partial F}{\partial \sigma} \cdot \left(\mathbf{D}^r d\varepsilon \right) + \frac{\partial F}{\partial \sigma} \cdot \left(\mathbf{C}^r d\dot{\varepsilon} \right)}{1 + \frac{1}{h_{rp}} \frac{\partial F}{\partial \sigma} \cdot \left(\mathbf{D}^r \frac{\partial G}{\partial \sigma} \right) + \frac{1}{h_{rp}} \frac{\partial F}{\partial \sigma} \cdot \left(\mathbf{C}^r \frac{\partial \dot{G}}{\partial \sigma} \right)} \tag{5.110}$$

Substituting the equation above in Eq. (5.102) gives the incremental elasto-visco-plastic constitutive law as

$$d\sigma = \mathbf{D}^{rp} d\varepsilon + \mathbf{C}^{rp} d\dot{\varepsilon} \tag{5.111}$$

where

$$\mathbf{D}^{rp} = \mathbf{D}^r - \frac{\mathbf{D}^r \frac{\partial G}{\partial \sigma} \otimes \frac{\partial F}{\partial \sigma} \mathbf{D}^r}{h_{rp} + \frac{\partial F}{\partial \sigma} \cdot \left(\mathbf{D}^r \frac{\partial G}{\partial \sigma} \right) + \frac{\partial F}{\partial \sigma} \cdot \left(\mathbf{C}^r \frac{\partial \dot{G}}{\partial \sigma} \right)} \tag{5.112}$$

$$\mathbf{C}^{rp} = \mathbf{C}^{r} - \frac{\mathbf{C}^{r}\dfrac{\partial \dot{G}}{\partial \sigma} \otimes \dfrac{\partial F}{\partial \sigma}\mathbf{C}^{r}}{h_{rp} + \dfrac{\partial F}{\partial \sigma}\cdot\left(\mathbf{D}^{r}\dfrac{\partial G}{\partial \sigma}\right) + \dfrac{\partial F}{\partial \sigma}\cdot\left(\mathbf{C}^{r}\dfrac{\partial \dot{G}}{\partial \sigma}\right)} \tag{5.113}$$

Figure 5.14 illustrates the elasto-visco-plastic model for 1D response.

5.3.3 Yield/failure criteria

Non-linear behavior requires the existence of yield functions. These yield functions are also called failure functions at the ultimate state when rocks rupture. It is common to use the Mohr–Coulomb yield criterion given by:

$$\tau = c + \sigma_n \tan\phi \text{ or } \sigma_1 = \sigma_c + q\sigma_3 \tag{5.114}$$

where c, ϕ and σ_c are cohesion, friction angle and uniaxial compressive strength, respectively. σ_c and q are related to cohesion and friction angle in the following form:

$$\sigma_c = \frac{2c\cos\phi}{1-\sin\phi} \text{ and } q = \frac{1+\sin\phi}{1-\sin\phi} \tag{5.115}$$

As the intermediate principal stress is indeterminate in the Mohr–Coulomb criterion and there is a corner-effect problem during the determination of the incremental elasto-plasticity tensor, the use of the Drucker–Prager criterion (Drucker and Prager 1952) is quite common in numerical analyses, which is given by

$$\alpha I_1 + \sqrt{J_2} = k \tag{5.116}$$

where

$$I_1 = \sigma_I + \sigma_{II} + \sigma_{III};\ J_2 = \frac{1}{6}\left((\sigma_I - \sigma_{II})^2 + (\sigma_{II} - \sigma_{III})^2 + (\sigma_{III} - \sigma_I)^2\right)$$

Nevertheless, it is possible to relate the Drucker–Prager yield criterion with the Mohr–Coulomb yield criterion. On π-plane, if the inner corners of the Mohr–Coulomb yield surface are assumed to coincide with the Drucker–Prager yield criterion, the following relations could be derived (Figure 5.15):

$$\alpha = \frac{2\sin\phi}{\sqrt{3}(3+\sin\phi)};\ k = \frac{6c\cos\phi}{\sqrt{3}(3+\sin\phi)} \tag{5.117}$$

where c,ϕ are cohesion and friction angle, respectively.

In rock mechanics, one of the recent yield criteria is Hoek–Brown's criterion (1980, 1997), which is written as

$$\sigma_1 = \sigma_3 + \sqrt{m\sigma_c\sigma_3 + s\sigma_c^2} \tag{5.118}$$

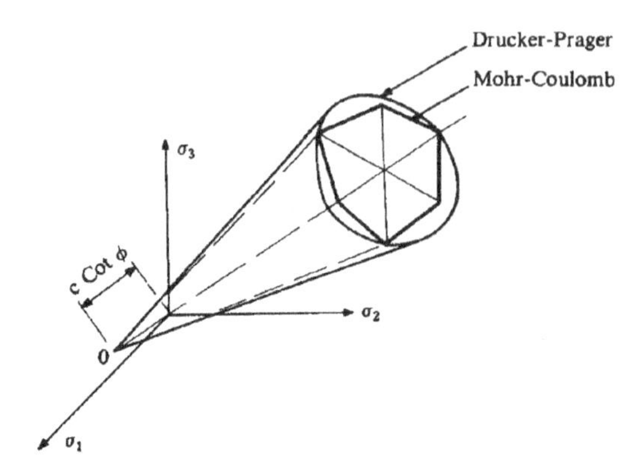

Figure 5.15 Illustration of yield criteria in principal stress space (from Owen and Hinton 1980).

where m and s denote coefficients. While the value of s is 1 for intact rock, the values of m and s change when they are used for rock mass.

Aydan (1995b) proposed a yield function for thermo-plasticity yielding of rock as given by

$$\sigma_1 = \sigma_3 + \left[S_\infty - (S_\infty - \sigma_c) e^{-b_1 \sigma_3} \right] e^{-b_2 T} \tag{5.119}$$

where S_∞ is the ultimate deviatoric strength while coefficients b_1, b_2 are empirical constants.

Several examples of applications of yield (failure) criteria to actual experimental results involving igneous, metamorphic and sedimentary rocks are described and compared with each other as well as with experimental results.

a. Sedimentary rocks

A series of uniaxial and triaxial compression and Brazilian tensile tests were carried out on Oya tuff, which is a well-known volcanic sedimentary rock in Japan (Seiki and Aydan 2003). Figure 5.16 shows the failure criteria of Mohr–Coulomb, Hoek and Brown and Aydan applied to experimental results. The best fits to experimental results were obtained for the Aydan criterion and Mohr–Coulomb criterion for $\sigma_3 > 0$. However, if the yield criterion is required to evaluate both uniaxial compressive strength and tensile strength, the criteria of Hoek–Brown and Mohr–Coulomb cannot evaluate the triaxial strength of rocks.

The next application is concerned with a very weak sandstone from Tono mine in Central Japan (Aydan et al. 2006). Figure 15.17 shows the fitted failure criteria of Mohr–Coulomb, Hoek and Brown and Aydan applied to the experimental results of sandstone of Tono. As seen from Figure 15.17, the best fits to experimental results were obtained for Aydan's criterion and Hoek and Brown's criterion. Nevertheless, the criterion of Hoek–Brown deviates from experimental results when the confining pressure is greater than 1 MPa.

A series of uniaxial and triaxial compression and Brazilian tests were carried out on samples from a limestone formation in which the Gökgöl karstic cave in

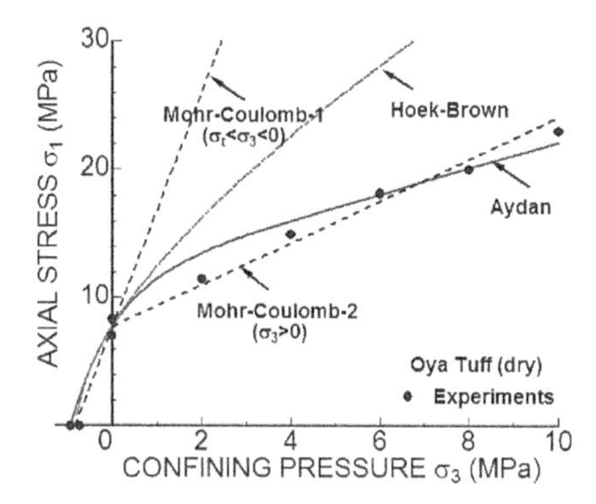

Figure 5.16 Comparisons of yield criteria for Oya tuff (dry).

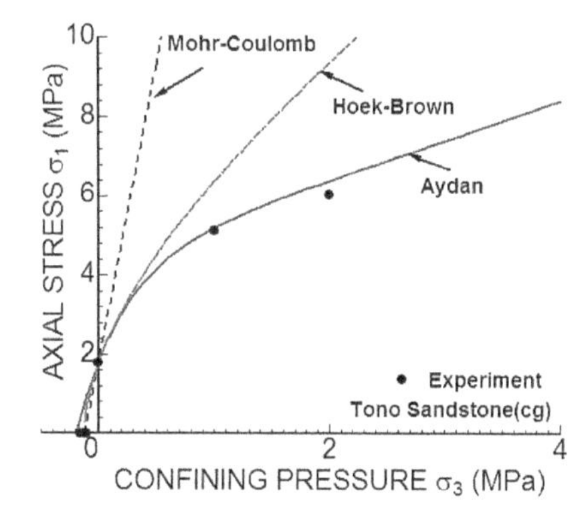

Figure 5.17 Comparisons of yield criteria for experimental results on a Tono sandstone.

Zonguldak province (Turkey) is located (Aydan et al. 2012). Triaxial compression experiments were carried out under confining pressures of up to 40 MPa. Figure 5.18 shows the experimental results together with several fitted failure criteria of bilinear Mohr–Coulomb, Hoek–Brown and Aydan. The best fits to experimental results were obtained from the applications of the bilinear Mohr–Coulomb criterion and Aydan's criterion. If the Hoek–Brown criterion is required to represent both the tensile strength and uniaxial compression strength, it is well fitted to the lower-bound values. However, the estimated curve by Aydan's criterion can better represent the bilinear Mohr–Coulomb criterion as well as all experimental results.

b. Igneous rocks

Inada granite is a well-known igneous hard rock and its uniaxial compressive strength is generally greater than 100 MPa. Figure 5.19 shows the fitted failure criteria of Mohr–Coulomb, Hoek and Brown and Aydan applied to experimental results carried out under confining pressures of up to 100 MPa. For confining pressures of up to 50 MPa, the best fits to experimental results are those of Aydan and Mohr–Coulomb. If the parameters of the Hoek–Brown criterion are

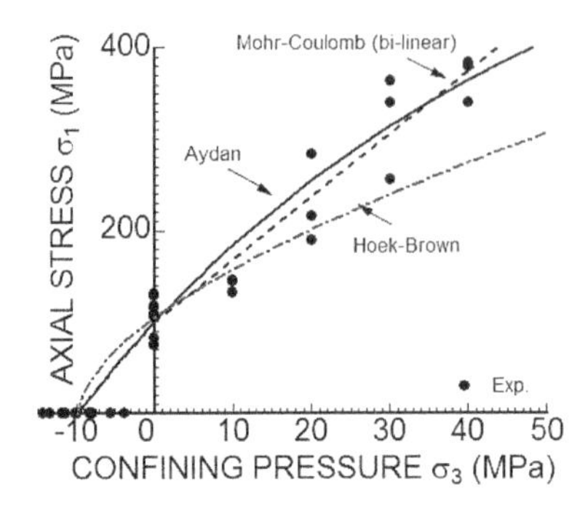

Figure 5.18 Comparisons of yield criteria for experimental results on a limestone.

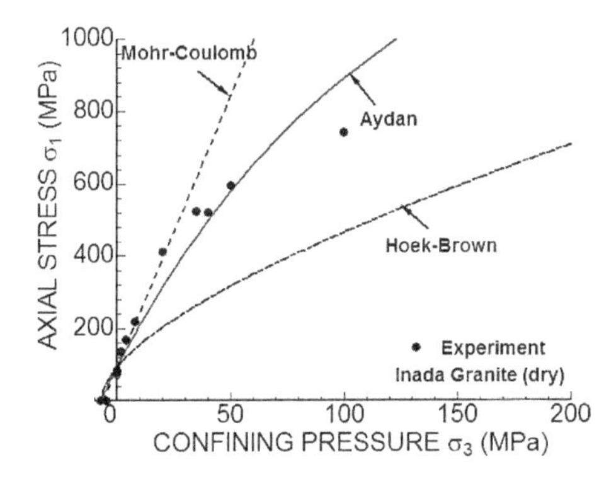

Figure 5.19 Comparisons of yield criteria for experimental results on an Inada granite.

determined to represent tensile strength and uniaxial compressive strength, the estimated triaxial strengths for high confining pressures are entirely different from experimental results.

Hirth and Tullis (1994) reported the results of triaxial experiments on quartz aggregates under very high confining pressures, which is almost six times its uniaxial compressive strength. The best fit to the experimental results is obtained for Aydan's criterion as seen in Figure 5.20. Up to a confining pressure of 1000 MPa, the estimation by the Mohr–Coulomb criterion is better than that by the Hoek–Brown criterion. Again, a very high discrepancy is observed among the estimations by the Hoek–Brown criterion and experimental results.

c. Metamorphic rocks

The yield (failure) criterion of metamorphic rocks must consider the anisotropy caused by the orientation of schistosity in relation to the applied principal stresses during experiments. Jaeger and Cook (1979) developed a procedure involving yield conditions for shearing along the schistosity plane and shearing through rock based on the Mohr–Coulomb yield (failure) criterion. There are some attempts to express the dependency of yield function to the schistosity orientation by several researchers (McLamore and Gray 1967; Donath 1964; Nasseri et al. 2003). When

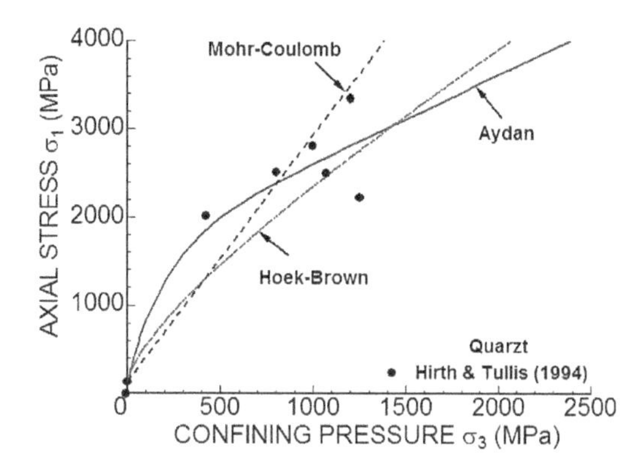

Figure 5.20 Comparisons of yield criteria for experimental results on quartz.

the yield (failure) criterion is evaluated, one should take into account the effects of characteristics of rocks as well as the schistosity orientation with respect to applied stresses. Nevertheless, the overall functional form of the yield criterion should be similar for a given orientation except for the specific values of their parameters.

Nasseri et al. (2003) reported extensive experimental research on the failure characteristics of Himalayan schists subjected to confining pressures of up to 100 MPa. They also reported the tensile strength characteristics of all rock types. Figure 5.21 shows the failure criteria of bilinear Mohr–Coulomb, Hoek–Brown and Aydan applied to experimental results of Himalayan chlorite schist for the orientation angle of 90°. In fitting relations of the criteria of Hoek–Brown and Aydan, the failure criteria are required to represent uniaxial compressive strength and tensile strength. We note that experimental results are well represented by the bilinear Mohr–Coulomb criterion. Once again, the Hoek–Brown criterion deviates from triaxial compressive experimental results if it is required to represent the tensile strength and compressive strength. Aydan's criterion provides the best continuous fit to experimental results as noted from Figure 5.21.

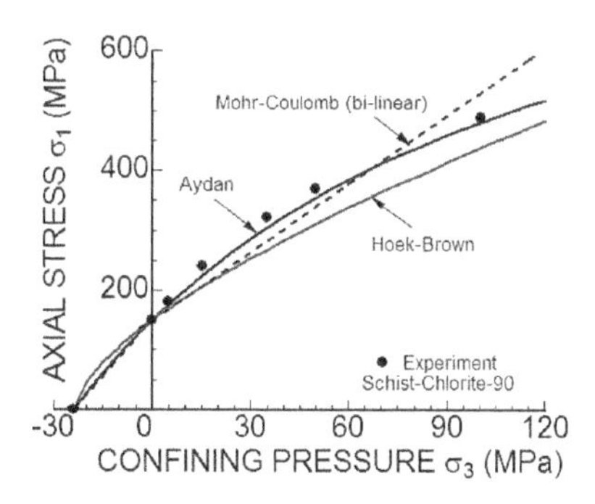

Figure 5.21 Comparisons of yield criteria for experimental results on chlorite schist.

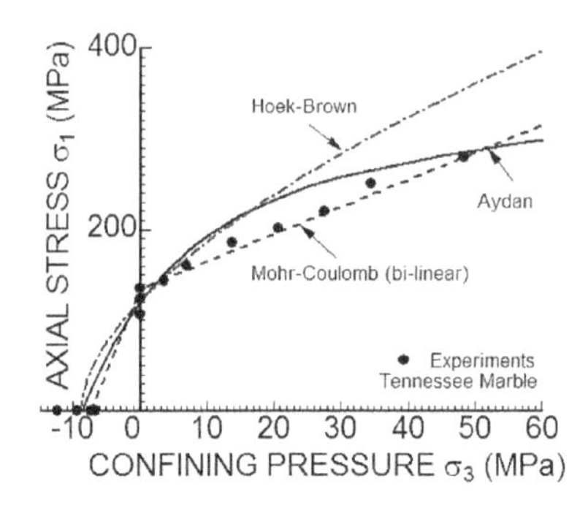

Figure 5.22 Comparisons of yield criteria for experimental results on a Tennessee marble.

Waversik and Fairhurst (1970) presented results of triaxial compressive tests on a Tennessee marble while Haimson and Fairhurst (1970) reported tensile and uniaxial compressive strength of the same rock. The failure criteria of bilinear Mohr–Coulomb, Hoek–Brown and Aydan are fitted to a combined set of experiments on a Tennessee marble with the requirement of representing its tensile and uniaxial compressive strength as shown in Figure 15.22. The overall tendency of fitted criteria to experimental results remains the same. Particularly, the requirement of representing both tensile and compressive strength by the criterion of Hoek–Brown results in different values of parameter m as also noted by Betournay et al. (1991). The value of m is roughly equal to the ratio of uniaxial compressive strength to tensile strength, which is known as the brittleness index.

d. Application to thermal triaxial compression experiments
In geomechanics, there is almost no yield (failure) criterion incorporating the effect of temperature on yield (failure) properties of rocks although there are some experimental studies (Hirth and Tullis 1994). Aydan's criterion is the only criterion known to incorporate the temperature and it was used to study the stress state of the earth (Aydan 1995b). Figure 5.23 shows the experimental results for three different values of ambient temperature reported by Hirth and Tullis (1994) while Figure 5.24 shows the reduction of strength with temperature for a given confining pressure of 1.17–1.2 GPa.
Aydan's yield (failure) criterion is applied to experimental results shown in Figures 5.23 and 5.24, and the results are shown in Figure 5.25.

5.4 Equivalent models for discontinua

Rock masses in nature contain numerous discontinuities in the form of cracks, joints, faults, bedding planes, and so on. Therefore, various continuum equivalent models of discontinuum have been proposed and used to assess the stability of rock tunnels since the beginning of 1970. Discontinuum is distinguished from continuum by the existence of contacts or interfaces between the discrete bodies that comprise the system.

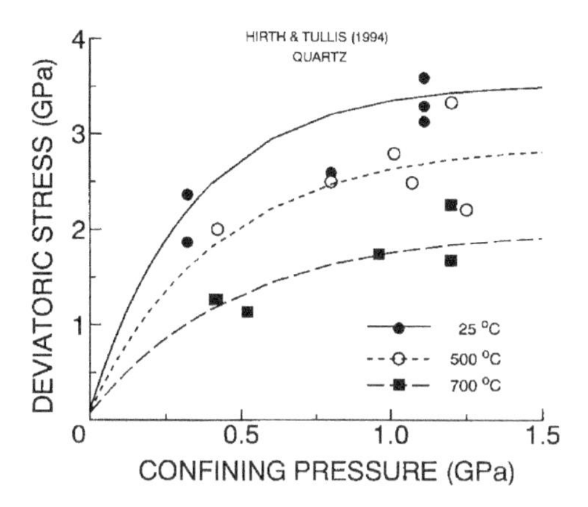

Figure 5.23 Experimental results of Hirth and Tullis (1994) on quartz for three different ambient temperatures.

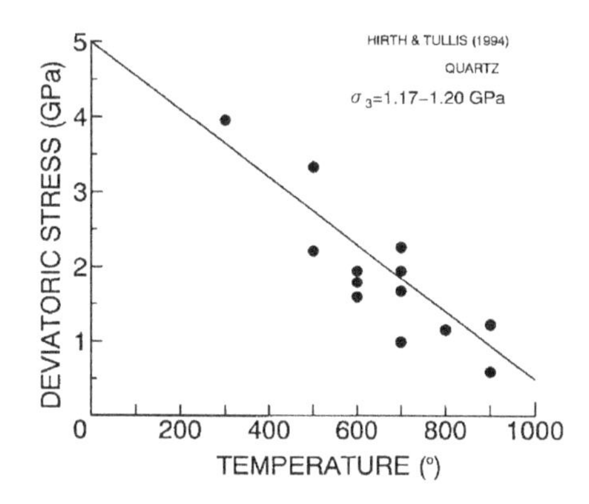

Figure 5.24 The reduction of deviatoric strength of quartz as a function of temperature for a confining pressure of 1.17–1.2 GPa.

Relative sliding or separation movements in such localized zones present an extremely difficult problem in mechanical modeling and numerical analysis. The formulation for representing contacts is very important when a system of interacting blocks is considered, and it has been receiving considerable interest among researchers. The main characteristics of the models are described in this section.

5.4.1 Equivalent elastic compliance model (Singh's model)

Singh's model (1973) is based on the theory proposed by Hill (1963) for composite materials, and the elastic constitutive law of the rock mass is obtained by making the following assumptions (Figure 5.26):

1. Discontinuities are distributed in sets in the rock mass.

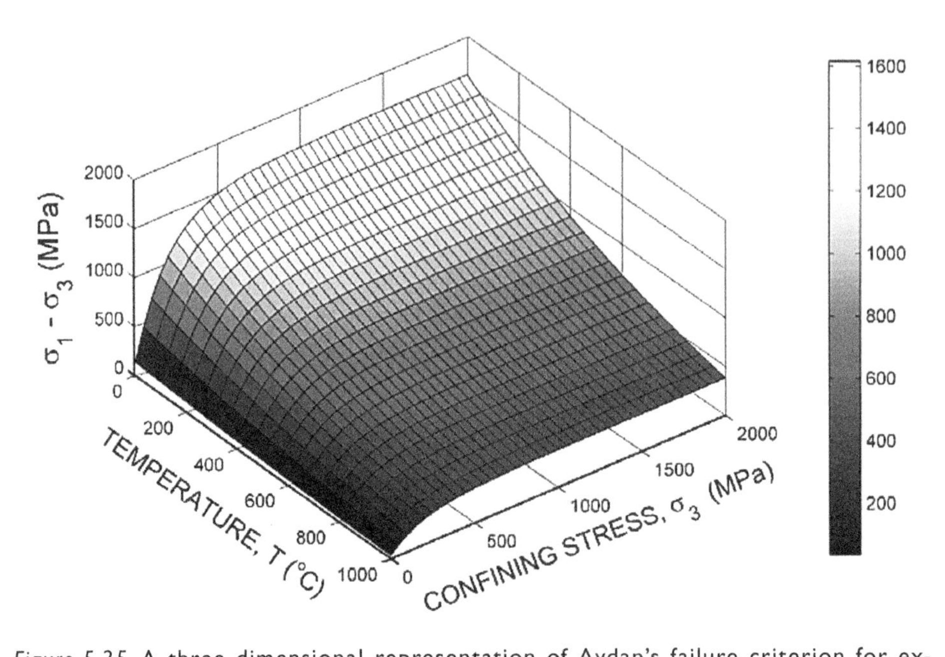

Figure 5.25 A three-dimensional representation of Aydan's failure criterion for experimental results of Hirth and Tullis (1994).

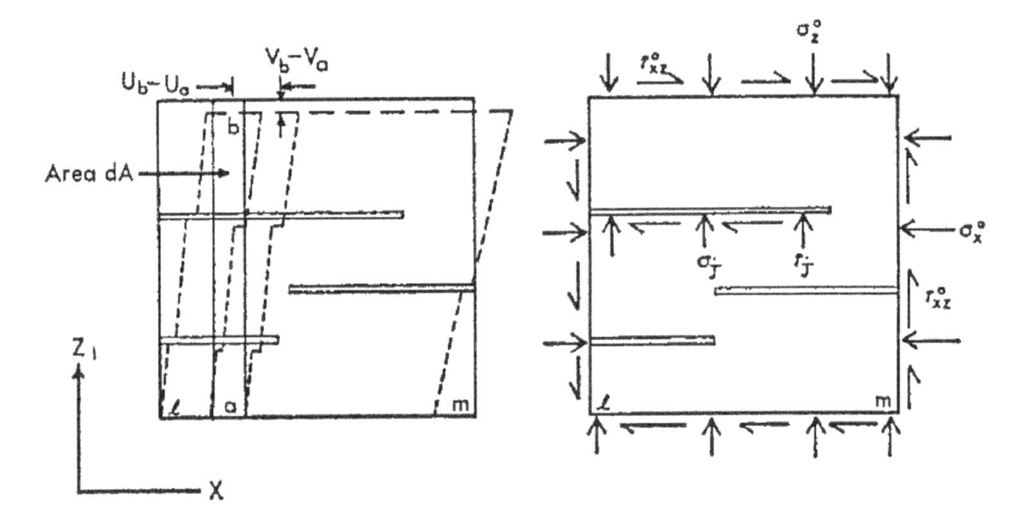

Figure 5.26 Singh's model.

2. The geometry of discontinuities (area and orientation) are known.
3. The constitutive law of discontinuities is expressed in terms of relative normal and shear displacements and applied shear and normal stresses, and the shear and normal stiffnesses are used to express the behavior of discontinuities.

4. The stress tensor acting on discontinuities is related to that acting on the representative volume through a tensor called a stress-concentration tensor.
5. The strain tensor of the representative element is a linear sum of the strain tensor of the intact rock and the additional strain tensor due to discontinuities.
6. The volume of discontinuities is assumed to be negligible as compared with that of the rock so that the stress tensor acting on the representative volume of the rock mass is the same as that on the intact rock.

The constitutive law derived in a local coordinate system is then transformed into that in the global coordinate system. The formulations given by Goodman (1976) are the simplified form of Singh's model. Application of these models to tunnels in jointed media with a cross-continuous pattern and intermittent pattern is given and compared with a discrete model. This model is the first equivalent model to be applied to rock engineering problems.

5.4.2 Crack tensor model (CTM)

This model proposed by Oda (1982) for rock masses follows the same steps of Singh's model to obtain the elastic constitutive law of the rock mass. The main differences are as follows:

1. Constitutive law is directly derived in a global coordinate system.
2. The geometry of discontinuities (area and orientation) are represented by a series of even order tensors (up to fourth-order tensors).
3. The additional strain tensor due to discontinuities is determined from a procedure utilizing an analogy to the theoretical solutions for penny shape or elliptical inclusions embedded in the elastic medium used in the linear elastic fracture mechanics (Figure 5.27).

The application of this model to tunnels, particularly branching tunnels, has been described in a recent paper by Oda et al. (1984, 1993).

5.4.3 Damage model

The damage model is based upon the theory originally proposed by Kachanov (1958) for creeping metals. It is elaborated by Murakami (1985) by introducing a second-order tensor called the damage tensor and it is applied to a rock mass by Kyoya (1989) and Kawamoto et al. (1988). Assumptions 1 and 2 of Singh are also the same in this model. However, this model differs from other models and it is based on the following additional assumptions:

1. The discontinuities are assumed to be not transmitting any stress across, which implies that the discontinuities have no stiffness at all.
2. The stresses are assumed to be acting only on the intact parts, which implies the parallel connection principle for the stress field. The average stress (Cauchy stress) is related to the stress (net stress or intensified stress) on the intact part through

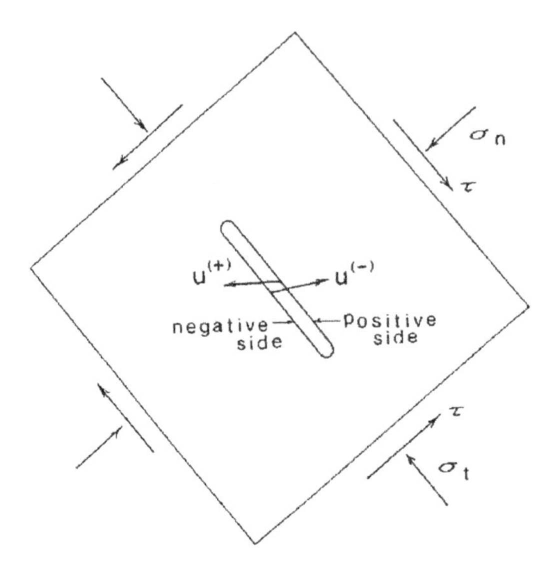

Figure 5.27 Crack tensor model (Oda et al. 1993).

the second-order damage tensor which represents a tensorial area reduction in the mass (Figure 5.28).

3. The strain tensor of the representative elementary volume is the same as that of the intact rock.
4. The constitutive law is introduced between the net stress tensor and the strain tensor.

It should be noted that this model could not directly be used for throughgoing discontinuity sets because of Assumption 1. Nevertheless, Kyoya (1989) introduced some coefficients for normal and shear responses to differentiate the behavior under tension and compression. The applications of this model to tunnels and underground caverns are described by Kyoya (1989) and Kawamoto et al. (1988).

5.4.4 Microstructure models

Aydan et al. (1992, 1996) proposed two models for discontinuous rock masses based on the microstructure theory of mechanics (Jones 1975). Although the first assumption of Singh is the same as that in this approach, this model differs from others. The fundamental differences are as follows:

1. Discontinuities have a finite volume which enables one to model a wide range of discontinuities from joints to faults or fractured zones.
2. The constitutive law of discontinuities is expressed in the conventional sense of mechanics. In other words, the constitutive law is expressed in terms of stresses and strains and it is uniquely defined.

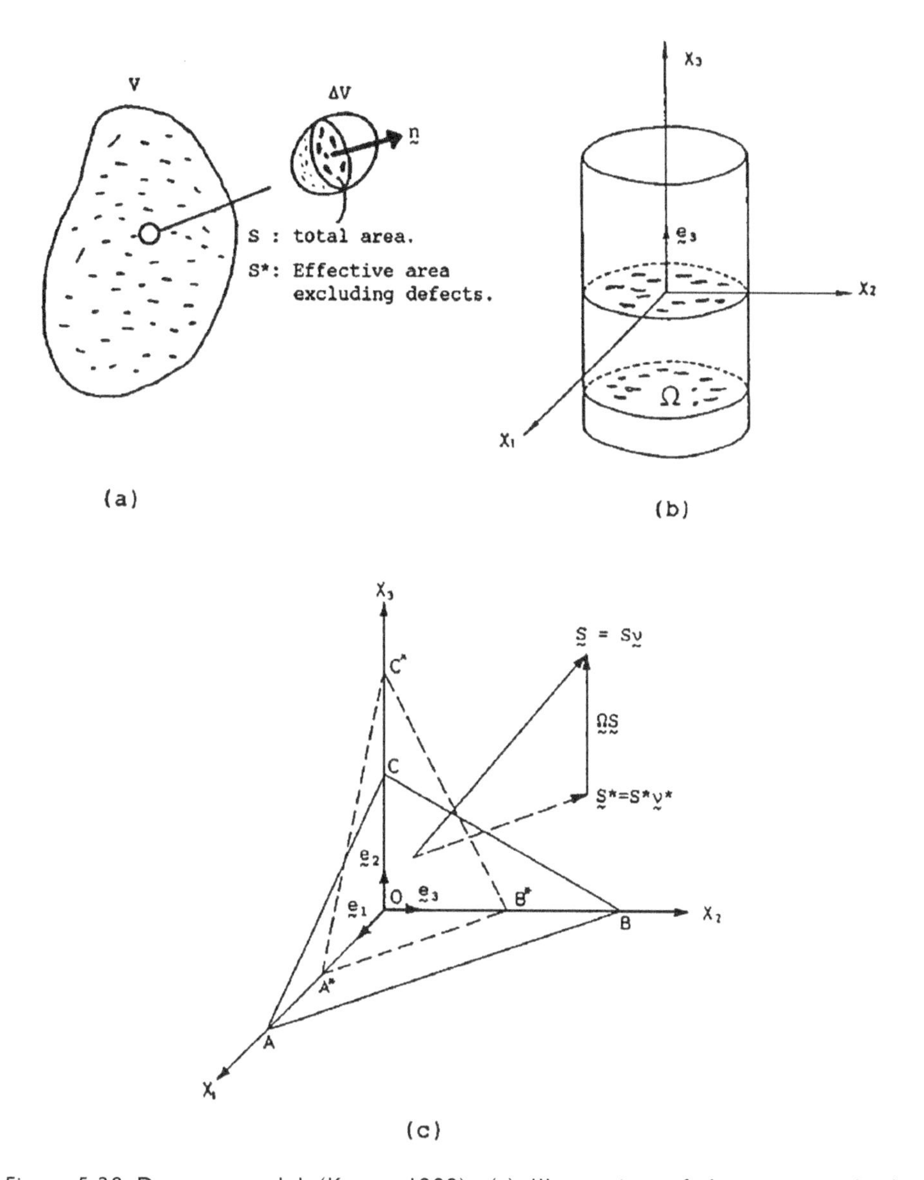

Figure 5.28 Damage model (Kyoya 1989). (a) Illustration of damage in a body, (b) illustration of damage in 1D, (c) Illustration of Cauchy stress and net-stress of damage model.

3. The constitutive law is not restricted to elasticity and it can be of any kind that can describe the mechanical response of discontinuities.
4. Stress and strain fields of each constituent are related to each other using two concepts, namely, Globally Series and Locally Parallel Model (GSLPM) and Globally Parallel and Locally Series Model (GPLSM) (Figure 5.29).

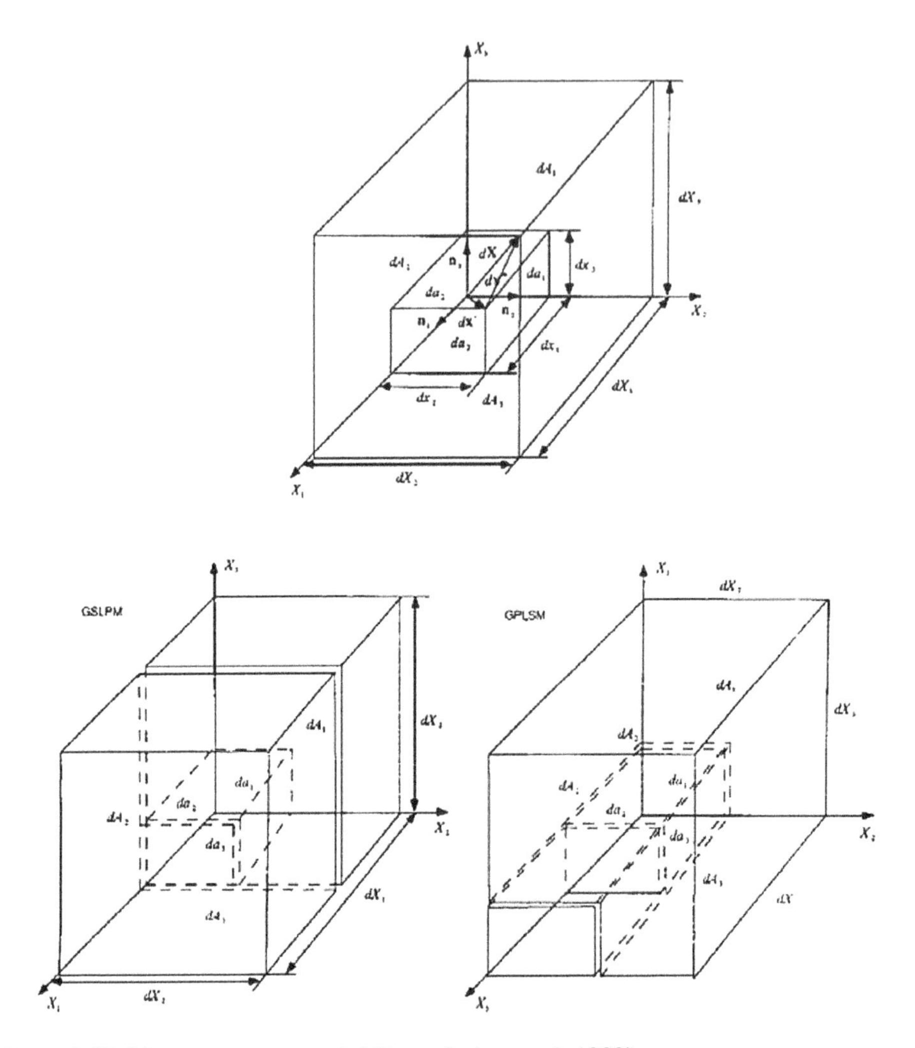

Figure 5.29 Microstructure model (from Aydan et al. 1992).

5.4.5 Homogenization technique

The homogenization technique was mainly used to obtain the equivalent characteristics of composites (Bakhvalov and Panasenko 1984) and it has been recently applied to soil (Auriault 1983) and rocks (Tokashiki et al. 1998; Tokashiki and Aydan 2008) (Figure 5.30). Assumptions 1 and 3 of the microstructure model also hold for this technique. Stress and strain fields of constituents are obtained from a perturbation of the displacement field. An influence tensor, which is a gradient of six vectorial functions called characteristic deformation functions, for a given representative elementary volume (unit cell) is used to establish relations between the homogenized elasticity tensor and those of its constituents. Except for very simple cases, the equivalent parameters are obtained using a numerical method such as FEM.

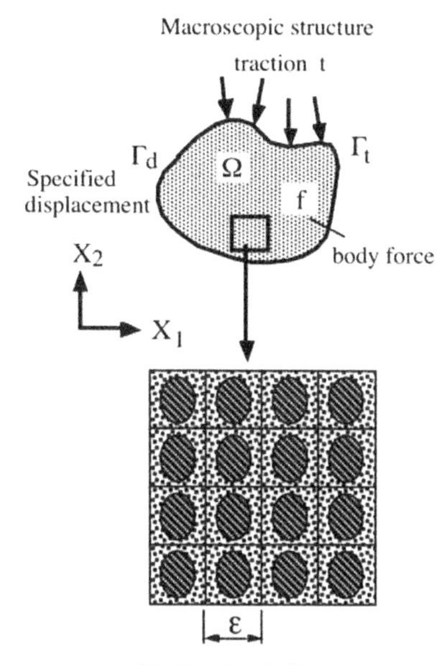

Figure 5.30 Homogenization model.

References

Auriault, J.L. 1983 *Homogenisation: Application to Porous Saturated Media*, Gdansk, Summer School on two-Phase Medium Mechanics.

Aydan, Ö. 1995a. Mechanical and numerical modelling of lateral spreading of liquefied soil. *Proceedings of International Symposium on Earthquake Geotechnical Engineering*, Tokyo, Balkema, 881–886.

Aydan, Ö. 1995b. The stress state of the earth and the earth's crust due to the gravitational pull. *The 35th US Rock Mechanics Symposium*, Lake Tahoe, 237–243.

Aydan, Ö. 2017. *Time Dependency in Rock Mechanics and Rock Engineering*, London, CRC Press, Taylor & Francis Group, 241 p.

Aydan, Ö., M. Daido, H. Tano, S. Nakama and H. Matsui 2006. The failure mechanism of around horizontal boreholes excavated in sedimentary rock. *50th US Rock Mechanics Symposium*, Paper No. 06–130.

Aydan, Ö., N. Tokashiki and M. Geniş 2012. Some considerations on yield (failure) criteria in rock mechanics ARMA 12-640. *Proceedings of 46th US Rock Mechanics/Geomechanics Symposium*, Chicago, 10 p, (on CD).

Aydan, Ö. and P. Nawrocki 1998. Rate-dependent deformability and strength characteristics of rocks. *International Symposium on the Geotechnics of Hard Soils-Soft Rocks*, Napoli, 1, 403–411.

Aydan, Ö, N. Tokashiki, T. Seiki and F. Ito 1992. Deformability and strength of discontinuous rock masses. *International Conference on Fractured and Jointed Rock Masses*, Lake Tahoe, 256–263.

Aydan, Ö, N. Tokashiki and T. Seiki 1996. Micro-structure models for porous rocks to jointed rock masses. *Proceedings of 3rd Asia-Pacific Conference on Computational Mechanics*, Seoul.

Bakhvalov, N. and G. Panasenko 1984. *Homogenization: Averaging Processes in Periodic media*, Dordrecht, Kluwer.

Betournay, M.C., B. Gorski, D. Labrie, R. Jackson and M. Gyenge 1991. New considerations in the determining of Hoek and Brown material constants. *Proceedings on 7th International Congress on Rock Mechanics*, Aachen, 1, 195–200.

Cristescu, N.D. and U. Hunsche 1998. *Time Effects in Rock Mechanics*, Chester, NY, John Wiley & Sons, ISBN 978-0-471-95517-5.

Donath, F.A. 1964. Strength variation and deformational behavior in anisotropic rock, in W.R. Judd, ed., *State of Stress in the Earth's Crust*, NY, Elsevier, 281–297.

Drucker, D.C. and W. Prager 1952. Soil mechanics and plastic analysis for limit design. *Quarterly of Applied Mathematics*, 10(2), 157–165.

Eringen, A.C. 1980. *Mechanics of Continua*, Huntington, NY, Robert E. Krieger Publishing Co., 606 p.

Goodman, R.E. 1976. *Methods of Geological Engineering in Discontinuous Rocks*, St. Paul, MN, West, Pub. Co., , 472 p.

Haimson, B.C. and C. Fairhurst 1970. Some bit penetration characteristics in pink Tennessee marble. *Proceedings on 12th US Rock Mechanics Symposium*, 547–559.

Hill, R. 1963. Elastic properties of reinforced solids: some theoretical principles. *Journal of the Mechanics and Physics of Solids*, 11, 357–372.

Hirth, G. and J. Tullis 1994. The brittle-plastic transition in experimentally deformed quartz aggregates. *Journal of Geophysical Research*, 99, 11731–11747.

Hoek E. and E.T. Brown 1980. Empirical strength criterion for rock masses. *Journal of the Geotechnical Engineering Division*, ASCE, 106(GT9), 1013–1035.

Hoek, E. and E.T. Brown 1997. Practical estimates of rock mass strength. *International Journal of Rock Mechanics and Mining Sciences*, 34(8), 1165–1186.

Jaeger, J.C. and N.G.W. Cook 1979. *Fundamentals of Rock Mechanics*. 3rd ed., London, Chapman & Hall, pp. 79 and 311.

Jones, R.M. 1975. *Mechanics of Composite Materials*, New York, Hemisphere Pub. Co.

Kachanov, L.M. 1958. *The Theory of Creep* (English transl. by A. J. Kennedy), Boston, MA, National Lending Library.

Kawamoto, T., Y. Ichikawa and T. Kyoya 1988. Deformation and fracturing behaviour of discontinuous rock mass and damage mechanics theory. *International Journal for Numerical and Analytical Methods in Geomechanics*, 12(1), 1–30.

Kyoya, T. 1989. A fundamental study on the application of damage mechanics to the evaluation of mechanical characteristics of discontinuous rock masses, Ph.D. Thesis, Nagoya University.

Mase, G. 1970. *Theory and Problems of Continuum Mechanics*, Schaum's Outline Series, New York, McGraw Hill Co., 221 p.

McLamore, R. and K.E. Gray 1967. The mechanical behaviour of anisotropic sedimentary rocks. *Transactions of the American Society of Mechanical Engineers*, 62–76.

Murakami, S. 1985. Anisotropic damage theory and its application to creep crack growth analysis. *Proceedings of International Conference Constitutive Laws for Engineering Materials: Theory and Applications*, Amsterdam, Elsevier, 535–551.

Nasseri, B.M.H., K.S. Rao and T. Ramamurthy 2003. Anisotropic strength and deformational behaviour of Himalayan Schists. *International Journal of Rock Mechanics and Mining Sciences*, 40(1), 3–23.

Oda, M. 1982. Fabric tensor for discontinuous geological materials. *Soils and Foundations*, 22(4), 96–108.

Oda, M., K. Suzuki and T. Maeshibu 1984. Elastic compliance for rock-like materials with random cracks. *Soils and Foundations*, 24(3), 27–34.

Oda, M., T. Yamabe, Y. Ishizuka, H. Kumasaka, H. Tada and K. Kimura 1993. Elastic stress and strain in jointed rock masses by means of crack tensor analysis. *Rock Mechanics and Rock Engineering*, 26(2), 89–112.

Owen, D.R.J. and E. Hinton 1980. *Finite Element in Plasticity: Theory and Practice*, Swansea, Pineridge Press Ltd.

Perzyna, P. 1966. Fundamental problems in viscoplasticity. *Advances in Applied Mechanics*, 9(2), 244–368.

Seiki, T. and Ö. Aydan 2003. Deterioration of Oya tuff and its mechanical property change as building stone. *Proceedings of International Symposium on Industrial Minerals and Building Stones*, Istanbul, Turkey, 329–336.

Singh, B. 1973. Continuum characterization of jointed rock masses. Part I*The constitutive equations. *International Journal of Rock Mechanics and Mining Sciences*, 10, 311–335.

Tokashiki, N. and Ö. Aydan 2008. Evaluation of deformability properties of rocks with overlapping inclusions by different averaging methods. *The 12th International Conference of International Association for Computer Methods and Advances in Geomechanics (IACMAG)*, Goa, India, 797–804.

Tokashiki, N., Ö. Aydan, T. Kyoya and K. Sugawara 1998. A comparative study on assessment of mechanical properties of porous and heterogeneous rocks by various averaging methods. *Proceedings of the Fourth European Conference on Numerical Methods in Geotechnical Engineering*, NUMGE98, October 13–16, Udine, Italy, Springer, 401–410.

Waversik, W.R. and C. Fairhurst 1970. A study of brittle rock fracture in laboratory compression experiments. *International Journal of Rock Mechanics and Mining Sciences*, 7, 561–575.

Chapter 6

Laboratory tests

Mechanical, seepage, heat and diffusion properties related to constitutive laws described in Chapter 5 require tests on materials (such as rocks, concrete, metals or rock mass) appropriate to physical and environmental conditions. In this chapter, the fundamental principles of available testing techniques used in the field of rock mechanics and rock engineering are explained.

6.1 Laboratory tests on mechanical properties

Laboratory tests on mechanical properties of rocks may be determined from uniaxial tensile and compression tests, triaxial compression experiments, three/four bending experiments, direct shear and Brazilian experiments. Some of these experiments are illustrated in Figure 6.1 (Aydan 2016a, 2017; Aydan et al. 2016a, b, c; Jaeger and Cook 1979; Feng 2016; Brown 1981; Ulusay and Hudson 2007; Ulusay 2015; Hudson et al. 1972; Waversik and Fairhurst 1970; Zimmerman et al. 1986). The fundamentals of these testing techniques are described in the following sub-sections.

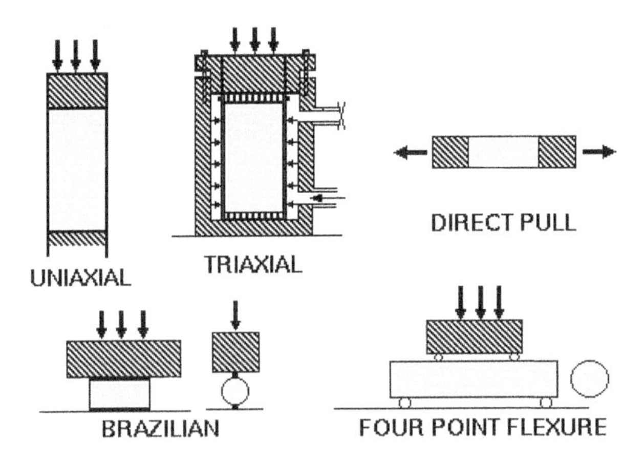

Figure 6.1 Illustration of some of the testing techniques for determining mechanical properties.

6.1.1 Uniaxial compression tests

Specimens from drill cores are prepared by cutting them to the specified length and are thereafter ground and measured. There are high requirements on the flatness of the end surfaces to obtain an even load distribution. The recommended ratio of height/diameter of the specimens is between 2 and 3. Strains and stress in uniaxial compression tests are defined as follows:

The axial strain is calculated from the equation

$$\varepsilon_a = \frac{\Delta l}{l_O} \tag{6.1}$$

where l_O is the original measured axial length and Δl is the change in measured axial length (defined to be positive for a decrease in length).

Diametrical strain can be determined either by measuring the changes in the diameter of the specimen or by measuring the circumferential strain. In the case of measuring changes in the diameter, the diametric strain is calculated from the equation below

$$\varepsilon_d = \frac{\Delta d}{d_O} \tag{6.2}$$

where d_0 is the original undeformed diameter of the specimen and Δd is the change in diameter (defined to be negative for an increase in diameter).

In the case of measuring the circumferential strain, ε_c, the circumference is $C = \pi d$; thus, the change in circumference is $\Delta C = \pi \Delta d$. Consequently, the circumferential strain, ε_c, is related to the diametric strain, ε_d, by

$$\varepsilon_c = \frac{\Delta C}{C} = \frac{\Delta d}{d_O} \tag{6.3}$$

so that $\varepsilon_c = \varepsilon_d$. where C and d_O are the original circumference and diameter of the specimen, respectively.

The compressive axial stress in the test specimen, σ_a, is calculated by dividing the compressive load P on the specimen by the initial cross-sectional area, A_O.

$$\sigma_a = \frac{P}{A_O} \tag{6.4}$$

where compressive stresses and strains are considered to be positive in this test procedure. For a given stress level, the volumetric strain, ε_v, is calculated from the equation

$$\varepsilon_v = \varepsilon_a + 2\varepsilon_d \tag{6.5}$$

The specimens are loaded axially up to failure or any other prescribed level whereby the specimen is deformed and the axial and the radial deformation can be measured using some equipment as shown in Figure 6.2.

Figure 6.2 A view of the experimental set-up and instrumentation in an uniaxial compression experiment at Nagoya University and the University of the Ryukyus.

There is a tremendous amount of studies on the stress and strain distributions induced in uniaxial compression experiments with the consideration of boundary conditions imposed in the experiments. Generally, the stiffness and Poisson's ratio of the platens and specimen are different from each other. Both theoretical and numerical analyses such as FEM indicate that the stress and strain are not uniform within the specimen. Particularly, the distributions are highly non-uniform near the end of the specimen as shown in Figure 6.3. The non-uniformity of the strain–stress distributions strongly depends upon the differences between stiffness and Poisson's ratio of platens and specimen and their geometry. Stresses and strain become uniform when the stiffness and Poisson's ratio of platens and specimen are the same, which is not the common case in many experimental studies. There is an entirely wrong interpretation of non-uniformity stress distributions caused by the differences of the stiffness and Poisson's ratio as the frictional effect. A frictional effect may come into action after a relative slip occurs between the platens and the ends of the specimen. In practice, Vaseline oil, Teflon sheets or brush type platens are used to deal with this issue.

6.1.2 Direct and indirect tensile strength tests (Brazilian tests)

Similar to the uniaxial compression tests, direct tensile tests are used to determine tensile strength and some deformability properties. The definitions of stress and strains are fundamentally the same except for the sign of stresses and strains. The specimens

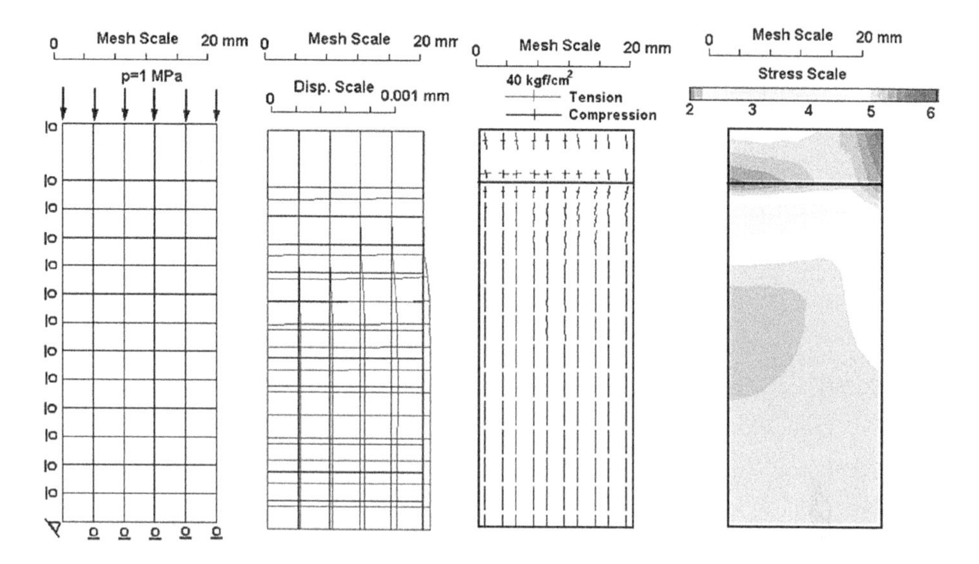

Figure 6.3 Finite element method simulation of the axisymmetric rock sample (1/4 of the specimen is used because of symmetry; from Aydan et al. 2016b).

are bonded to the loading platens or dog-shaped specimens are used to obtain the strain–stress relations.

As the preparation and procedures are quite cumbersome, indirect tensile stress tests are generally preferred. The common procedure is to load a solid or hollow cylindrical specimen under compression, which results in tensile stresses within the samples. A cylindrical specimen is loaded diametrically across the circular cross-section. This testing technique is known as the Brazilian tensile strength test. The loading causes tensile stresses perpendicular to the loading direction, which result in a tensile failure. Tensile stress induced in a solid cylinder of rock is theoretically given by the following equation:

$$\sigma_t = \frac{2F}{\pi D t} \tag{6.6}$$

where F, D and t are applied load, diameter and thickness of a rock sample, respectively. The nominal strain of the Brazilian tensile test sample may be given as (see Hondros 1959; Jaeger and Cook 1979 for details)

$$\varepsilon_t = 2\left[1 - \frac{\pi}{4}(1-\upsilon)\right]\frac{\sigma_t}{E} \text{ with } \varepsilon_t = \frac{\delta}{D} \tag{6.7}$$

For most rocks, the formula given above may be simplified to the following form:

$$\varepsilon_t = 0.82\frac{\sigma_t}{E} \tag{6.8}$$

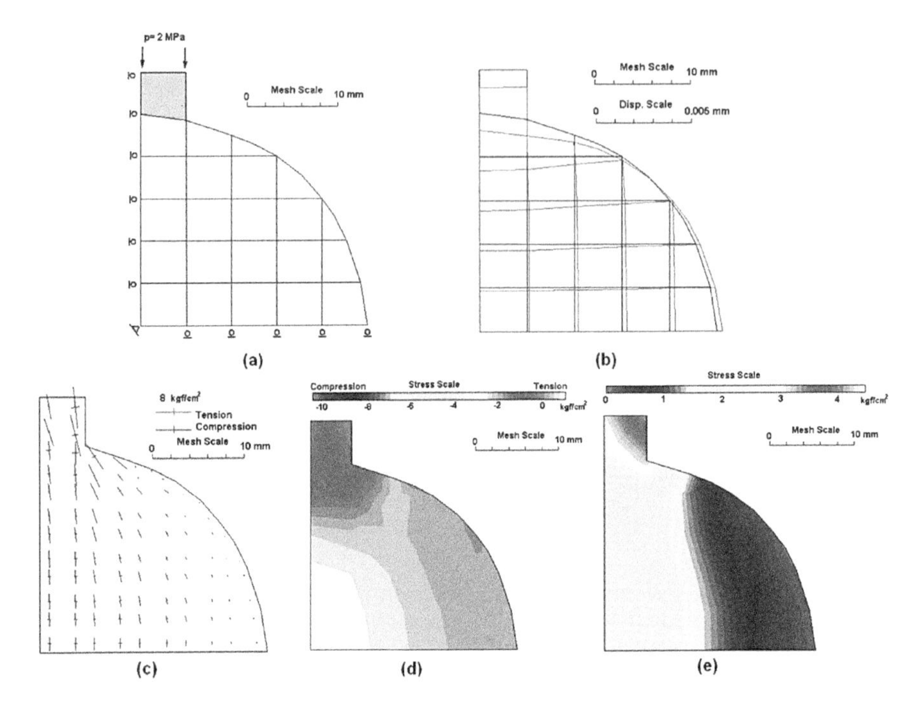

Figure 6.4 Boundary conditions and computed stress distributions. (a) FEM mesh and boundary conditions; (b) deformed configuration; (c) principal stresses; and (d) maximum principal stress contours; (e) maximum shear stress contours.

A plane stress finite element analysis was carried out for the Brazilian test. The properties of the platen were assumed to be those of the aluminum with an elastic modulus of 70 GPa. Uniform compressive pressure with an intensity of 20 kgf/cm^2 was applied on the platens and boundary conditions are shown in Figure 6.4.

The maximum tensile stress occurs near the center of the sample and its value is 1.08 kgf/cm^2. This is slightly greater than the theoretical estimation of 0.8 kgf/cm^2. This is probably due to the slight difference in the application of load boundary conditions. The computed radial displacement of the sample just below the platen was about 0.001 mm, which is almost equal to that estimated from Eq. (6.8). Therefore, it is possible to determine the elastic modulus besides the tensile strength of rocks. Furthermore, the strain response in experiments should be similar to those of the uniaxial compression experiments provided that deformability characteristics remain the same under both tension and compression.

Bending tests are also used to determine the tensile strength and deformability of rocks. The maximum tensile stress and maximum flexural strain of rock beam under three-point bending configuration with a concentrated load (F) and rectangular prismatic shape (b, t) and maximum deflection (δ) may be given in the following form:

$$\sigma_f = \frac{3FL}{2bt^2}; \varepsilon_f = \frac{6\delta t}{L^2}; \frac{\sigma_f}{\varepsilon_f} = \frac{F}{\delta} \cdot \frac{1}{4b}\left(\frac{L}{t}\right)^3 \tag{6.9}$$

Figure 6.5 Stress distribution in a beam subjected to a three-point bending condition.

Four-point bending is generally recommended due to uniform stress distributions in the area between two applied load points. Figure 6.5 illustrates stress distribution in a three-point bending test in a photo-elastic test.

6.1.3 Triaxial compression tests

Specimens from drill cores are prepared by cutting them to the specified length and are thereafter ground and measured to obtain the required flatness of the end surfaces. The recommended ratio of height/diameter of the specimens is between 2 and 3. A membrane is mounted on the surface of the specimen to seal the specimen from the surrounding pressure media. Deformation measurement equipment is mounded on the specimen and the specimen is inserted into the pressure cell whereupon the cell is closed and filled with oil. Hydrostatic pressure is applied in the first step. The specimen is then further loaded by increasing the axial load under constant or increasing cell pressure up to failure or any other predefined load level. A test setup and equipment, triaxial cell and instrumentation used in Nagoya University are shown in Figure 6.6.

6.1.4 Post-failure behavior in uniaxial and triaxial compression tests

The post-failure characteristics of rocks are of quite importance in rock engineering when failure of rock could not be prevented. For this purpose, servo-control testing devices were developed for investigating post-failure characteristics of rocks (e.g. Rummel and Fairhurst 1970; Waversik and Fairhurst 1970; Hudson et al. 1972; Kawamoto et al. 1980). The servo-control devices try to provide sufficient support to measure the intrinsic properties of rocks during the post-peak response. Such a support system is provided through the loading system, which utilizes either servo-controlled oil-based jacks or wedge-like solid support. Figure 6.7 shows the principle of the device, which is based on a wedge-like solid support concept at Nagoya University and Figure 6.8 illustrates how the wedge-like support system is activated during the deformation process.

Figure 6.9a shows general strain–stress responses for a typical rock sample while Figure 6.9b illustrates the brittle and ductile behavior of rocks. The stress drops rapidly when the rock exhibits a brittle behavior. On the other hand, the stress gradually decreases in the post-failure regime of ductile rocks. Figure 6.10 shows the true response of the triaxial behavior of a Ryukyu limestone under different confining pressures.

Figure 6.6 Triaxial compression device, cell, and an instrumented sample at Nagoya University Rock Mechanics Laboratory.

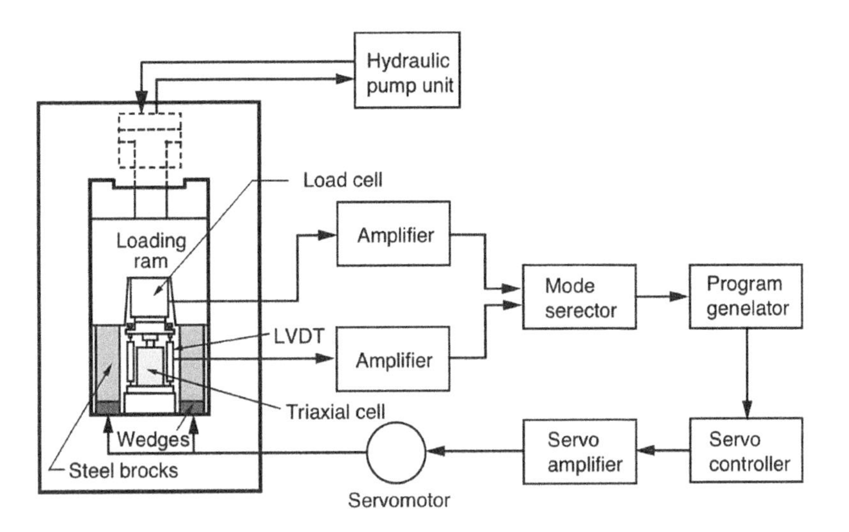

Figure 6.7 Servo-control testing device at Nagoya University.

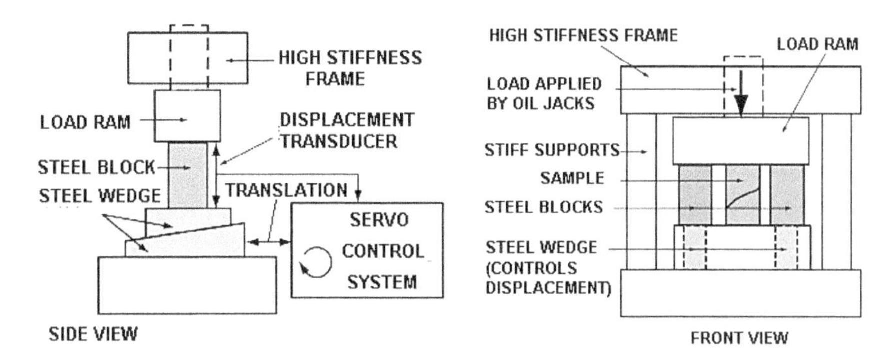

Figure 6.8 Illustration of the wedge-like solid support activation.

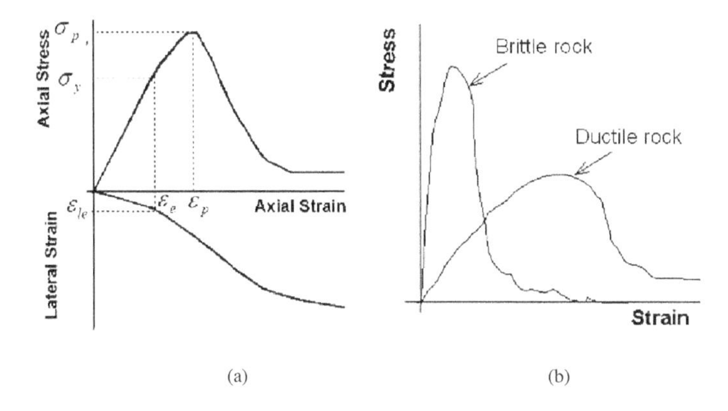

Figure 6.9 (a) Complete strain–stress relation of rocks and (b) illustration of the brittle and ductile behavior.

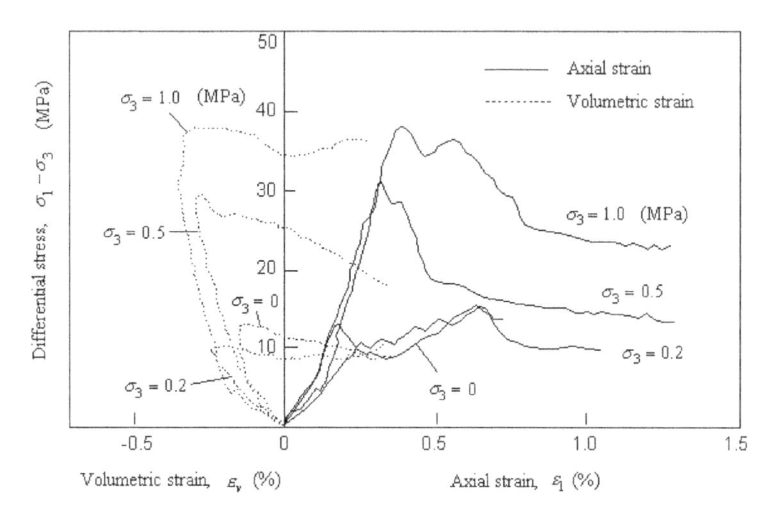

Figure 6.10 Strain–stress relation of the Ryukyu limestone under different confining pressures.

Figure 6.11 Views of a granite sample before and after testing.

Figure 6.12 Actual and X-ray CT images of a granite sample in the post-failure stage.

Figure 6.11 shows views of a granite sample tested by the Nagoya University servo-control testing machine. As noted in Figure 6.11, a macroscopic fracture zone is observed in the final post-failure stage. Figure 6.12 shows an actual image and an X-Ray CT tomographic image of another granite sample. As noted from the figure, a macroscopic fracture zone and a zone of fine cracks are seen in the X-ray CT tomographic image (Aydan et al. 2016a).

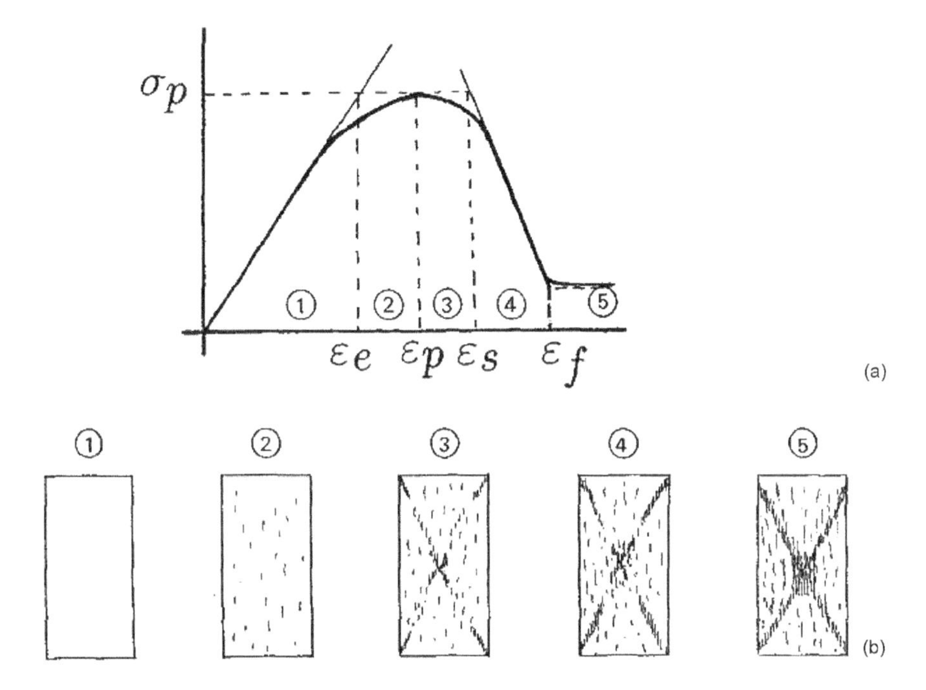

Figure 6.13 Idealized five different fracturing situations in a given rock sample during the complete strain–stress response (from Aydan et al. 1992, 1993), (a) strain-stress relation; (b) idealized fracture states.

Figure 6.13 illustrates the idealized five different fracturing situations in a given rock sample during the complete strain–stress response (Aydan et al. 1993). During stages 4 and 5, microscopic fractures coalesce into macroscopic shear bands. It should be noted that there is an argument if the strain–stress response in stages 4 and 5 should be considered as a part of constitutive law or not.

6.1.5 Direct shear tests

Direct shear test devices can also be used to obtain the shear strength properties of intact rock or rock discontinuities. There are different types of direct shear test devices. Figure 6.14 shows a shear testing machine named OA-DSTM designed and built originally in 1991 for direct shear testing under three loading conditions, namely, conventional direct shear loading, direct shear creep loading and direct shear cyclic loading at Nagoya University (Aydan et al. 1990, 1994, 2016c). Both shear and normal load on the shearing plane are designed to be 200 kN and the system is displacement controlled. Furthermore, the direct shear box was vertical to eliminate dead loads on the shearing plane. The normal load is first imposed on the sample and then shear loading is applied through a vertical jack. However, the system was needed to be upgraded for operational purposes as well as dealing with dynamic loading conditions. Aydan et al. (2016c) have recently upgraded the shear-testing machine by adding the dynamic shear loading option (Figure 6.15).

Figure 6.14 View of original shear-testing machine.

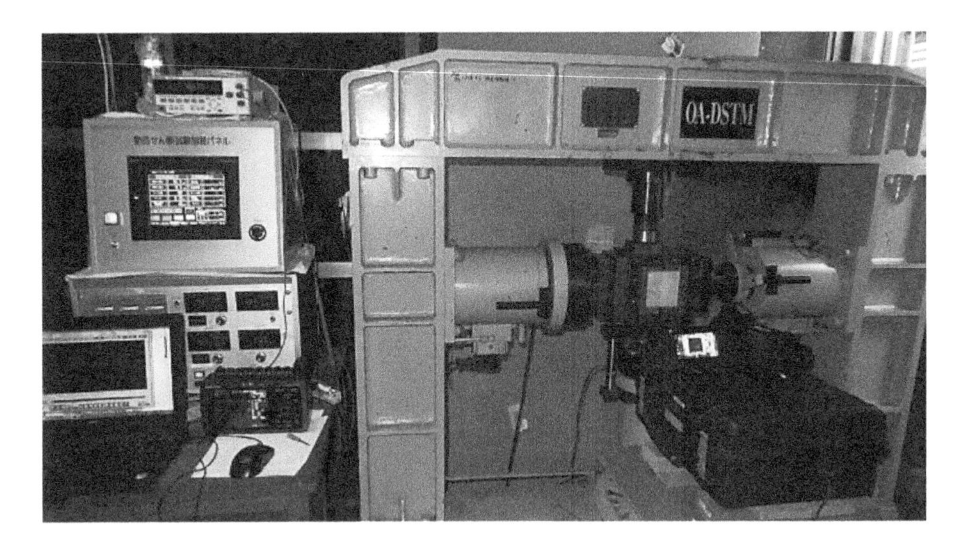

Figure 6.15 A view of the dynamic shear testing machine OA-DSTM.

The size of direct shear samples can be 100 × 100 × 100 mm or 150 × 75 × 75 mm. The original design size was 150 × 75 × 75 mm to eliminate the rotational effects on the sample. This can be achieved if the ratio of sample length over sample height is greater than 2. The shear load and displacement, normal load are directly recorded to computers using the outputs from the system. The outputs are real-time values of shear displacement in mm, and shear and normal loads in kN. Several examples of various direct shear tests on rock discontinuities and soft rocks are explained. Rock discontinuities are planar (polished and saw-cut of marbles). Soft rocks are sandy Ryukyu limestone (Awa-ishi), coral-stone and Oya tuff.

a. Conventional direct shear testing

Figure 6.16 shows the responses measured during the direct shear experiment on the shear response of coral-stone (honeycomb-like coral limestone). Once peak load exceeded, the deformation rate increases as noted from the figure.

An example of conventional direct shear test results on a polished planar surface of the marble is shown in Figure 6.17a. As seen from the figure, the yielding friction coefficient is about 0.27 (15°) and hardening is observed. When the friction coefficient exceeds 0.52 (27.5°), the relative shear displacement starts to increase. The ultimate friction angle coefficient is about 0.62 (31.8°) while the residual friction coefficient is 0.58 (30.1°).

Figure 6.17b shows the view of the interface after direct shear testing. As noted from the figure, the contacts on the planar surfaces were quite small despite that the surfaces were polished and planar. Furthermore, striations occurred on the surface parallel to the sliding direction, which would be commonly observed on the fault surface and slickenside surfaces. In addition, this may also imply that it is practically very difficult to prepare exact planar surfaces, which results in full contact of surfaces of blocks.

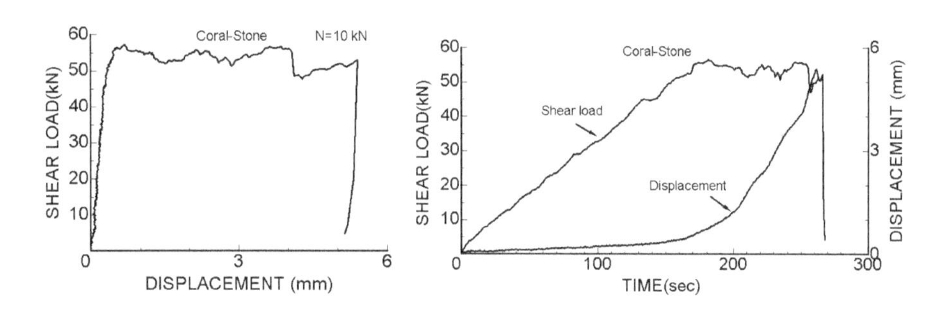

Figure 6.16 Shear displacement-shear load relation of coral-stone.

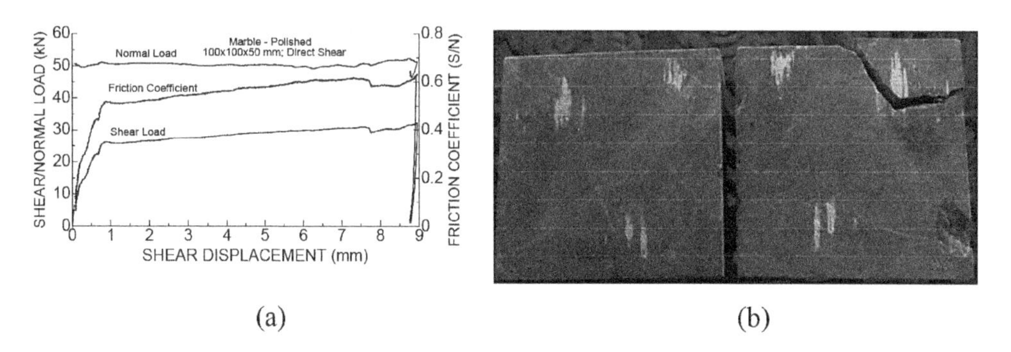

Figure 6.17 (a) Relative displacement, shear, normal and friction coefficient of the polished interface between blocks and (b) views of the sheared surfaces.

Table 6.1 Friction angles of interfaces of marble blocks

Condition	*Tilting test*	*Direct shear test (NS: 5 MPa)*			
		Initial	*Flow*	*Peak*	*Residual*
Polished	16–19	15	27	31.8	30.1
Saw-cut	28–35	23	28	31.0	29.0

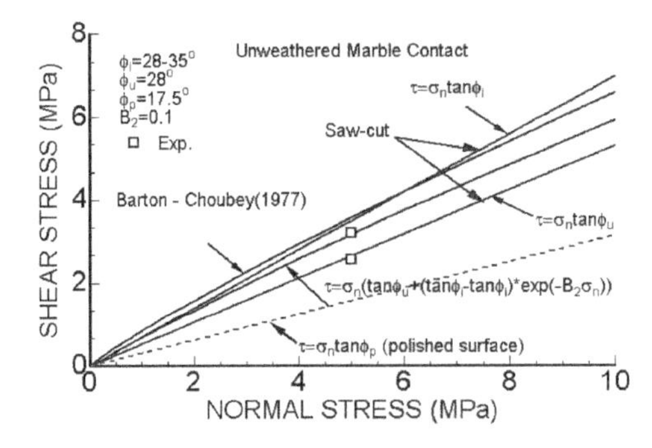

Figure 6.18 Plot of experimental results for polished marble contacts together with some failure criteria for rock discontinuities.

Tilting tests are carried out on the original blocks of polished and saw-cut surfaces. The results are given in Table 6.1 together with those from the obtained direct shear experiment under a normal stress of 5 MPa. From the comparison of the table, the friction angle of polished surfaces seems to be the non-representative frictional property of planar discontinuities. On the other hand, the friction angle saw-cut surfaces are closer to the intrinsic friction of planar discontinuities. Nevertheless, it must be noted that the traces of saws on the surfaces would cause the friction angle to be directional (Aydan et al. 1996).

Direct shear test results are plotted together with the results from tilting experiments in Figure 6.18. In the same figure, some failure criteria for rock discontinuities (i.e. Barton and Choubey 1977; Aydan 2008; Aydan et al. 1996) are also plotted. It is interesting to note that the shear strength of polished surface under high normal pressure is within the bounds obtained from the friction angle tests determined from tilting tests. The best fit to experimental results is obtained from the failure criterion of Aydan (Aydan 2008; Aydan et al. 1966). The failure criterion of Barton and Choubey (1977) is close to the upper bound strength envelope.

b. Multi-stage direct shear testing

A multi-stage (multi-step) direct shear test on a saw-cut surface of a sandy Ryukyu limestone sample, which consists of two blocks with dimensions of 150 × 75 × 37.55 mm, was carried out. The initial normal load was about 17 kN and increased to 30, 40, 50, 60 and 70 kN during the experiment. Figure 6.19 shows the shear displacement and shear load responses during the experiment. As noted from the figure, the relative slip occurs between blocks at a constant rate after each increase of normal and shear loads. This experiment is likely to yield the shear strength of the interface two blocks under different normal stress levels.

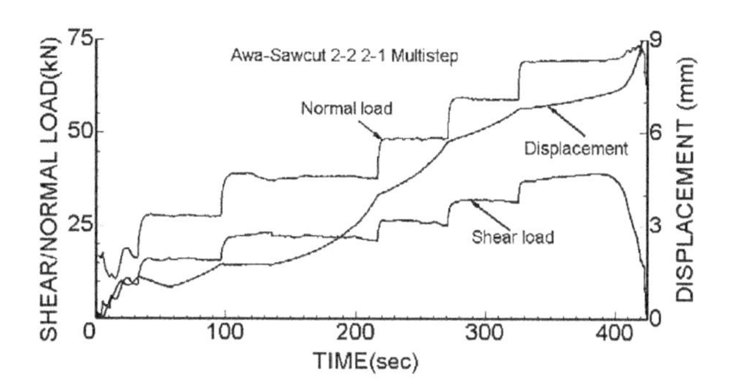

Figure 6.19 Shear stress–shear load response of the interface of sandy limestone blocks during the multi-stage (step) direct shear experiment.

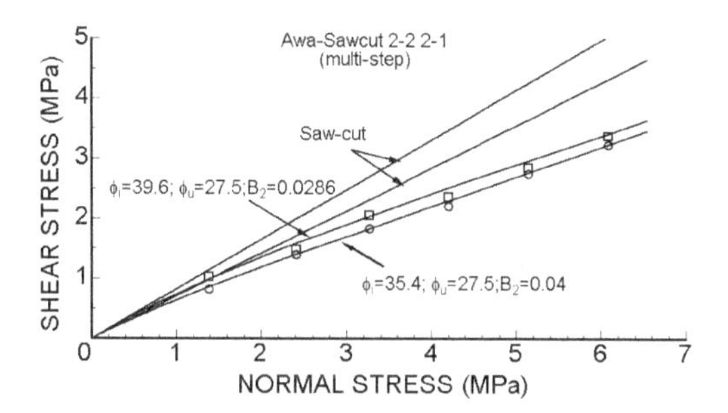

Figure 6.20 Comparison of the shear strength envelope for the interface of sandy limestone blocks with experimental results from the multi-stage(step) direct shear experiment.

Figure 6.20 shows the peak and residual levels of shear stress for each level of normal stress increment. Tilting tests were carried out on the same interface and the apparent friction angles ranged between 35.4° and 39.6°. Tilting test results and direct shear test results are plotted in Figure 6.20 together with shear strength envelopes using the shear strength failure criterion of Aydan (Aydan 2008; Aydan et al. 1966).

As noted from the figure, the friction angles obtained from tilting tests are very close to the initial part of the shear strength envelopes. However, the friction angle becomes smaller as the normal stress level increases. In other words, the friction angle obtained from tilting tests on saw-cut surfaces cannot be equivalent to the basic friction angle of planar discontinuities and interfaces of rocks. The basic friction angle of the planar interface of sandy limestone blocks is obtained as 27.5° for the range of given normal stress levels.

6.1.6 Tilting tests

The tilting test technique is one of the cheapest techniques to determine the frictional properties of rock discontinuities and interfaces under different environmental conditions (Barton and Choubey 1977; Aydan et al. 1995; Aydan et al. 2016b). This technique can be used to determine the apparent friction angle of discontinuities (rough or planar) under low-stress levels. It gives the maximum apparent friction angle of discontinuities, which would be one of the most important parameters to determine the shear strength criteria of rock discontinuities as well as various contacts. Therefore, the data for determining the parameters of the shear strength criteria for rock discontinuities should utilize both tilting test and direct shear experiment as described in the previous sub-section.

a. Theory of tilting tests
   Let us assume that a block is put upon a base block with an inclination α as illustrated in Figure 6.21a. The dynamic force equilibrium equations for the block can be easily written as follows:
   For the s-direction,

$$W\sin\alpha - S = m\frac{d^2s}{dt^2} \tag{6.10}$$

For the n-direction,

$$W\cos\alpha - N = m\frac{d^2n}{dt^2} \tag{6.11}$$

Let us further assume that the following frictional law holds at the initiation and during the motion of the block (Aydan and Ulusay 2002) as illustrated in Figure 6.21b:
   At the initiation of sliding,

$$\frac{S}{N} = \tan\phi_s \tag{6.12}$$

During motion,

$$\frac{S}{N} = \tan\phi_d \tag{6.13}$$

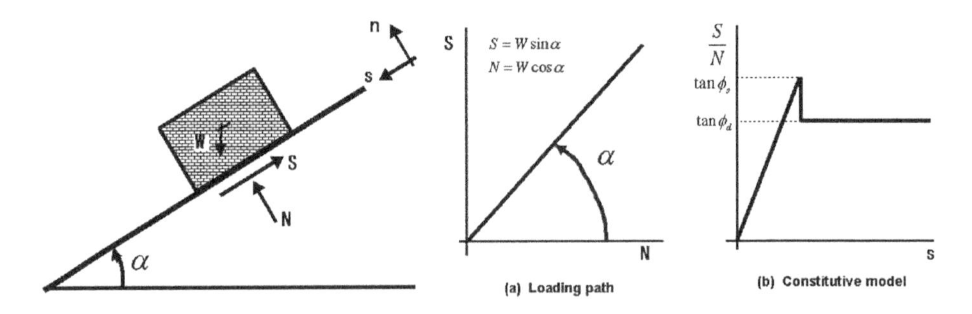

Figure 6.21 (a) Mechanical model for tilting experiments and (b) loading path in tilting experiments and constitutive relation.

At the initiation of sliding, the inertia terms are zero so that the following relation is obtained:

$$\tan\alpha = \tan\phi_s \tag{6.14}$$

The above relation implies that the angle of inclination (rotation) at the initiation of sliding should correspond to the static friction angle of the discontinuity.

If the normal inertia term is negligible during the motion and the frictional resistance is reduced to dynamic friction instantaneously, one can easily obtain the following relations for the motion of the block:

$$\frac{d^2s}{dt^2} = A \tag{6.15}$$

where $A = g(\sin\alpha - \cos\alpha\tan\phi_d)$.

The integration of differential Eq. (6.15) will yield the following:

$$s = A\frac{t^2}{2} + C_1 t + C_2 \tag{6.16}$$

As the following conditions hold at the initiation of sliding

$$s = 0 \text{ and } v = 0 \text{ at } t = T_s \tag{6.17}$$

Equation (6.16) takes the following form:

$$s = \frac{A}{2}(t - T_s)^2 \tag{6.18}$$

Coefficient A can be obtained either from a given displacement s_n at a given time t_n with the condition, that is,

$$t_n > T_s$$

$$A = 2\frac{s_n}{(t_n - T_s)^2} \tag{6.19}$$

or from the application of the least square technique to measured displacement response as follows:

$$A = 2\frac{\sum_{i=1}^{n} s_i (t_i - T_s)^2}{\sum_{i=1}^{n} (t_i - T)_s^4} \tag{6.20}$$

Once constant A is determined, the dynamic friction angle is obtained from the following relation:

$$\phi_d = \tan^{-1}\left(\tan\alpha - \frac{1}{\cos\alpha}\frac{A}{g}\right) \tag{6.21}$$

b. Tilting device and set-up

An experimental device consists of a tilting device operated manually (Aydan et al. 2019). During experiments, the displacement of the block and rotation of the base are measured through laser displacement transducers produced by KEYENCE while the acceleration responses parallel and perpendicular to the shear movement are measured by a three-component accelerometer (TOKYO SOKKI) attached to the upper block and WE7000 (YOKOGAWA) data acquisition system. The measured displacement and accelerations are recorded onto laptop computers. The weight of the accelerometer is about 98 gf. Figure 6.22 shows the experimental set-up.

A series of tilting tests are carried out on some discontinuities. Responses of some of these experiments are described as examples (Aydan et al. 2019). The measured responses during a tilting test on a rough discontinuity plane are shown in Figure 6.23 as an example. Figure 6.24 shows views of the tilting test on a rough discontinuity plane of granite. As noted from the responses of inclination (rotation) angle, relative displacement, and acceleration shown in Figure 6.24, fairly consistent results are observed. The static and dynamic friction coefficients of the interface were calculated from the measured displacement response and weight of the upper block using the tilting testing equipment shown in Figure 6.23. The static and dynamic friction angles were estimated to be 32.3°–37.6° and 30.3°–35.6°, respectively.

Similarly, experimental results on saw-cut discontinuity planes of Ryukyu limestone samples are shown in Figure 6.25. The static and dynamic friction angles of the interface were calculated from measured displacement responses explained in the previous section and they were estimated to be 28.8°–29.6° and 24.3°–29.2°, respectively.

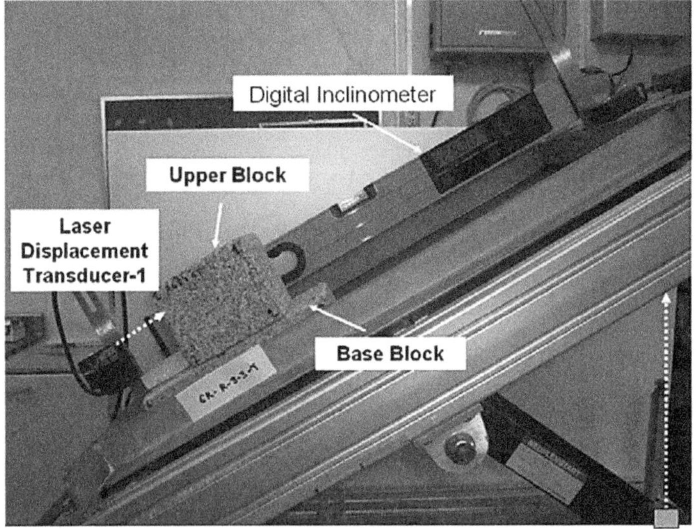

Figure 6.22 A view of the experimental set-up for the tilting device.

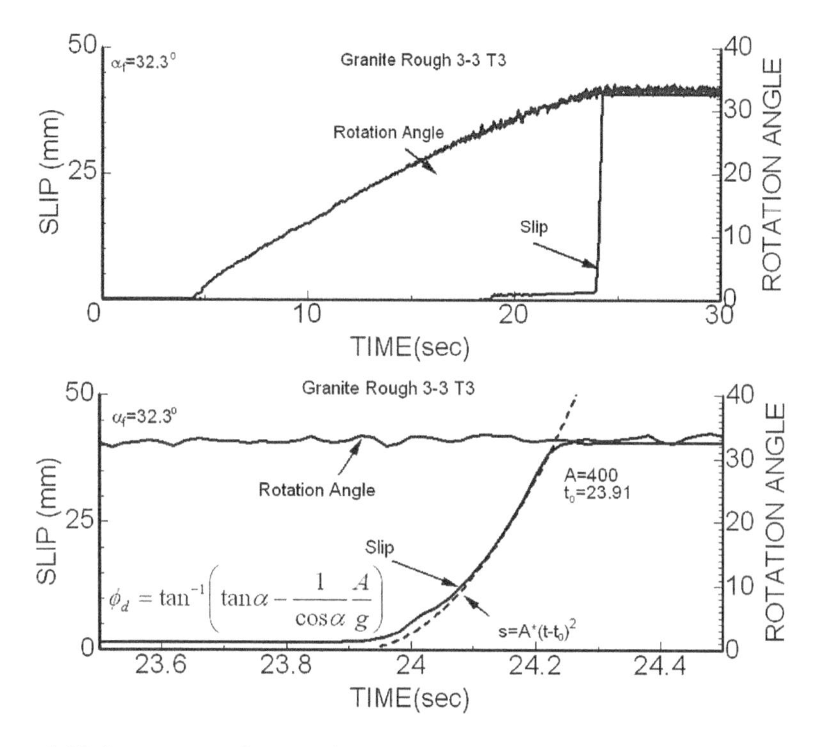

Figure 6.23 Responses of rough discontinuity of granite during a tilting test.

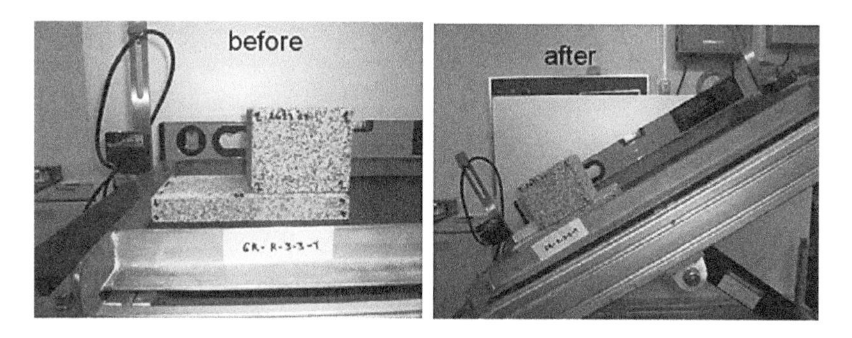

Figure 6.24 Views of the tilting experiment on a rough discontinuity plane of granite.

6.1.7 Experimental techniques for creep tests

The methods for creep tests described herein are concerned with creep characteristics of rocks under the indirect tensile stress regime of Brazilian test, uniaxial and triaxial compression tests and direct shear tests with the consideration of available creep testing techniques used in the rock mechanics field as well as other disciplines of engineering under laboratory conditions (Ishizuka et al. 1993; Ito and Akagi 2001; Ito et al. 2008; Aydan et al. 2016b; Aydan 2017; Cristescu and Hunsche 1998).

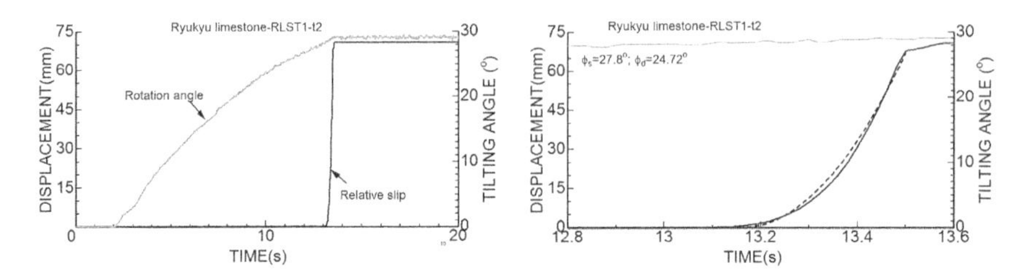

Figure 6.25 Responses of saw-cut discontinuity planes of Ryukyu limestone samples during a tilting test.

a. Apparatuses

Apparatuses for creep tests can be of cantilever-type or load/displacement-controlled type (Figure 6.26). Although the details of each testing machine may differ, the required features of apparatuses for creep tests are described herein.

Cantilever-type apparatus has been used in creep tests since early times. It is practically the most suitable apparatus for creep tests, as the load level can be easily kept constant in time-space (Figure 6.26a). The severest restrictions of this type of apparatus are the level of applicable load, which depends upon the length of the cantilever arm and its oscillations during the application of the load. Cantilever-type apparatus utilizing multi-arm lever overcome load limit restrictions and up to 500 kN load can be applied onto samples. The oscillation problem is also technically dealt with.

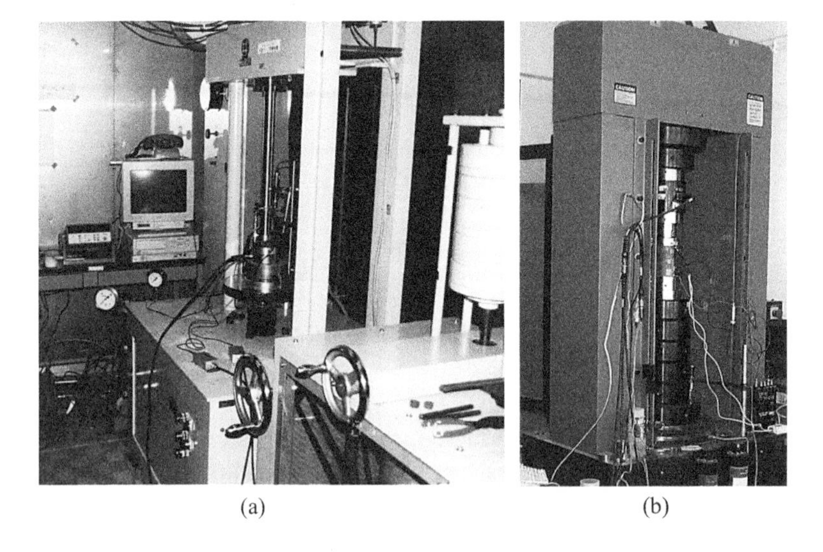

(a) (b)

Figure 6.26 Views of some of creep testing apparatuses. (a) Cantilever-type creep testing apparatus and (b) servo-controlled testing rig.

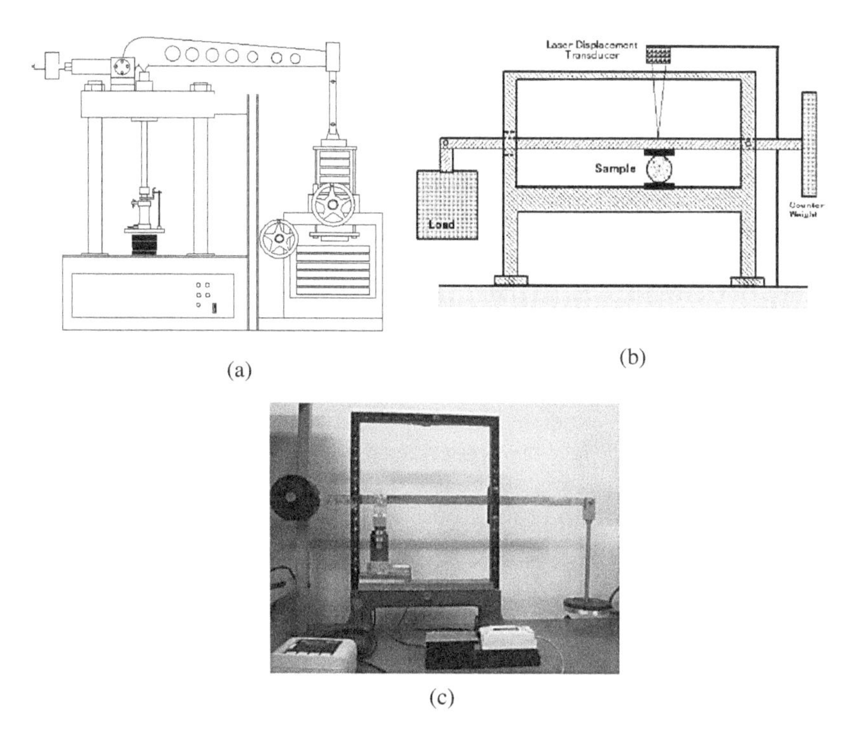

Figure 6.27 Schematic illustration of cantilever-type apparatuses for creep tests. (a) Uniaxial creep device (UR); (b) Brazilian creep device (UR) and (c) impression creep test device (UR).

The load is applied onto samples by attaching deadweights to the lever, which may be done manually for low-stress creep tests or mechanically for high-stress creep tests. In triaxial experiments, special load cells are required and the confining pressure is generally provided through oil pressure. The utmost care must be taken for keeping the confining pressure constant with the consideration of continuous power-supply for the compressor of the confining pressure system as well as the monitoring system.

b. Brazilian creep tests (Figure 6.27b)
The loading jigs and procedure used in the suggested method (SM) for Brazilian tests by ISRM should be followed unless the size of the samples differs from the conventional size. The displacement should be measured continuously or periodically as suggested in the SM. The load application rate may be higher than that used in the SM. Once the load reaches the designated load level, it should be kept constant thereafter. If experiments are required to be carried out under saturated conditions, the jigs and sample should be put in a water-filled special cell.

c. Uniaxial compression creep tests
The displacement is measured continuously or periodically as suggested in the SM. The load application rate may be higher than that used in the SM. Once

the load reaches the designated load level, it should be kept constant thereafter. If experiments are required to be carried out under saturated conditions, the sample should be put in a water-filled special cell.

d. Triaxial compression creep tests
The displacement is measured continuously or periodically as suggested in the SM. The load application rate may be higher than that used in the SM. Once the load reaches the designated load level, it should be kept constant thereafter. If experiments are required to be carried out under saturated conditions, the sample should be put in a water-filled special cell.

e. Impression creep test as an index test
The impression creep test technique utilizes the indenter, which is a cylinder with a flat end. The indenter makes a shallow impression on the surface of the specimen and it is therefore named "impression creep". There may be two different loading schemes during the experiments, namely, direct application of the dead weight (Figure 6.28) or load by a cantilever frame (Figure 6.27c). The potential of the use of this technique for the creep characteristics of rocks was explored by Mousavi et al. 2008 and Rassouli et al. 2010. The critical issue with this technique is the definition of strain and stress, which can be related to those in conventional creep experiments.

f. Direct shear test device
The servo-control shear-testing device shown in Figure 6.15 can also be used for creep tests on rock discontinuities.

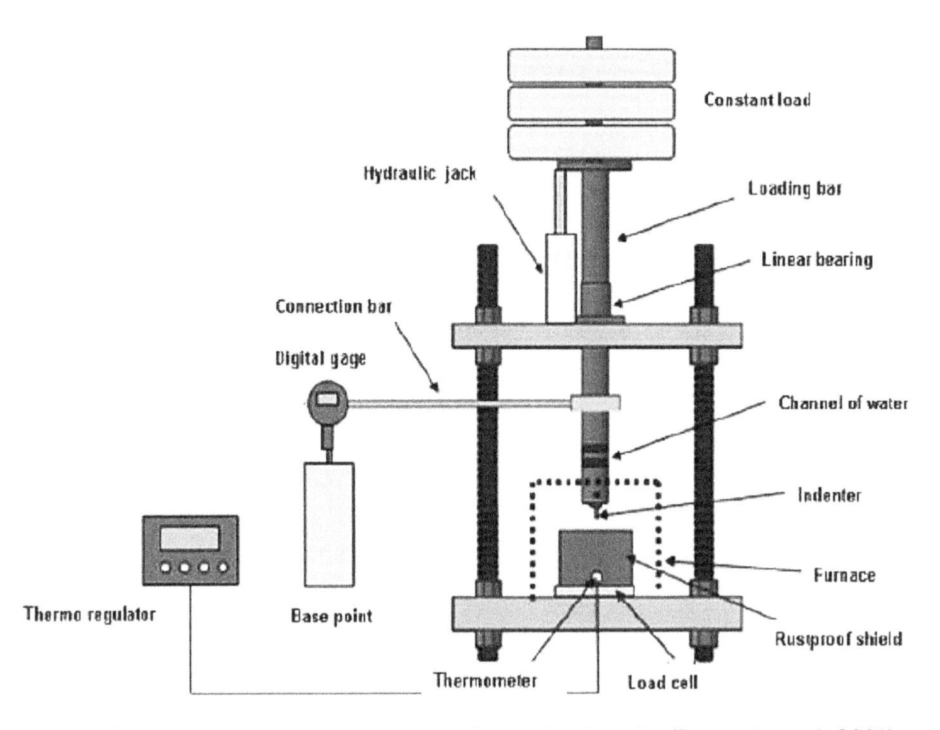

Figure 6.28 Impression creep apparatus utilizing deadweight (Rassouli et al. 2001).

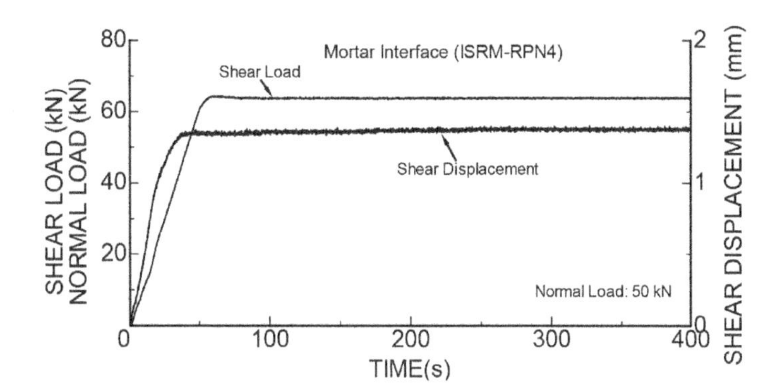

Figure 6.29 Direct shear creep test on a mortar interface with a surface roughness profile number 4.

An example of a direct creep test on a mortar sample having a surface roughness profile number (RPN4) of ISRM is shown in Figure 6.29. As noted from experimental responses, the testing device can keep shear load constant on the sample, which is the major problem when servo-control testing machines are used. Although the duration of the test is short, the creep behavior of the interface indicates almost a linear response.

6.2 Thermal properties of rocks and their measurements

Thermal properties such as specific heat, heating or cooling coefficient and thermal conductivity are important to assess the heat transport through solids as noted from Eqs. (4.30, 4.70 and 4.110). There are many techniques to measure thermal properties such as specific heat coefficient, thermal conductivity, thermal diffusion and thermal expansion coefficient. The details of such techniques can be found in various publications and textbooks (i.e. Crank 1956; Carslaw and Jaeger 1959; Somerton 1992; Jumikis 1983; Popov et al. 1999). Specific heat coefficient is commonly measured using the calorimeter tests.

The earlier and common technique for thermal conductivity measurement is the divided bar technique (Birch 1950) based on the steady-state heat flow assumption and it is illustrated in Figure 6.30. This technique utilizes reference materials with well-known thermal properties.

There are also techniques for measuring thermal conductivity utilizing transient heat flow (i.e. Carslaw and Jaeger 1959; Popov et al. 1999; Sass et al. 1984). These techniques utilize line or plane sources and temperature variations are measured using either a contact sensor or an infrared camera. These techniques utilize the analytical solutions developed by Crank (1956) and Carslaw and Jaeger (1959).

An experimental technique using a device similar to a calorimeter-type apparatus is described in this section to measure thermal properties of rock materials from a single experiment and its applications are given (Aydan 2017). Let us consider a solid (s) is enveloped by fluid (i.e. water (w)) as illustrated in Figure 6.31. It is assumed that solid and fluid have different thermal properties and temperature.

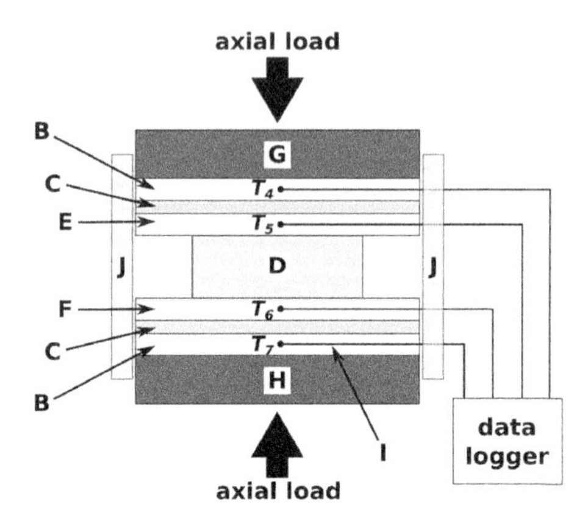

Figure 6.30 The key components of a divided-bar apparatus (Popov et al. 1999). A – pivot point; B – brass disks; C – reference material; D – rock specimen; E – hot plate; F – cold plate; G – heat source (concealed Peltier device); H – heat sink; I – holes for the insertion of temperature sensors and J – thermal insulation.

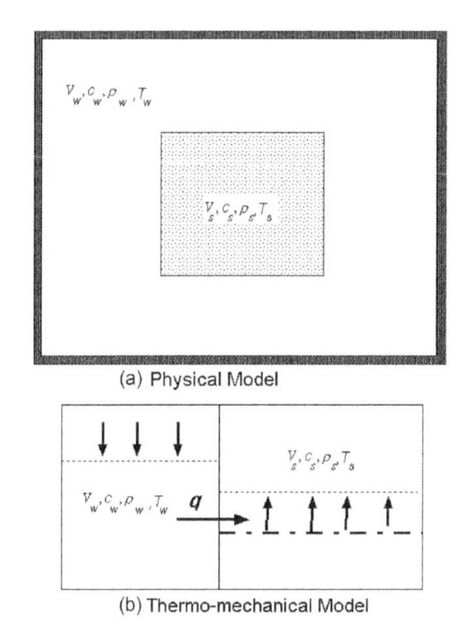

Figure 6.31 An illustration of a (a) physical and (b) thermo-mechanical model.

For theoretical modeling, the following parameters are defined as follows: Q: heat; ρ: density; k: thermal conductivity; c: specific heat coefficient; T: temperature; m: mass; h: cooling coefficient; V: volume; A_s: surface area of solid and λ: thermal expansion coefficient. The heat of a body is given in the following form:

$$Q = m \cdot c \cdot T = \rho V \cdot c \cdot T \tag{6.22}$$

Its unit is Joule ($J = N \cdot m$).

Assuming that mass and specific heat coefficient are constant, the heat rate (heat flux) is given in the following form:

$$\frac{dQ}{dt} = q = m \cdot c \cdot \frac{dT}{dt} \tag{6.23}$$

The Newton cooling law is written as:

$$q = h \cdot A_s \cdot \Delta T \tag{6.24}$$

where ΔT is the temperature difference between a solid and enveloping fluid and its unit is Watt ($W = J / s$).

In this particular model, the temperature of the surrounding fluid is assumed to be higher than the solid enveloped by the fluid. Furthermore, there is no heat flow from the system outward. In other words, it is thermally isolated. The heat flux from the fluid can be given as

$$q_W = -\rho_W \cdot c_W \cdot V_W \frac{dT_W}{dt} \tag{6.25}$$

The heat from the fluid into the solid should be equal with the use of the Newton cooling law as written below:

$$-\rho_W \cdot c_W V_W \frac{\partial T_W}{\partial t} = h \cdot A_s \left(T_W - T_s\right) \tag{6.26}$$

Similarly, the heat change of the solid should be equal to that supplied from the fluid as given by

$$\rho_S \cdot c_S \cdot V_S \frac{\partial T_s}{\partial t} = h \cdot A_s \left(T_W - T_s\right) \tag{6.27}$$

It should be noted that the sign of heat flux is +. Rewriting Eq. (6.26) yields the following:

$$T_s = T_W + \frac{\rho_W c_W v_W}{h \cdot A_S} \cdot \frac{\partial T_W}{\partial t} \tag{6.28}$$

If the derivation of Eq. (6.26) with respect to time is inserted into Eq. (6.25), one easily gets the following:

$$\frac{\partial^2 T_W}{\partial t^2} + \alpha \frac{\partial T_W}{\partial t} = 0 \tag{6.29}$$

where

$$\alpha = h \cdot A_s \frac{\rho_W \cdot c_W \cdot V_W + \rho_S \cdot c_S \cdot V_S}{\rho_W \cdot c_W \cdot V_W \cdot \rho_S \cdot c_S \cdot V_S} \tag{6.30}$$

The solution of Eq. (6.29) would be obtained as follows:

$$T_w = C_1 + C_2 e^{-\alpha t} \tag{6.31}$$

Integral coefficients C_1 and C_2 of Eq. (6.31) are obtained from the following conditions:

$$T_w = T_i \text{ at } t = 0 \quad \text{and } T_1 = T_f \text{ at } t = \infty \tag{6.32}$$

as

$$C_1 = T_f; C_2 = T_i - T_f \tag{6.33}$$

By using integral constants given by Eq. (6.33), Eq. (6.27) becomes

$$T_w = T_f + (T_i - T_f)e^{-\alpha t}; + \frac{\partial T_w}{\partial t} = -\alpha (T_i - T_f)e^{-\alpha t} \tag{6.34}$$

The average temperature of solid is obtained by inserting Eq. (6.34) into Eq. (6.28) as

$$T_s = T_f - \frac{\rho_w \cdot c_w \cdot V_w}{\rho_s \cdot C_s \cdot V_s}(T_i - T_f)e^{-\alpha t} \tag{6.35}$$

As $T_s = T_o$ at the time $t = 0$, Eq. (6.35) can be rewritten as

$$\frac{T_f - T_o}{T_i - T_f} = \frac{\rho_w \cdot c_w \cdot V_w}{\rho_s \cdot c_s \cdot V_s} \tag{6.36}$$

Inserting Eq. (6.36) into Eq. (6.35) yields

$$T_s = T_f - (T_f - T_o)e^{-\alpha t} \tag{6.37}$$

The temperature difference between solid and enveloping fluid can be obtained from Eqs. (6.40) and (6.42) as

$$\Delta T_{ws} = (T_w - T_s) = (T_i - T_o)e^{-\alpha t} \tag{6.38}$$

Therefore, if the values of T_o, T_f, T_i, ρ_w, ρ_s, V_w, V_s and c_w are known, the specific heat coefficient of solid can be easily obtained. For example, the specific heat coefficient of water is 4.1783–4.2174 J/g/K for a temperature range of 0°–90°C. As the thermal properties of water remain almost constant for the given temperature range, the water would be used as the fluid in the experimental set-up. After obtaining the specific heat coefficient, the coefficient α is obtained from Eqs. (6.34), (6.37) or (6.38) using the curve fitting technique to the experimental response. Then using the value of the coefficient α, the value of the cooling coefficient is obtained from Eq. (6.30).

For determining the thermal conductivity coefficient (k), the following approach is used. The Fourier law for the one-dimensional situation can be written as

$$q = -kA\frac{\partial T}{\partial x} \tag{6.39}$$

Assuming the specimen has a length (L) and using Newton's cooling law, we may write the following relationship:

$$hA\Delta T = kA\frac{\Delta T}{L} \tag{6.40}$$

Equation (6.40) can be rewritten and the following relation holds between cooling coefficient and thermal conductivity

$$k = h \cdot L \tag{6.41}$$

The characteristics length of a solid sample can be obtained from the volume of the solid from the following relationship:

$$L = \sqrt[3]{V_s} \tag{6.42}$$

Linear thermal expansion coefficient (λ) is defined as:

$$\lambda = \frac{1}{L} \cdot \frac{dL}{dT} \tag{6.43}$$

where L is the length of the sample and $\frac{dL}{dT}$ is the variation of the length of the sample with respect to temperature variation and it is determined under the unstrained condition or 100 gf load on the sample. If the variation of length of the sample at the equilibrium state with respect to the initial length before the commencement of the experiment is measured, it would be straightforward to obtain a linear expansion coefficient. Similarly, width or diametrical changes can be also measured and thermal expansion coefficients can be evaluated from the variation of side length or diameter for a given temperature difference.

The technique described in this sub-section is unique and quite practical considering the laborship in other techniques. The device for determining the thermal properties of geo-materials consists of a thermostated cell equipped with temperature sensors. The fundamental features of this device are illustrated in Figure 6.32. In the experiments, the temperatures of the sample, water, air and thermostat are measured. The method utilizes the thermal properties of water, whose properties remain to be the same up to 90°C, to infer the thermal properties of geo-material substances. If the continuous measurements of temperatures are available, one can easily infer the thermal properties from the following equations:

The specific heat of a geo-material is given by

$$c_s = \frac{\rho_w \cdot c_w \cdot V_w}{\rho_s \cdot V_s} \frac{T_i - T_f}{T_f - T_o} \tag{6.44}$$

where ρ_w is the density of water, c_w is the specific heat coefficient of water, V_w is the volume of water, ρ_s is the density of sample, c_s is the specific heat coefficient of sample, V_s is the volume of sample, T_i is the initial temperature of water, T_o is the initial temperature of sample and T_f is the equilibrium temperature.

Heat conduction coefficient (α) is obtained from fitting experimental results to the following equation:

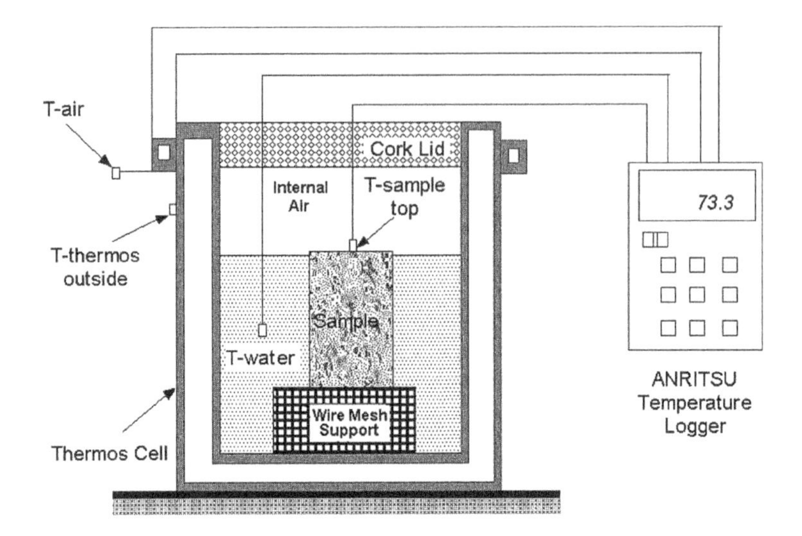

Figure 6.32 Illustration of the experimental set-up.

$$\Delta T_{WS} = (T_W - T_s) = (T_i - T_o)e^{-\alpha t} \tag{6.45}$$

If the heat conduction coefficient (α) is determined, then Newton's cooling coefficient is determined from the following equation:

$$h = \frac{\alpha}{A_s} \cdot \frac{\rho_W \cdot c_W \cdot V_W \cdot \rho_S \cdot c_S \cdot V_S}{\rho_W \cdot c_W \cdot V_W + \rho_S \cdot c_S \cdot V_S} \tag{6.46}$$

Finally, the thermal conductivity coefficient is obtained from the following equation:

$$k = h \cdot L \tag{6.47}$$

where L is the characteristic sample side length.

If the sample temperature can be measured, it will be very easy to determine the specific heat coefficient of sample and subsequent properties. This is possible for granular materials since the temperature sensor can be embedded in the center of samples. However, it is quite difficult to determine the equilibrium temperature T_f for solid samples. Therefore, the following procedure is followed for this purpose:

Step 1: Determine the heat conduction coefficient (α)
Step 2: Plot the following equation in time-space

$$T_s = T_W - (T_i - T_o)e^{-\alpha t} \tag{6.48}$$

Step 3: Determine the peak value from Eq. (6.48) and assign it as the equilibrium temperature T_f.
Step 4: Then proceed to determine the rest of the thermal properties using the procedure described above.

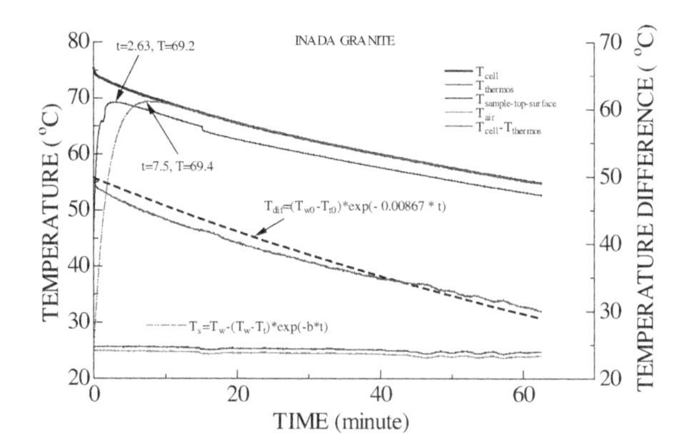

Figure 6.33 Application of the procedure to an Inada granite sample.

Figure 6.33 shows the application of the method to a cylindrical Inada granite sample. The temperature of the sample at the top was also measured. As noted in the figure, the sample temperature at the top of the sample achieves the peak value before the computed response. Since the temperature of the sample averaged over the total volume of the sample in the proposed temperature, the computed sample temperature achieves its peak value later than that at the top surface of the sample.

Thermal conductivity could be measured using the TK04 system described by Blum (1997). This system employs a single-needle probe (Von Herzen and Maxwell 1959), heated continuously, in a half-space configuration for hard rock. The needle probe is a thin metal tube that contains a thermistor and a heater wire. The needle is assumed to approximate an infinitely long, continuous medium, and the temperature near the line source is measured as a function of time. If it is assumed that the sediment or rock sample to be measured can be represented as solids in a fluid medium, it is then possible to determine a relationship between thermal diffusivity and thermal conductivity. With this assumption, the change in temperature of the probe as a function of time is given to a good approximation by Von Herzen and Maxwell 1959

$$T(t) = \frac{q}{4k} \ln\left(4\frac{\alpha_t}{Ba^2}\right) \tag{6.49}$$

where T is the temperature [°C], q is the heat input per unit time per unit length [W/m], k is the thermal conductivity of the sediment or rock sample [W/m·°C], t is the time after the initiation of the heat [s], α is the thermal diffusivity of the sample [m^2/s], B is a constant (1.7811) and a is the probe radius [m].

This relationship is valid when t is large compared with a^2/α. A plot of T vs. $\ln(t)$ yields a straight line, the slope of which determines k (Von Herzen and Maxwell 1959).

6.3 Tests for seepage parameters

Some of the theoretical backgrounds of permeability tests are also explained in Section 7.3 of Chapter 7. Longitudinal flow-type permeability tests are described in this subsection.

6.3.1 Falling head tests

When the rock is quite permeable, falling head tests, which utilize the dead weight of the fluid, are also used for determining the permeability of rocks and discontinuities. In this subsection, analytical solutions for falling head tests for longitudinal flow are given.

The experimental set-up used for this kind of test is shown in Figure 6.34 (Aydan et al. 1997). As seen from the figure, two manometers having cross-sections a are assumed to be attached to both ends of the sample. During a test, the change of pressure and velocity of flow can be measured through these manometers. The level h_2 of water at the lower tank is assumed to be constant in the following formulation. When an experiment starts, the flow rate inside the pipe can be given as:

$$v_p = -a\frac{\partial h_1}{\partial t} \tag{6.50}$$

where h_1 is the level of water inside the manometer (1). At a given time, the flow rate through the cross-section area (A) of the specimen is given by

$$v_t = \overline{v}A \tag{6.51}$$

It is assumed that the flow rate through the specimen should be equal to the flow rate of the pipe. Then, the pressure gradient in a specimen can be given in the following form:

$$\frac{\partial p}{\partial x} \approx -\rho g\frac{(h_1 - h_2)}{L} \tag{6.52}$$

where ρ is the density, g is the gravitational acceleration. Substituting Eq. (6.52) into Eq. (6.51) and equalizing the resulting equation to Eq. (6.50) yield the following differential equation for the change of water height h_1:

$$\frac{\partial h_1}{h_1 - h_2} = -\frac{kA\rho g}{La\eta}\partial t \tag{6.53}$$

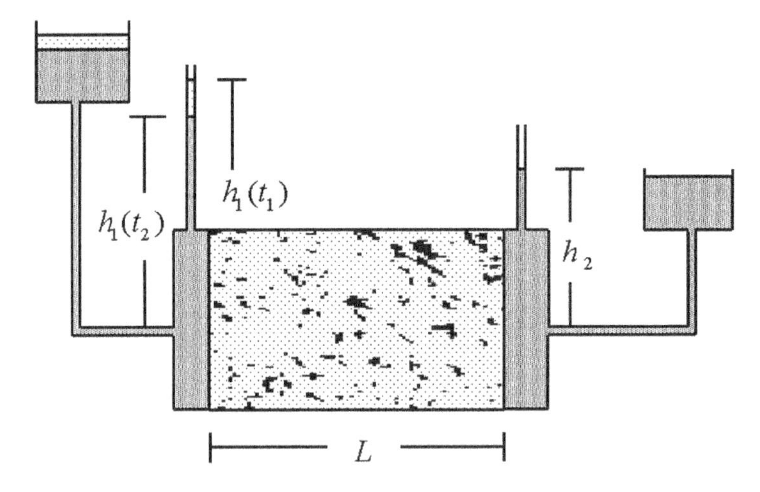

Figure 6.34 Illustration of the longitudinal falling head test.

where L is the sample length. The solution of the above differential equation is:

$$h_1 = h_2 + Ce^{-\alpha t} \tag{6.54}$$

where

$$\alpha = \frac{kA}{La}\frac{\rho g}{\eta}$$

If initial conditions are given by

$$h_1 = h_{10} \text{ at } t = 0$$

where h_{10} is the water height at manometer 1 at $t = 0$. Thus, the integration coefficient C is obtained as follows:

$$C = h_{10} - h_2 \tag{6.55}$$

Inserting the above integration coefficient into Eq. (6.54) yields

$$-\alpha t = \ln\left(\frac{\Delta h}{\Delta h_o}\right) \tag{6.56}$$

where

$$\Delta h = h_1 - h_2, \quad \Delta h_o = h_{10} - h_2$$

If α is substituted into the above equation, the following expression for permeability is obtained

$$k = \frac{La}{A}\frac{\ln\left(\dfrac{h_{10} - h_2}{h_1 - h_2}\right)}{t}\frac{\eta}{\rho g} \tag{6.57}$$

6.3.2 *Transient pulse test method*

Brace et al. (1968) proposed a transient pulse method for longitudinal flow tests. In this method, the following assumptions are made (Aydan 1998; Aydan and Üçpırtı 1997):

- fluid flow obeys Darcy's law,
- change of fluid density inside pores with respect to time is negligible,
- the volumes of reservoirs (V_1, V_2) are constant and
- the relation between pressure and volumetric strain of fluid is linear.

Permeability is obtained from pressure changes, which are applied to the ends of a specimen, with respect to time (Figure 6.35). During experiments, the flow rate is not measured. The volumetric strain of fluid inside reservoirs V_1 and V_2 can be written as follows:

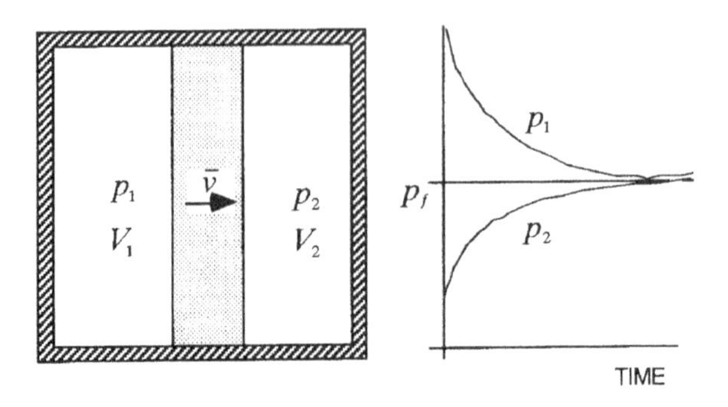

Figure 6.35 Illustration of transient pulse-method for longitudinal flow.

$$\varepsilon_V^1 \approx \frac{\Delta V_1}{V_1}, \quad \varepsilon_V^2 \approx \frac{\Delta V_2}{V_2} \tag{6.58}$$

Similarly, for the volumetric strain rate of fluid, the following relations can also be written as:

$$\dot{\varepsilon}_V^1 \approx \frac{\Delta \dot{V}_1}{V_1}, \quad \dot{\varepsilon}_V^2 \approx \frac{\Delta \dot{V}_2}{V_2} \tag{6.59}$$

or

$$\Delta \dot{V}_1 = \dot{\varepsilon}_V^1 V_1, \quad \Delta \dot{V}_2 = \dot{\varepsilon}_V^2 V_2 \tag{6.60}$$

If the following relations exist between the volumetric strain of fluid and pressure in reservoirs:

$$\varepsilon_V^1 = -c_f p_1, \quad \varepsilon_V^2 = -c_f p_2 \tag{6.61}$$

and the compressibility coefficient (c_f) is constant, for volumetric strain rate, the following relations can be also written

$$\dot{\varepsilon}_V^1 = -c_f \dot{p}_1, \quad \dot{\varepsilon}_V^2 = -c_f \dot{p}_2 \tag{6.62}$$

Flow rates may be defined as

$$v_{t1} = \Delta \dot{V}_1, \quad v_{t2} = \Delta \dot{V}_2 \tag{6.63}$$

Using Eqs. (6.59), (6.62) and (6.63), flow rates can be rewritten in the following form:

$$v_{t1} = -c_f V_1 \frac{\partial p_1}{\partial t}, \quad v_{t2} = -c_f V_2 \frac{\partial p_2}{\partial t} \tag{6.64}$$

Introducing the following boundary conditions

$$p = p_1 \ at \quad x = 0, \quad p = p_2 \quad at \quad x = L$$

and using Darcy's law, the following relations can be obtained for flow rates

$$v_{t1} = -\frac{kA}{\eta}\left(\frac{dp_1}{dx}\right)_{x=0}, \quad v_{t2} = -\frac{kA}{\eta}\left(\frac{dp_2}{dx}\right)_{x=L} \tag{6.65}$$

where A is the cross-section area of the sample and L is the length of the sample. Pressure gradients in the above equations are as follows:

$$\frac{dp_1}{dx} \approx -\frac{(p_1 - p_2)}{L}, \quad \frac{dp_2}{dx} \approx -\frac{(p_2 - p_1)}{L}$$

Inserting the above equation into Eq. (6.65) and equating the resulting equation into Eq. (6.64) yield the following set of equations

$$\frac{\partial p_1}{\partial t} = -\beta\frac{1}{V_1}(p_1 - p_2) \tag{6.66}$$

$$\frac{\partial p_2}{\partial t} = \beta\frac{1}{V_2}(p_1 - p_2) \tag{6.67}$$

where

$$\beta = \frac{kA}{c_f \eta L}$$

Brace et al. (1968) solved similar equations by using the Laplace transformation technique. Herein, the method of elimination will be used for solving the set of equations above (see Kreyszig (1983) for the method). Equation (6.66) can be rearranged as follows:

$$p_2 = p_1 + \frac{V_1}{\beta}\frac{\partial p_1}{\partial t} \tag{6.68}$$

Taking the time derivative of the above equation, the following expression is obtained:

$$\frac{\partial p_2}{\partial t} = \frac{\partial p_1}{\partial t} + \frac{V_1}{\beta}\frac{\partial^2 p_1}{\partial t^2} \tag{6.69}$$

Substituting Eqs. (6.68) and (6.69) into Eq. (6.67) and rearranging the resulting equation yield the following homogeneous differential equation:

$$\frac{\partial^2 p_1}{\partial t^2} + \alpha\frac{\partial p_1}{\partial t} = 0 \tag{6.70}$$

where

$$\alpha = \beta\frac{V_1 + V_2}{V_1 V_2}$$

The general solution of this differential equation is

$$p_1 = C_1 + C_2 e^{-\alpha t} \tag{6.71}$$

Introducing the following initial conditions

$$p_1 = p_i \; at \quad t = 0, \quad p_1 = p_f \quad at \quad t = \infty$$

where p_i is the applied initial pressure at Reservoir 1 (V_1), p_f is the final pressure at the end of the test, yields the integration constants C_1 and C_2 as:

$$C_1 = p_f, \quad C_2 = p_i - p_f \tag{6.72}$$

Inserting these integration constants into Eq. (6.71) gives

$$p_1 = p_f + \left(p_i - p_f\right)e^{-\alpha t} \tag{6.73}$$

Taking the time derivative of the above equation

$$\frac{\partial p_1}{\partial t} = -\left(p_i - p_f\right)\beta \frac{V_1 + V_2}{V_1 V_2} e^{-\alpha t} \tag{6.74}$$

Substituting Eqs. (6.73) and (6.74) into Eq. (6.66) and rearranging yield the following equation:

$$p_2 = p_f - \left(p_i - p_f\right)\frac{V_1}{V_2} e^{-\alpha t} \tag{6.75}$$

For the following initial condition for p_2

$$p_2 = p_0 \text{ at } t = 0$$

Equation (6.75) takes the following form

$$\left(p_i - p_f\right) = \left(p_f - p_0\right)\frac{V_2}{V_1} \tag{6.76}$$

The above equation can be rewritten in a different way for $p_i - p_0$ as follows:

$$\left(p_i - p_f\right) = \left(p_i - p_0\right)\frac{V_2}{V_1 + V_2} \tag{6.77}$$

Inserting this equation into Eq. (6.73) and rearranging yield the following:

$$-\alpha t = \ln\left(\frac{p_1 - p_f}{p_i - p_0}\frac{V_1 + V_2}{V_2}\right) \tag{6.78}$$

where

$$\alpha = \frac{kA}{c_f L \eta}\frac{V_1 + V_2}{V_1 V_2}$$

From the above equations, one gets the following equation to compute permeability

$$k = \frac{\eta c_f L}{A}\frac{V_1 V_2}{V_1 + V_2}\ln\left(\frac{\Delta p_o}{\Delta p}\frac{V_2}{V_1 + V_2}\right)\frac{1}{t} \tag{6.79}$$

where $\Delta p = p_1 - p_f, \Delta p_o = p_i - p_o$. When gas is used as a permeation fluid, p_1 and p_2 are replaced with $U_1\left(= p_1^2\right)$ and $\left(= p_2^2\right)$, and permeability can be calculated using the same relation given above.

If the volume of Reservoir 2 (V_2) is much greater than the volume of Reservoir 1 (V_1), $\left(V_2 >> V_1\right)$ (for instance, the outer side of specimen is open to air) p_0 and p_f given in the above equation will be equal to atmospheric pressure (p_a). For this particular case, Eq. (6.79) takes the following form:

$$k = \frac{\eta c_f L V_1}{A} \ln\left(\frac{p_i - p_a}{p_1 - p_a}\right)\frac{1}{t} \tag{6.80}$$

For different values of α, the relations between normalized pressure change and time and the natural logarithm of normalized pressure change and time for transient pulse tests were computed and are shown in Figure 6.36. As seen in Figure 6.36, there is a linear relationship between the natural logarithm of normalized pressure change and time. In both figures, time is taken as a unitless parameter. However, the unit depends on the description of the problem. For instance, it can be a year, day or second.

Theoretical and experimental curves for a transient pulse test on a rock salt (halite) specimen are shown in Figure 6.37. As seen from this figure, there is a slight difference between the curve obtained from the test and the curve from the theory, particularly in the initial stages. To obtain the permeability value, the linear part of normalized pressure change and time relation is generally used to compute permeability for the selected range.

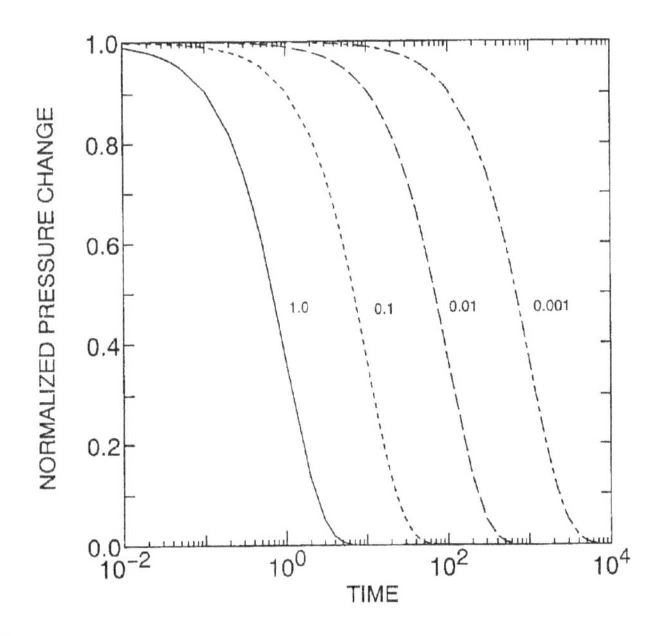

Figure 6.36 Computed time–pressure relations.

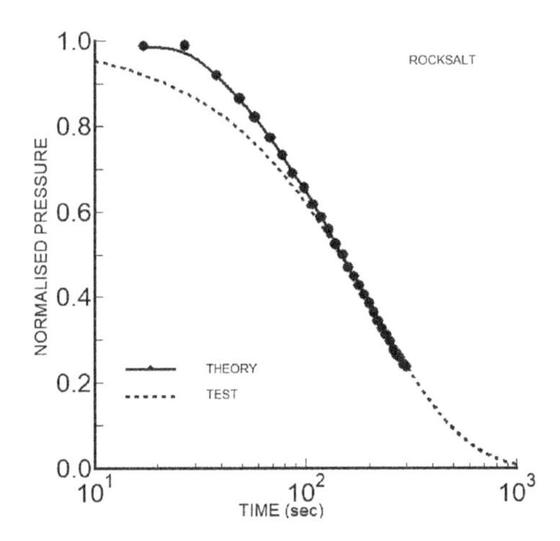

Figure 6.37 An experimental result on a rock salt (from Aydan and Ucpirti 1997).

6.4 Tests for diffusion parameters

Diffusion experiments are concerned with the concentration variation of a given substance through rocks and rock masses. There are different techniques and they can utilize one-dimensional, axisymmetric or equivalent volumetric models to interpret the experimental results. A testing technique for an equivalent volumetric model is described by Aydan (2003) and also in Section 7.5 in Chapter 7. The most difficult aspect in the diffusion experiments is how to assess the amount of the concentration of a given substance in time. The tests can be categorized as destructive tests by slicing the sample and checking the concentration or non-destructive testing techniques such as electrical resistivity or the X-ray CT scanning technique (Kano et al. 2004; Sato and Aydan 2014; Aydan et al. 2016a). However, these non-destructive tests require some correlations between the amount of concentration of the substance and measured parameters such as electrical resistivity (or conductivity) or CT value (or number) in the X-ray CT scanning technique. Figure 6.38 shows an experimental set-up for measuring the resistivity and amount of water loss during the drying process of saturated Oya tuff and related electrical resistivity changes (Kano et al. 2004). The diffusion coefficient of water migration in this particular experiment can be evaluated using the procedure described in Section 7.5.

Oya tuff has been extracted from the Oya town of Utsunomiya City and it is commercially known as Oya stone. Figure 6.39 shows X-ray CT scan images and CT value distribution with height at time intervals of 1, 2, 3 and 4 hours reported by Sato and Aydan (2014). As expected, water migrates upward gradually. Although the water absorption front is clearly observed in CT scanning images, the front is not straight. This may be due to inclusions of highly absorptive nodules such as clays in the sample. Additional experiments were carried out by Aydan et al. (2016a). Figure 6.40 shows the water migration in the sample at different time steps. The sample was immersed in water up to a height of 40 mm from the bottom. The rock started to absorb water from all sides and becomes saturated after 180 minutes.

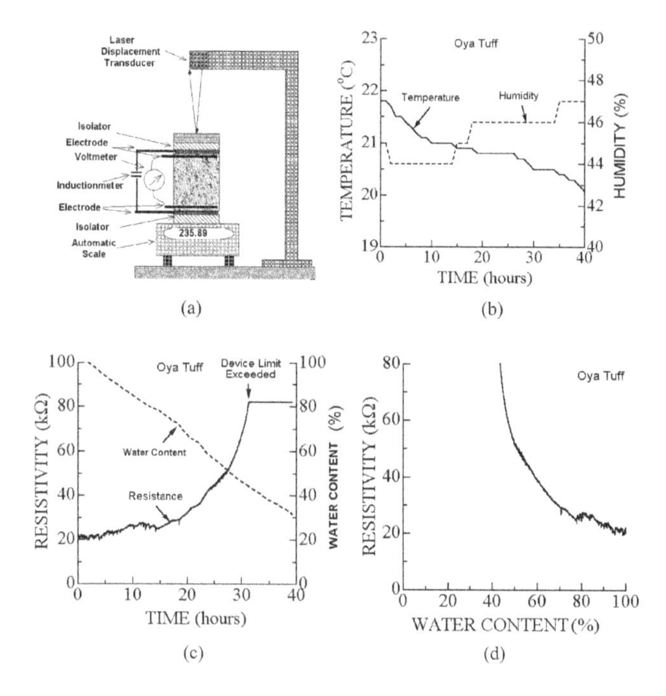

Figure 6.38 An experimental set-up for measuring water-migration during the drying process of Oya tuff (from Kano et al. 2004). (a) Experimental set-up; (b) environmental conditions; (c) time variation of water content and resistivity and (d) water content versus resistivity.

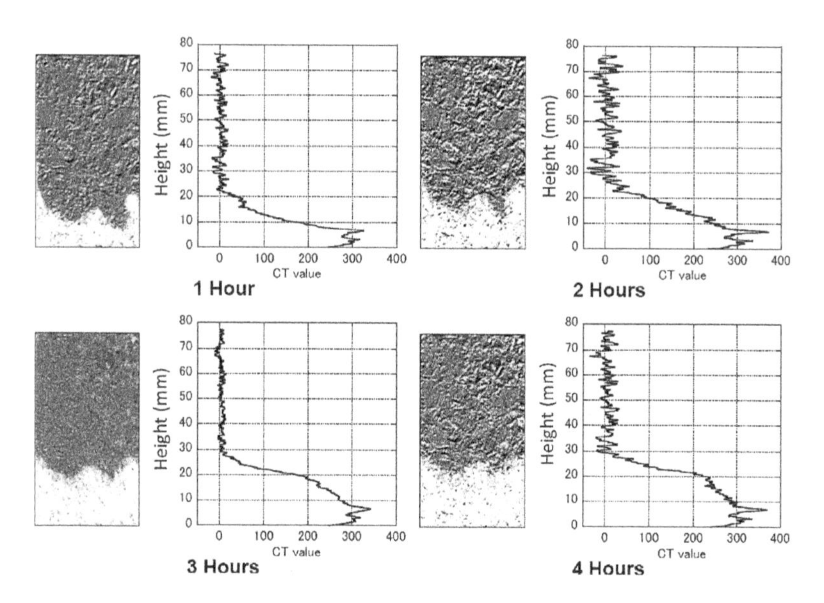

Figure 6.39 X-ray CT scan images and CT value distribution with height at different time intervals in the Oya tuff sample (from Sato and Aydan 2014).

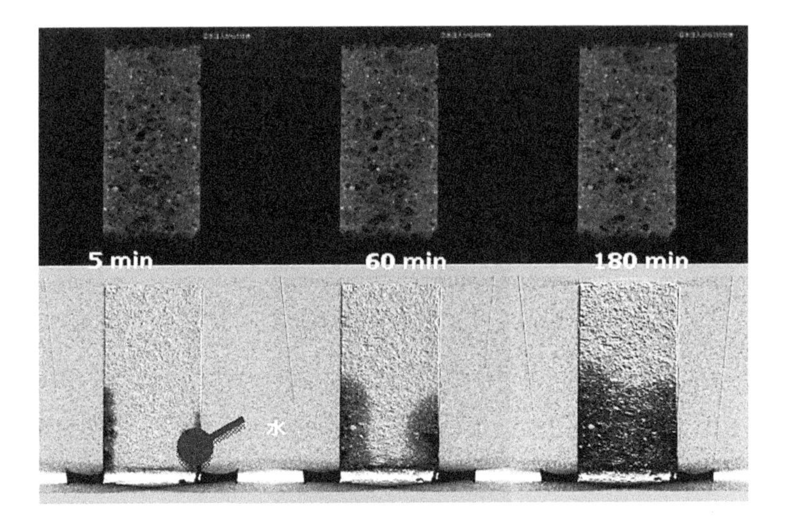

Figure 6.40 Water migration process in Oya tuff (from Aydan et al. 2016a).

The governing equation for one-dimensional diffusion of a given substance can be written as:

$$\frac{\partial \theta}{\partial t} = D \frac{\partial^2 \theta}{\partial x^2} \tag{6.81}$$

The utilization of the method of separating variable for the solution of partial differential Eq. (6.81) takes the following form

$$\theta = \exp(-D\lambda^2 t)(A \sin \lambda x + B \cos \lambda x) \tag{6.82}$$

where A and B are constants of integration. The most general solution would be the sum of Eq. (6.82) as given below (e.g. Crank 1956; Carslaw and Jaeger 1959; Farlow 1982)

$$\theta = \sum_{i=0}^{\infty} \exp\left(-D\lambda_i^2 t\right)\left(A_i \sin \lambda_i x + B \ \cos \lambda_i x\right) \tag{6.83}$$

where A_i, B_i and λ_i are determined from the boundary and initial conditions.

Let's assume that the boundary and initial conditions are given as follows:

Boundary conditions

$$\begin{aligned} \theta(0,t) &= C_1(t) \\ \theta(L,t) &= C_2(t) \end{aligned} \tag{6.84a}$$

with $C_2 > C_1$.

Initial conditions

$$\theta(x,0) = 0.0 \tag{6.84b}$$

The boundary conditions imply that when time goes to infinity, the final distribution of concentration should take the following form:

$$\theta(x,\infty) = C_1 + (C_2 - C_1)\frac{x}{L} \tag{6.85}$$

Utilization of the method of separating variables and Fourier series for imposing the boundary and initial conditions yields the following particular solution as follows

$$\theta = C_1 + C_2\left(\frac{x}{L} + \sum_{i=1}^{n}\left(\frac{2}{i\pi}(-1)^i \sin\frac{i\pi x}{L}\exp\left(-D\left(\frac{i\pi}{L}\right)^2 t\right)\right)\right) \tag{6.86}$$

As noted from Eq. (6.86), the exact solution is a sum of series and the major issue is "what should be the value of n" for an acceptable solution for computations. For $n = 2$, 10 and 100, Figure 6.41 shows the plots of solution (6.86) for the following conditions:

$$D = 0.01; C_1 = 0.0; C_2 = 1.0$$

As noted in Figure 6.41, if the value of n is greater than 10, the solutions become almost the same. However, when the value of n is 1 or 2, the results become unacceptable.

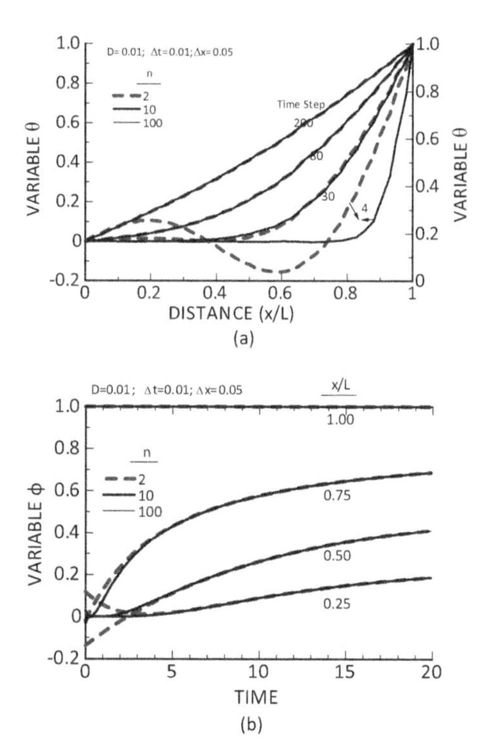

Figure 6.41 Effect of the number of terms on variations of variables at the selected time and distances. (a) Distribution of concentration at selected times and (b) variation of concentration at selected locations.

References

Aydan, Ö. 1998. Finite element analysis of transient pulse method tests for permeability measurements. *The 4th European Conference on Numerical Methods in Geotechnical Engineering-NUMGE98*, Udine, 719–727.

Aydan, Ö. 2003. The moisture migration characteristics of clay-bearing geo-materials and the variations of their physical and mechanical properties with water content. *2nd Asian Conference on Saturated Soils (UNSAT-ASIA 2003)*, 383–388.

Aydan, Ö. 2008. New directions of rock mechanics and rock engineering: Geomechanics and Geoengineering. *5th Asian Rock Mechanics Symposium (ARMS5)*, Tehran, 3–21.

Aydan, Ö. 2016a. Issues on rock dynamics and future directions. Keynote, ARMS2016, Bali.

Aydan, Ö. 2016b. Considerations on friction angles of planar rock surfaces with different surface morphologies from tilting and direct shear tests. ARMS2016, Bali, Indonesia.

Aydan, Ö. 2017. *Time Dependency in Rock Mechanics and Rock Engineering*, London, CRC Press, Taylor & Francis Group, 241 p.

Aydan, Ö. and H. Üçpırtı 1997. The theory of permeability measurement by transient pulse test and experiments. *Journal of the School of Marine Science and Technology*, 43, 45–66.

Aydan, Ö. and R. Ulusay 2002. Back analysis of a seismically induced highway embankment during the 1999 Düzce earthquake. *Environmental Geology*, 42, 621–631.

Aydan, Ö., Y. Ichikawa, S. Ebisu, S. Komura and A. Watanabe 1990. Studies on interfaces and discontinuities and an incremental elasto-plastic constitutive law. *International Conference Rock Joints*, ISRM, 595–601.

Aydan, Ö., T. Akagi, T. Ito and T. Kawamoto 1992. Deformation behavior of tunnels in squeezing rocks and its prediction. *Journal of Japan Society of Civil Engineers, Geotechnical Division*, 448/III-19, 73–82.

Aydan, Ö., T. Akagi and T. Kawamoto 1993. Squeezing potential of rocks around tunnels; theory and prediction. *Rock Mechanics and Rock Engineering*, 26(2), 137–163.

Aydan, Ö., T. Akagi, H. Okuda and T. Kawamoto 1994. The cyclic shear behavior of interfaces of rock anchors and its effect on the long-term behavior of rock anchors. *International Symposium on New Developments in Rock Mechanics and Rock Engineering*, Shenyang, 15–22.

Aydan, Ö., Y. Shimizu and T. Kawamoto 1995. A portable system for in-situ characterization of surface morphology and frictional properties of rock discontinuities. Field measurements in geomechanics. *4th International Symposium*, 463–470.

Aydan, Ö., Y. Shimizu and T. Kawamoto 1996. The anisotropy of surface morphology and shear strength characteristics of rock discontinuities and its evaluation. *NARMS'96*, 1391–1398.

Aydan, Ö., H. Ücpirti and N. Turk 1997. Theory of laboratory methods for measuring permeability of rocks and tests (in Turkish). *Bulletin of Rock Mechanics*, Ankara, 13, 19–36.

Aydan, Ö., N. Tokashiki and M. Edahiro 2016a. Utilization of X-ray CT scanning technique in rock mechanics applications. ARMS2016, Bali.

Aydan, Ö., T. Ito and F. Rassouli 2016b. *Chapter 11: Tests on Creep Characteristics of Rocks*, CRC Press, 333–364.

Aydan, Ö., N. Tokashiki, J. Tomiyama, N. Iwata, K. Adachi and Y. Takahashi 2016c. The development of a servo-control testing machine for dynamic shear testing of rock discontinuities and soft rocks, EUROCK2016, Ürgüp, 791–796.

Aydan, Ö., Y. Ohta, N. Iwata and R. Kiyota 2019. The evaluation of static and dynamic frictional properties of rock discontinuities from tilting and stick-slip tests. *Proceedings of 46th Japan Rock Mechanics Symposium*, Iwate, 105–110.

Barton, N.R. and V. Choubey 1977. The shear strength of rock joints in theory and practice. *Rock Mechanics*, 10, 1–54.

Birch, F. 1950. Flow of heat in the Front Range. Colorado. Bull Geol Soc Am, 61, 567–630.

Blum, P. 1997. Physical properties handbook. *ODP Technical Note*, 26. doi: 10.2973/odp.tn.26.1997.

Brace, W.F., J.B. Walsh and W.T. Frangos 1968. Permeability of granite under high pressure. *Journal of Geophysical Research*, 73(6), 2225–2236.

Brown, E.T. 1981. *Suggested Methods for Rock Characterization, Testing, Monitoring*, Oxford, Pergamon Press.

Carslaw, H. S. and J.C. Jaeger 1959. *Conduction of Heat in Solids*. 2nd ed., Oxford, Clarendon Press.

Crank, J. 1956. *Mathematics of Diffusion*, Oxford, Clarendon Press.

Cristescu, N.D. and U. Hunsche 1998. *Time Effects in Rock Mechanics*, New York, John Wiley & Sons.

Farlow, S.J. 1982. *Partial Differential Equations for Scientists and Engineers*, New York, Wiley.

Feng, X.T. 2016. *Laboratory and Field Testing*, ISRM Book Series, London, CRC Press, 2, 631 p.

Hondros, G. 1959. The evaluation of Poisson's ratio and the modulus of materials of low tensile resistance by the Brazilian (indirect tensile) tests with particular reference to concrete. *Australian Journal of Applied Sciences*, 10, 243–268.

Hudson, J.A., S.L. Crouch and C. Fairhurst 1972. Soft, stiff and servo-controlled testing machines. A review with reference to rock failure. *Engineering Geology*, 6, 155–189.

Ishizuka, Y., H. Koyama and S. Komura 1993. Effect of strain rate on strength and frequency dependence of fatigue failure of rocks. *Proceedings of Assessment and Prevention of Failure Phenomena in Rock Engineering*, 321–327.

Ito, T. and T. Akagi 2001. Methods to predict the time of creep failure. *Proceedings of the 31st Symposium on Rock Mechanics of Japan*, 77–81 (in Japanese).

Ito, T., Ö. Aydan, R. Ulusay and Ö. Kaşmer 2008. Creep characteristics of tuff in the vicinity of Zelve antique settlement in Cappadocia region of Turkey. *Proceedings of 5th Asian Rock Mechanics Symposium (ARMS5)*, Tehran, 337–344.

Jaeger, J.C. and N.G.W. Cook 1979. *Fundamentals of Rock Mechanics*. 3rd ed., London, Chapman & Hall, pp. 79 and 311.

Jumikis, A.R. 1983. Rock Mechanics, Trans tech Publications, 613 p.

Kano, K., T. Doi, M. Daido and Ö. Aydan 2004. The development of electrical resistivity technique for real-time monitoring and measuring water-migration and its characteristics of soft rocks. *3rd Asian Rock Mechanics Symposium*, Kyoto, 2, 851–854.

Kawamoto, T., N. Tokashiki and Y. Ishizuka 1980. On uniaxial compression test of rock-like materials using a new type of high stiff testing machine Japan. *Material Science Journal (Zairyo)*, 30(322), 517–523.

Kreyszig, E. 1983. *Advanced Engineering Mathematics*, New York, John Wiley & Sons.

Mousavi, M., M. Jafari and F.S. Rassouli 2008. The impression creep test as new method for creep measuring the soft rocks, in A. Majdi and A. Ghazvinian, eds., *ARMS 5th International Symposium*, Tehran, November 24–26. Iranian Society for Rock Mechanics, Iran, 407–413.

Popov, Y., D. Pribnow, J. Sass, C. Williams and H. Burkhardt 1999. Characterisation of rock thermal conductivity by high-resolution optical scanning. *Geothermics*, 28, 253–276.

Rassouli, F.S., M. Moosavi and M.H. Mehranpour 2010. The effects of different boundary conditions on creep behavior of soft rocks. The 44th U.S. Rock mechanics Symposium & 5th U.S. Canada symposium, Salt Lake City, Utah.

Rummel, F. and G. Fairhurst 1970. Determination of the post-failure behaviour of brittle rock using a servo-controlled testing machine. *Rock Mechanics*, 2, 189–204.

Sass, J., C. Stone and R. Munroe 1984. Thermal conductivity determinations on solid rock – a comparison between a steady-state divided-bar apparatus and a commercial transient line-source device. *Journal of Volcanology and Geothermal Research*, 20(1–2), 145–153.

Sato, A. and Ö. Aydan 2014. An X-ray CT imaging of water absorption process of soft rocks. *Unsaturated Soils: Research and Applications – Proceedings of the 6th International Conference on Unsaturated Soils*, UNSAT, Taylor Francis-Balkema, 675–678.

Somerton, W.H. 1992. *Thermal Properties and Temperature-Related Behavior of Rock/Fluid Systems. Developments in Petroleum Science*, 37, Amsterdam, Elsevier Science Publishers B.V., 257 p.

Ulusay, R. 2015. *The ISRM Suggested Methods for Rock Characterization, Testing, Monitoring, 2007–2014*, Vienna, Springer.

Ulusay, R. and J.A. Hudson 2007. The complete ISRM suggested methods for rock characterization, testing, monitoring, 1974–2007, Ankara.

Von Herzen, R. and A.E. Maxwell 1959. The measurement of thermal conductivity of deep-sea sediments by a needle-probe method. *Journal of Geophysical Research*, 64(10), 1557–1563.

Waversik, W.R. and C. Fairhurst 1970. A study of brittle rock fracture in laboratory compression experiments. *International Journal of Rock Mechanics and Mining Sciences*, 7, 561–575.

Zimmerman, R.W., W.H. Somerton and M.S. King 1986. Compressibility of porous rocks. *Journal of Geophysical Research*, 91(B12), 12765–12777.

Chapter 7

Methods for exact (closed-form) solutions

The solutions of governing equations of coupled or uncoupled motion, mass transportation and energy transport phenomena require certain methods, which may be exact or approximate. When the resulting equations including initial and/or boundary conditions are simple to solve, the exact (closed-from) solutions are preferred (e.g. Kirsch 1898; Jaeger and Cook 1979; Eringen 1961; Fenner 1938; Galin 1946; Savin 1961; Timoshenko and Goodier 1951, 1970; Detournay 1983, 1986; Aydan 1989, 1994a, b, 1995a, b, 1997, 2008, 2017, 2018; Verruijt 1970). As the resulting equations and initial and/or boundary conditions are generally complex in many rock engineering problems, the use of approximate (numerical) methods such as finite difference, finite element or boundary element methods becomes necessary.

7.1 Basic approaches

The fundamental governing equations presented in Chapter 4 are either ordinary differential or partial differential equations for mass, momentum and energy conservations laws with chosen constitutive laws. The closed-form (analytical) methods attempt to solve the equations in their original form and they can be categorized into intuitive, complex variable or separation of variable techniques.

7.1.1 Intuitive function methods

This method is fundamentally based on choosing a function intuitively, which satisfies boundary/initial conditions. For example, the well-known Kirsch's solutions for circular opening under biaxial far-field stresses are based on this approach and utilize Airy's stress function. The readers are advised to refer several textbooks such as that of Jaeger and Cook (1979) on this aspect.

7.1.2 Solution by separating variables

This method assumes that the solution consists of the convolution of several functions of the independent variables (e.g. Kreyszig 1983). The insertion of these functions to partial differential equations results in separated ordinary differential equations. Then the integral coefficients are determined so that the boundary and/or initial conditions are satisfied. Particularly, the determination of integral coefficients may be quite cumbersome. This method is also known as Fourier's method.

7.1.3 Complex variable method

This method is a very powerful tool to obtain solutions to many problems in elasticity. The method was originally developed by Kolosov (1909), and it is further expanded by several Soviet Union (SSCB) mathematicians. For example, Muskhelishvili (1962) provided a comprehensive textbook on this solution method. Similar textbooks by Green and Zerna (1968), and England (1971) can be found in the literature. The well-known textbook by Kreyszig (1983) also describes this method. The method is quite powerful to solve the partial differential equations subjected to very complex far-field boundary stress conditions for anisotropic elastic materials. Several applications of this method are described in the textbook by Jaeger and Cook (1979). Gerçek (1988, 1996) provided the application of the method for stress concentration around cavities having different shapes under biaxial far-field stresses, in which the integral constants are obtained numerically. The textbook by Verruijt (1970) describes the utilization of this method in seepage problems.

7.2 Closed-form solutions for solids

In this section, several specific applications of the closed-form solutions are described.

7.2.1 Visco-elastic rock sample subjected to uniaxial loading

Aydan (1997) proposed a method to model the dynamic response of rock samples during loading. In this subsection, this method and several examples of its application to some typical situations are presented.

a. Theoretical formulation

Let us consider a sample under uniaxial loading as shown in Figure 7.1a. The force equilibrium of such a sample can be written in the following form (Figure 7.1b):

$$\sigma = D_r\varepsilon + C_r\dot{\varepsilon} + \rho H\ddot{u} \quad (7.1)$$

where D_r, C_r, ρ and H are elastic modulus, viscosity coefficient, density and sample height, respectively. If acceleration $\ddot{u}$ is uniform over the sample and its strain ε is defined as

$$\varepsilon = \frac{u}{H} \quad (7.2)$$

Equation (7.1) becomes

$$\sigma = D_r\varepsilon + D_r\dot{\varepsilon} + \rho H^2\ddot{\varepsilon} \quad (7.3)$$

Let us assume that stress is applied onto the sample in the following form (Figure 7.2)

for $0 \leq t \leq T_0$

$$\sigma = \frac{\sigma_0}{T_0}t \quad (7.4)$$

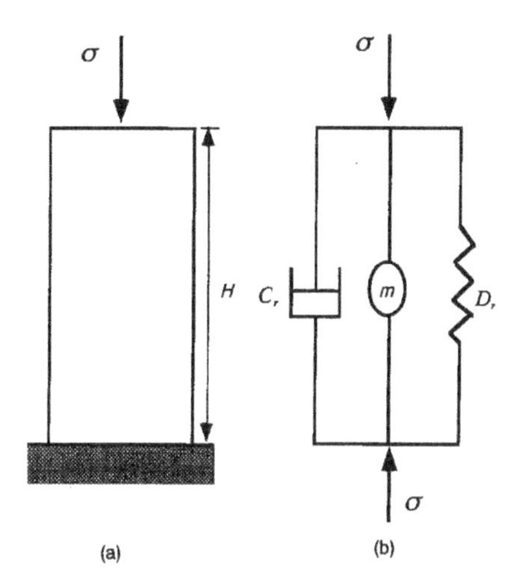

Figure 7.1 (a) Uniaxial compression test and (b) its mechanical model.

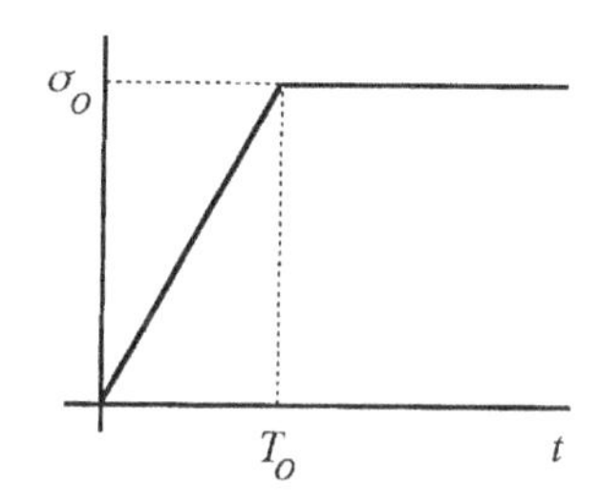

Figure 7.2 Time-history of uniaxial loading.

for $t \geq T_0$

$$\sigma = \sigma_0 \tag{7.5}$$

The solutions to this ordinary differential equation are:

Case 1: roots are real

$$\varepsilon = C_1 e^{\lambda_1 t} + C_2 e^{\lambda_2 t} + \varepsilon_p \tag{7.6}$$

where

$$\lambda_1 = \frac{1}{2\rho H^2}\left(-C_r + \sqrt{C_r^2 - 4D_r\rho H^2}\right), \lambda_2 = \frac{1}{2\rho H^2}\left(-C_r - \sqrt{C_r^2 - 4D_r\rho H^2}\right)$$

for $0 \leq t \leq T_0$

$$\varepsilon_p = \frac{\sigma_0}{\rho H^2 T_0}\frac{1}{\lambda_1^2\lambda_2^2}\left[\lambda_1\lambda_2 t + (\lambda_1 + \lambda_2)\right]$$

for $0 \le t \le T_0$

$$\varepsilon_p = \frac{\sigma_0}{\rho H^2} \frac{1}{\lambda_1 \lambda_2}$$

Case 2: roots are the same

$$\varepsilon = [C_1 + C_2 t] e^{\lambda t} + \varepsilon_p \tag{7.7}$$

where

$$\lambda = -\frac{C_r}{2\rho H^2}$$

for $0 \le t \le T_0$

$$\varepsilon_p = \frac{1}{\lambda^3} \frac{\sigma_0}{\rho H^2 T_0}$$

for $t \ge T_0$

$$\varepsilon = \frac{1}{\lambda^2} \frac{\sigma_0}{\rho H^2}$$

Case 3: roots are complex

$$\varepsilon = e^{pt} [A \cos qt + B \sin qt] + \varepsilon_p \tag{7.8}$$

where

$$p = -\frac{C_r}{2\rho H^2}, \quad q = \sqrt{4 D_r \rho H^2 - C_r^2}$$

for $0 \le t \le T_0$

$$\varepsilon_p = \frac{1}{\left(p^2 + q^2\right)^2} \frac{\sigma_0}{\rho H^2 T_0} \left[\left(p^2 + q^2\right) t + 2p\right]$$

for $t \ge T_0$

$$\varepsilon_p = \frac{1}{p^2 + q^2} \frac{\sigma_0}{\rho H^2}$$

Integration constants C_1 and C_2 can be determined from the following initial conditions:

for $0 \leq t < T_0$

$$\begin{aligned} \varepsilon &= 0 \quad \text{at} \quad t = 0 \\ \dot{\varepsilon} &= 0 \quad \text{at} \quad t = 0 \end{aligned} \tag{7.9}$$

for $t \geq T_0$

$$\begin{aligned} \varepsilon &= \varepsilon_0 \quad \text{at} \quad t = T_0 \\ \dot{\varepsilon} &= \dot{\varepsilon}_0 \quad \text{at} \quad t = T_0 \end{aligned} \tag{7.10}$$

Integration constants can be easily obtained for the above conditions for each case. However, their specific forms are not presented as they are too lengthy.

b. Applications

Several applications of the theoretical relations derived in the previous section are given herein to investigate the effects of viscosity coefficient, elasticity coefficient, loading rate and sample height.

1. *The effect of viscosity coefficient*: Figure 7.3 shows the effect of the viscosity coefficient on the deformation responses of a sample. It is of great interest that when rock is elastic, an oscillating behavior must be observed. Furthermore, the stress–strain relation is not linear and it also oscillates as the applied stress

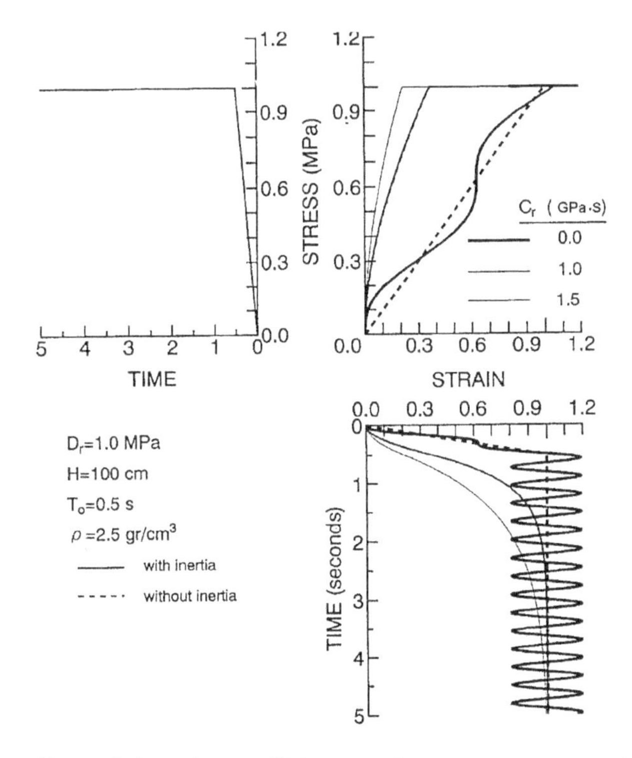

Figure 7.3 The effect of viscosity coefficient on dynamics response of a sample subjected to uniaxial compression.

is linearly increased. However, this oscillating behavior is suppressed as the viscosity coefficient increases.

2. *The effect of elasticity coefficient*: Figure 7.4 shows the effect of elasticity coefficient on the deformation responses of a sample with a viscosity coefficient of 0 GPa · s. The amplitudes of the oscillating part and stationary part of strain decrease as the value of the elasticity coefficient increases. Nevertheless, the oscillating behavior is apparent for each case.
3. *The effect of loading rate*: Figure 7.5 shows the effect of loading rate on the deformation responses of a sample with a viscosity coefficient of 0 GPa · s. While the amplitude of the stationary part of strain remains the same, the amplitude of the oscillating part of strain decreases as the loading rate decreases. Although the oscillating behavior could not be suppressed, the effect of oscillation tends to become smaller.
4. *The effect of sample height*: Figure 7.6 shows the effect of sample height on the deformation responses of a sample with a viscosity coefficient of 0 GPa · s. The amplitudes of the oscillating part and stationary part of strain remain the same while the period of oscillations becomes larger as the value of sample height increases.

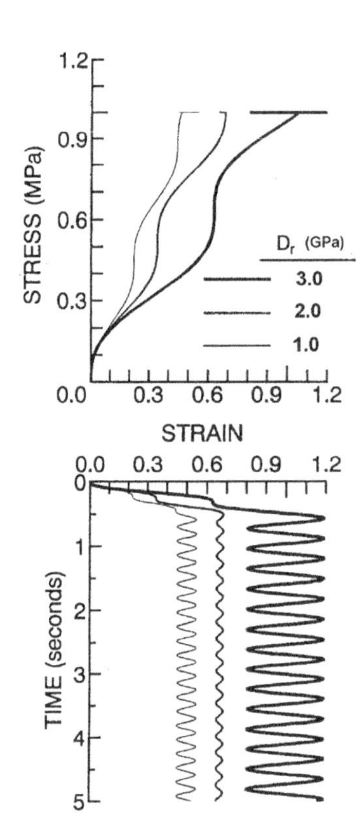

Figure 7.4 The effect of the elastic coefficient on the dynamics response of a sample subjected to uniaxial compression.

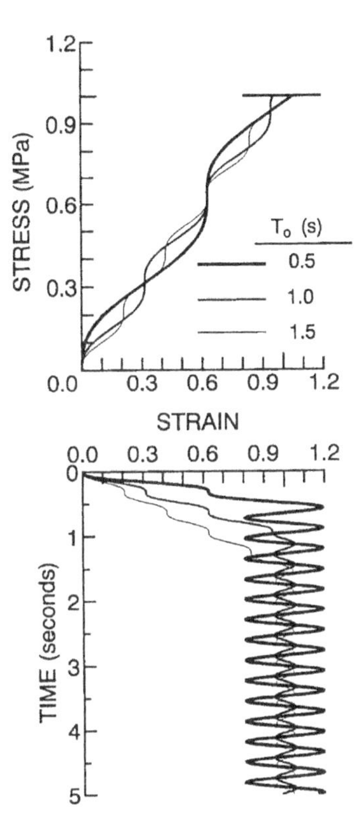

Figure 7.5 The effect of loading rate on dynamics response of a sample subjected to uniaxial compression.

7.2.2 Visco-elastic layer on an incline

A semi-infinite slab on an incline is considered and it is assumed to be subjected to instantaneous gravitational loading (Figure 7.7a). The original formulation was developed by Aydan (1994a, b, 1995a, b) and it is adopted herein for assessing the dynamic response of a semi-infinite layer on an incline, which is a very close-situation to the slope stability assessment of Terzaghi (1960).

a. Formulation

Let us assume that the deformation is purely due to shearing under gravitational loading and the slab behaves in a visco-elastic manner of Kelvin–Voigt type given by (Figure 7.7b):

$$\tau = G\gamma + \eta\dot{\gamma} \tag{7.11}$$

where G is the elastic shear modulus and η is the viscous shear modulus. This model is known as the Voigt–Kelvin model (Eringen 1980). When $G = 0$, then it simply corresponds to a Newtonian fluid. On the other hand, when $\eta = 0$, it corresponds to a Hookean solid.

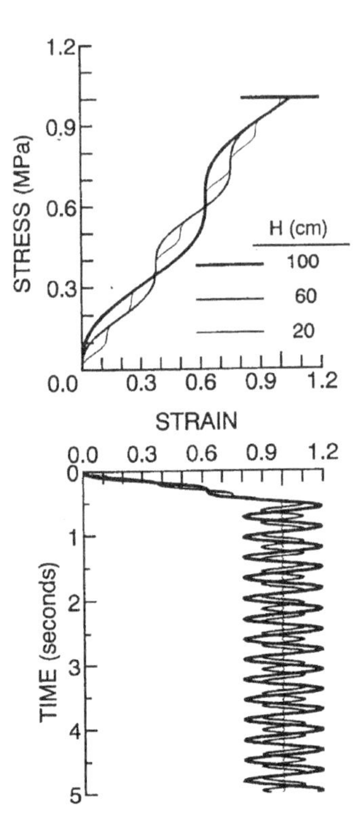

Figure 7.6 The effect of sample height on the dynamics response of a sample subjected to uniaxial compression.

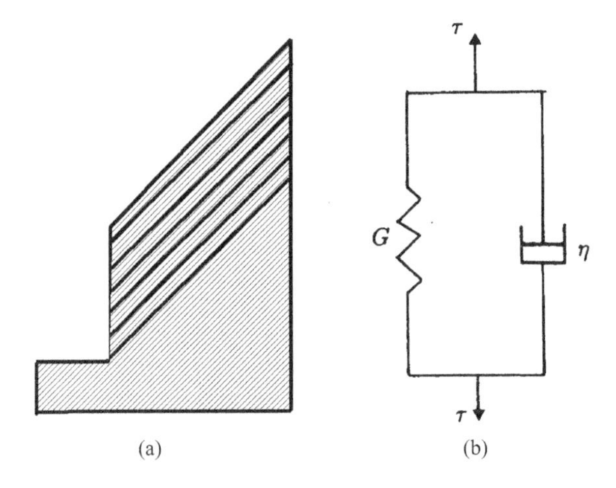

Figure 7.7 (a) An illustration of semi-infinite slope (modified from Terzaghi 1960) and (b) constitutive model.

Let us consider an infinitesimal element within an inclined infinitely long layer as illustrated in Figure 7.8. The governing equation takes the following form by considering the equilibrium of the element by applying Newton's second Law (Eringen 1980):

$$\frac{\partial \tau}{\partial y} - \frac{\partial p}{\partial x} + \rho g \sin\alpha = \rho \ddot{u} \tag{7.12}$$

If the thickness of the liquified layer does not vary with x and the medium consists of the same material, then $\partial p/\partial x = 0$ and the above equation becomes:

$$\frac{\partial \tau}{\partial y} + \rho g \sin\alpha = \rho \ddot{u} \tag{7.13}$$

Assuming that the shear strain and shear strain rate can be defined as

$$\gamma = \frac{\partial u}{\partial y}, \quad \dot{\gamma} = \frac{\partial \dot{u}}{\partial y} \tag{7.14}$$

and introducing the constitutive law given by Eq. (7.11) into Eq. (7.13) yield the following partial differential equation

$$\rho \frac{\partial^2 u}{\partial t^2} + \eta \frac{\partial^2}{\partial y^2}\left(\frac{\partial u}{\partial t}\right) + G \frac{\partial^2 u}{\partial y^2} = \rho g \sin\alpha \tag{7.15}$$

Let us assume that the solution of this partial differential equation using the separation of variable technique is given as (i.e. Kreyszig 1983; Zachmanoglou and Thoe 1986)

$$u(y,t) = Y(y) \cdot T(t) \tag{7.16}$$

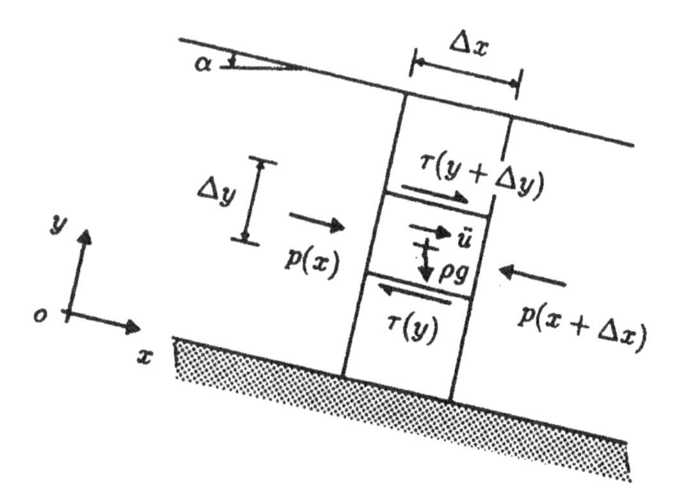

Figure 7.8 Mechanical model for shearing of a semi-infinite slope.

As a particular case based on an intuitive approach, $Y(y)$ is assumed to be of the following form by considering an earlier solution of the equilibrium equation without an inertial term for a semi-infinite slab with free-surface boundary conditions (Figure 7.9a):

$$Y(y) = y\left(H - \frac{y}{2}\right) \tag{7.17}$$

By introducing this relation into Eq. (7.15), we obtain

$$\rho\frac{\partial^2 T}{\partial t^2}y\left(H - \frac{y}{2}\right) + \eta\frac{\partial T}{\partial t} + GT = \rho g \sin\alpha \tag{7.18}$$

Integrating the above equation with respect to y for bounds $y=0$ and $y=H$ results in the following second-order non-homogeneous ordinary differential equation

$$\frac{\partial^2 T}{\partial t^2} - \frac{3\eta}{\rho H^2}\frac{\partial T}{\partial t} + \frac{3G}{\rho H^2}T = \frac{3g\sin\alpha}{H^2} \tag{7.19}$$

The solutions of this differential equation are:

Case 1: roots are real

$$T = C_1 e^{-\lambda_1 t} + C_2 e^{-\lambda_2 t} + \frac{1}{\lambda_1\lambda_2}\frac{3g\sin\alpha}{H^2} \tag{7.20}$$

where

$$\lambda_1 = \frac{1}{2\rho H^2}\left(-3\eta + \sqrt{9\eta^2 - 12G\rho H^2}\right), \quad \lambda_2 = \frac{1}{2\rho H^2}\left(-3\eta - \sqrt{9\eta^2 - 12G\rho H^2}\right)$$

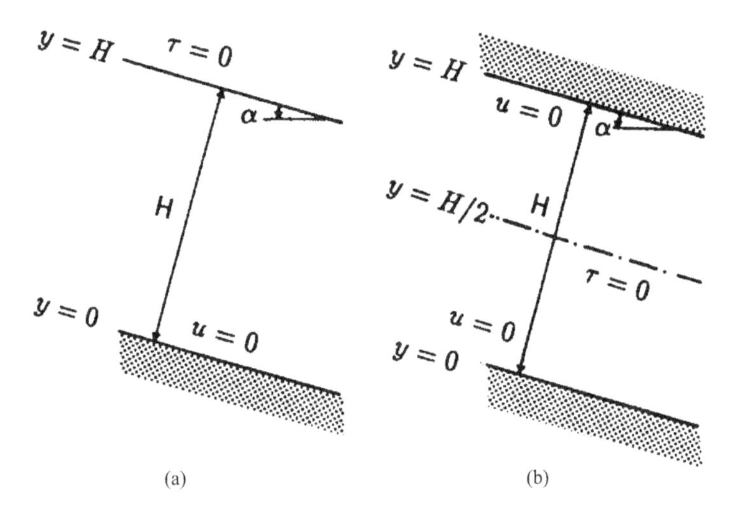

Figure 7.9 Boundary conditions. (a) Free surface condition and (b) constrained surface condition.

Case 2: roots are the same

$$T = [C_1 + C_2 t]e^{-\lambda t} + \frac{1}{\lambda^2}\frac{3g\sin\alpha}{H^2} \tag{7.21}$$

where

$$\lambda = -\frac{3\eta}{2\rho H^2}$$

Case 3: roots are complex

$$T = e^{pt}[A\cos qt + B\sin qt] + \frac{1}{p^2+q^2}\frac{3g\sin\alpha}{H^2} \tag{7.22}$$

where

$$p = -\frac{3\eta}{2\rho H^2}, \quad q = \sqrt{12G\rho H^2 - 9\eta^2}$$

Integration constants C_1 and C_2 can be determined from the following initial conditions:

$$\begin{aligned} u(y,t) &= 0 \quad \text{at} \quad t = 0 \\ \dot{u}(y,t) &= 0 \quad \text{at} \quad t = 0 \end{aligned} \tag{7.23}$$

For the above initial conditions, the integration constants for each case are

Case 1: roots are real

$$C_1 = -\frac{1}{\lambda_1(\lambda_2 - \lambda_1)}\frac{3g\sin\alpha}{H^2}, \quad C_2 = -\frac{1}{\lambda_2(\lambda_2 - \lambda_1)}\frac{3g\sin\alpha}{H^2} \tag{7.24}$$

Case 2: roots are the same

$$C_1 = -\frac{1}{\lambda^2}\frac{3g\sin\alpha}{H^2}, \quad C_2 = \frac{1}{\lambda^2}\frac{3g\sin\alpha}{H^2} \tag{7.25}$$

Case 3: roots are complex

$$C_1 = -\frac{1}{p^2+q^2}\frac{3g\sin\alpha}{H^2}, \quad C_2 = \frac{p}{q}\cdot\frac{1}{p^2+q^2}\frac{3g\sin\alpha}{H^2} \tag{7.26}$$

Integration constants for constrained boundary conditions can be obtained similarly. This situation may be quite relevant to the response of soft-layers sandwiched between two relatively rigid layers in underground excavations and trap-door experiments used particularly for underground openings in soil (Terzaghi 1946).

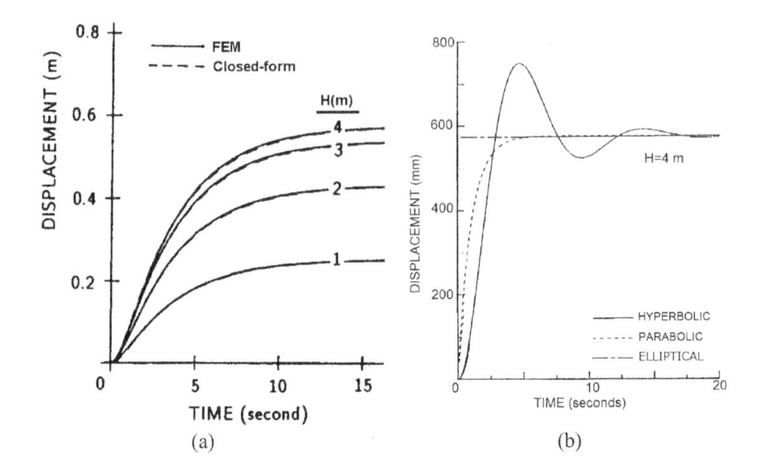

Figure 7.10 (a) Dynamic response of a 4-m thick semi-infinite slab under instantaneously applied gravitation load; (b) comparison of responses obtained for hyperbolic, parabolic and elliptical forms of the differential equation for a 4-m thick slab under instantaneous gravitational loading.

b. Applications
   Figure 7.10 compares the solutions obtained from the closed-form solution and FEM for a 4-m thick semi-infinite slab. Using the solutions presented in this section, one may compare the expected responses under different circumstances. Such a comparison has been already done by Aydan 1994a, b, 1995a, b. The negligence of the inertia component in Eq. (7.11) results in a parabolic partial differential equation. If the viscous effect is neglected, the resulting equation would result in a differential equation of the elliptical form. Figure 7.10b compares the responses obtained for three situations of the differential equation. As noted from the figure, all solutions converge to the solution obtained from the elliptical form (static case). The inertia component implies that displacement, as well as resulting stress and strain responses, would be greater than those of the elliptical form.

7.2.3 One-dimensional bar embedded in rock

The equation of motion for the axial responses of rockbolts and rock anchors together with the consideration of inertia component including mass proportional damping can be written in the following form (Figure 7.11)

$$\rho\frac{\partial^2 u_b}{\partial t^2}+h_a^*\frac{\partial u_b}{\partial t^2}=\frac{\partial \sigma}{\partial x}+\frac{2}{r_b}\tau_b \tag{7.27}$$

The analytical solutions for Eq. (7.27) are extremely difficult for given constitutive laws, boundary and initial conditions. However, it is possible to obtain solutions for simple cases, which may be useful for the interpretation of results of site-investigations. Equation (7.27) may be reduced to the following form by omitting the effect of damping and interaction with surrounding rock as

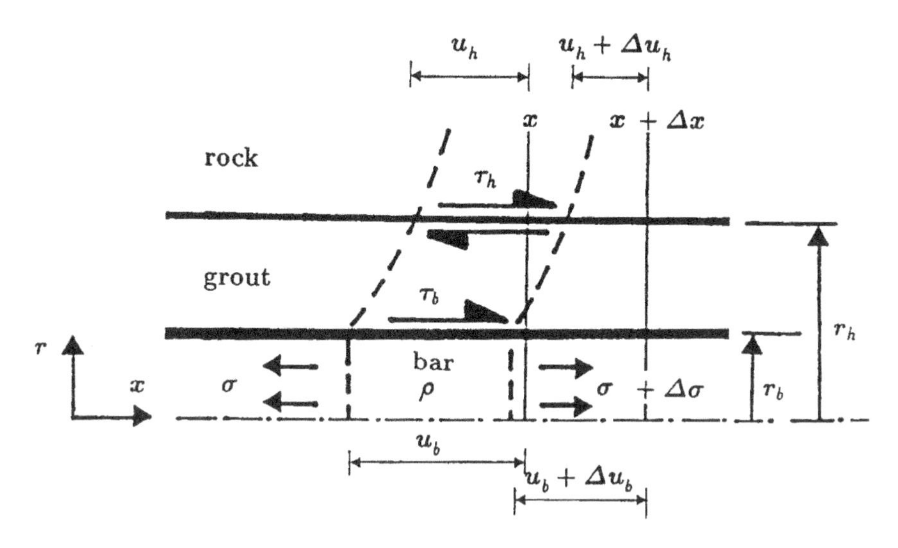

Figure 7.11 Modeling of the dynamic axial response of tendons.

$$\frac{\partial^2 u_b}{\partial t^2} = V_p^2 \frac{\partial^2 u_b}{\partial x^2} \tag{7.28}$$

where

$$V_p = \sqrt{\frac{E_b}{\rho}}$$

The general solution of partial differential Eq. (7.28), which is also known as the D'Alambert solution, may be given as

$$u_b = h\left(x - V_p t\right) + H\left(x + V_p t\right) \tag{7.29}$$

For a very simple situation, the solution may be given as follows:

$$u_b = A \sin \frac{2\pi}{L}\left(x \pm V_p t\right) \tag{7.30}$$

where L is the tendon length. Thus, the Eigenvalues of the tendon may be obtained as follows

$$f_p = n \frac{1}{2L} V_p,\ n = 1,2,3 \tag{7.31}$$

Similarly, the Eigenvalues of traverse vibration of the tendon under a given pre-stress may be obtained as follows:

$$f_T = n \frac{1}{2L} V_T,\ n = 1,2,3 \tag{7.32}$$

where

$$V_T = \sqrt{\frac{\sigma_o}{\rho}}$$

7.2.4 Circular cavity in the elastic rock under a far-field hydrostatic stress

An analytical solution is herein presented in the case of a circular underground opening excavated in a hydrostatic state of stress (Aydan et al. 1992, 1993). To start the derivations, the following conditions are set. Then, the governing equations for bolted and unbolted sections are (Figure 7.12):

$$\frac{d\sigma_r}{dr} + \frac{\sigma_r - \sigma_\theta}{r} = 0 \tag{7.33}$$

where r is the distance from the opening center, σ_r is the radial stress and σ_θ is the tangential stress.

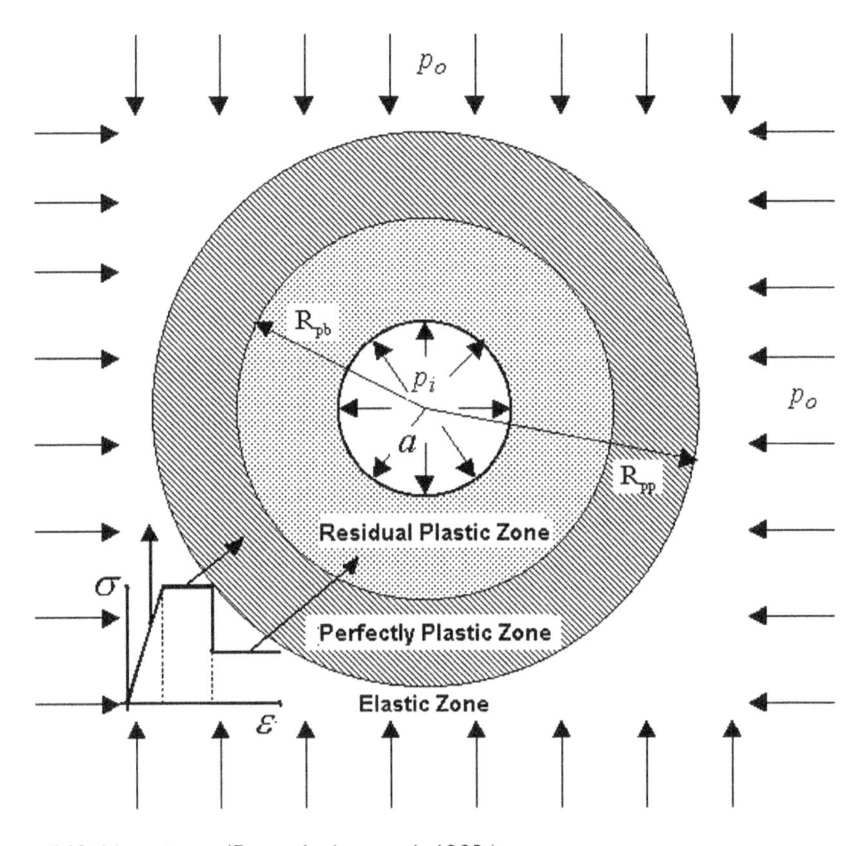

Figure 7.12 Notations. (From Aydan et al. 1993.)

The constitutive law between stresses and strains of rock is of the following form:

$$\left\{\begin{array}{c}\varepsilon_r\\ \varepsilon_\theta\end{array}\right\}=\frac{1-\nu_r^2}{E_r}\left[\begin{array}{cc}1 & -\dfrac{\nu_r}{1-\nu_r}\\ -\dfrac{\nu_r}{1-\nu_r} & 1\end{array}\right]\left\{\begin{array}{c}\sigma_r\\ \sigma_\theta\end{array}\right\} \text{or} \left\{\begin{array}{c}\sigma_r\\ \sigma_\theta\end{array}\right\}$$

$$=\frac{E_r(1-\nu_r)}{(1+\nu_r)(1-2\nu_r)}\left[\begin{array}{cc}1 & -\dfrac{\nu_r}{1-\nu_r}\\ -\dfrac{\nu_r}{1-\nu_r} & 1\end{array}\right]\left\{\begin{array}{c}\varepsilon_r\\ \varepsilon_\theta\end{array}\right\} \tag{7.34}$$

where

E_r: elastic modulus of rock, ε_r: radial strain; ε_θ: tangential strain in u

$$\varepsilon_r=\frac{du}{dr},\quad \varepsilon_\theta=\frac{u}{r} \tag{7.35}$$

The stresses and displacements in the bolted and unbolted sections can be obtained by solving the governing Eqs. (7.33) together with the constitutive law (7.34) and the boundary conditions. Substituting the constitutive law (7.34) in governing Eqs. (7.33) together with the relations (7.35) results in the following differential equations:

$$r^2\frac{d^2u}{dr^2}+r\frac{du}{dr}-u=0 \tag{7.36}$$

The general solutions of the above differential equations are of the following forms:

$$u=A_1 r+A_2\frac{1}{r} \tag{7.37}$$

By introducing the following boundary conditions for each section

- Unbolted section

  $\sigma_r=P_b$ at $r=b$

  $\sigma_r=\sigma_0$ at $r=\infty$

  integration constants A_1 and A_2 are obtained as

$$A_1=\frac{(1+\nu_r)(1-2\nu_r)}{E_r}\sigma_0,\quad A_2=\frac{(1+\nu_r)}{E_r}b^2(\sigma_0-P_b).$$

By inserting the above constants in Eq. (7.37), the following expressions can be obtained after some manipulations:

$$u = \frac{1+\nu_r}{E_r}\left[(1-\nu)_r \sigma_0 r + (\sigma_0 - P_b)\frac{b^2}{r^2}\right] \tag{7.38}$$

The above displacement fields are for an initially unstressed body. As the initial displacement field has already taken place in rock mass before the excavation of openings, this initial displacement field has to be subtracted from the above expressions. The initial displacement field is the state when P_b was equal to the far-field stress σ_0. Inserting these identities in the above expressions yields

$$u_0 = \frac{1+\nu_r}{E_r}(1-\nu_r)\sigma_0 r \tag{7.39}$$

Finally, one obtains the following expressions for displacement fields due to the excavation of opening

$$u_e = \frac{1+\nu_r}{E_r}(\sigma_0 - P_b)\frac{b^2}{r^2} \tag{7.40}$$

where subscript e denotes excavation. Variations in displacement fields will, in turn, bring about variations in radial and tangential stresses. These variations are obtained from the above relations together with Eq. (7.35) and the constitutive law (7.34) as

$$\sigma_{re} = -(\sigma_0 - P_b)\frac{b^2}{r^2} \tag{7.41}$$

$$\sigma_{\theta e} = (\sigma_0 - P_b)\frac{b^2}{r^2} \tag{7.42}$$

The stress state in the surrounding medium is, therefore, defined as the sum of variation in the post-excavation state and the pre-excavation stress state given as

$$\sigma_r = \sigma_0 + \sigma_{re} = \sigma_0 - (\sigma_0 - P_b)\frac{b^2}{r^2} \tag{7.43}$$

$$\sigma_\theta = \sigma_0 + \overset{*}{\sigma}_{\theta e} = \sigma_0 + (\sigma_0 - P_b)\frac{b^2}{r^2} \tag{7.44}$$

Note: The above relations for stresses can also be directly obtained from the governing Eq. (7.33) with the use of the following identities

$$\sigma_r + \sigma_\theta = 2\sigma_0 \tag{7.45}$$

and the same boundary conditions.

7.2.5 Unified analytical solutions for circular/spherical cavity in an elasto-plastic rock

a. General solution

Constitutive laws

A generalized form of the constitutive law between stresses and strains of rock in the elastic region for the radially symmetric problem (cylindrical and spherical openings) can be given as:

$$\begin{Bmatrix} \sigma_r \\ \sigma_\theta \end{Bmatrix} = \begin{bmatrix} \lambda + 2\mu & n\lambda \\ \lambda & n\lambda + 2\mu \end{bmatrix} \begin{Bmatrix} \varepsilon_r \\ \varepsilon_\theta \end{Bmatrix} \tag{7.46}$$

where n is the shape coefficient and has a value of 1 for cylindrical opening and 2 for spherical opening; σ_r is the radial stress; σ_θ is the tangential stress; ε_r is the radial strain; ε_θ is the tangential strain and λ and μ are Lame constants and given as:

$$\lambda = \frac{Ev}{(1+v)\ (1-2v)}; \quad \mu = \frac{E}{2(1+v)} \tag{7.47}$$

where E is the elastic modulus of rock and v is the Poisson ratio of rock.

Equilibrium equation

When the problem is radially symmetric, the momentum law for the static case takes the following form:

$$\frac{d\sigma_r}{dr} + n\frac{\sigma_r - \sigma_\theta}{r} = 0 \tag{7.48}$$

where r is the distance from the opening center.

Compatibility condition

The compatibility condition between strain components for radially symmetric openings is given as:

$$\frac{d\varepsilon_\theta}{dr} + \frac{\varepsilon_\theta - \varepsilon_r}{r} = 0 \tag{7.49}$$

Relations between strain components and radial displacement (u) are:

$$\varepsilon_r = \frac{du}{dr}, \quad \varepsilon_\theta = \frac{u}{r} \tag{7.50}$$

Behavior of rock material

An elastic-perfect-residual plastic model as shown in Figure 7.13 approximates the behavior of rock. Although it is possible to consider the strain-softening behavior, it is

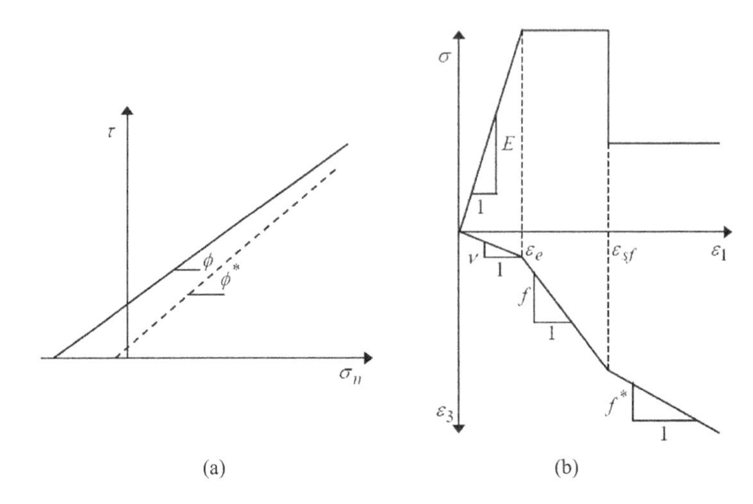

Figure 7.13 Mechanical models for rock mass. (a) Yield criterion and (b) strain–stress behavior.

extremely difficult to obtain closed-form solutions and numerical techniques would be necessary. Rock was assumed to obey the Mohr–Coulomb yield criterion. Although it is possible to derive solutions for the Hoek–Brown criterion, it is not intentionally done as the generalized Hoek–Brown criterion violates the Euler theorem used in the classical theory of plasticity for constitutive modeling of rocks. Failure zones about radially symmetric openings excavated in rock mass for elastic-perfect-residual plastic behavior and yield functions for each region are illustrated in Figure 7.14 and are given by:

$$\sigma_1 = q\sigma_3 + \sigma_c, \quad q = \frac{1+\sin\varphi}{1-\sin\varphi} \quad \text{perfectly plastic region} \tag{7.51a}$$

$$\sigma_1 = q^*\sigma_3 + \sigma_c^*, \quad q^* = \frac{1+\sin\varphi^*}{1-\sin\varphi^*}, \quad \text{residual plastic region} \tag{7.51b}$$

where σ_1 is the maximum principal stress; σ_3 is the minimum principal stress; σ_c is the uniaxial compressive strength of intact rock; σ_c^* is the uniaxial compressive strength of broken rock; φ is the internal friction angle of intact rock and φ^* is the internal friction angle of broken rock. Relations between total radial and tangential strains in plastic regimes are assumed to be of the following form

$$\varepsilon_r = -f\ \varepsilon_\theta \quad \text{for a perfectly plastic region} \tag{7.52a}$$

$$\varepsilon_r = -f^*\varepsilon_\theta \quad \text{for a residual plastic region} \tag{7.52b}$$

where f and f^* are physical constants obtained from tests. These constants may be interpreted as plastic Poisson's ratios (Aydan et al. 1995a, b, 1996).

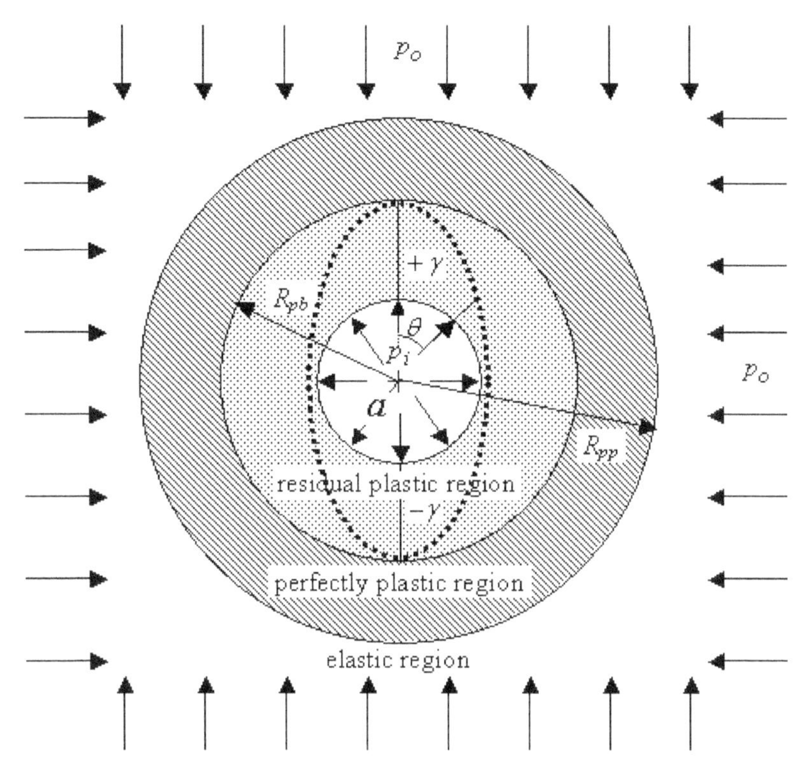

Figure 7.14 States about opening and notations (gravity is considered in the dotted zone).

i. Stress and Strain Field Around Opening

   1. Residual Plastic Zone ($a \le r \le R_{pb}$)
      Inserting the yield criterion Eq. (7.52b) into the governing Eq. (7.48) with $\sigma_3 = \sigma_r$ and $\sigma_1 = \sigma_\theta$ yields

$$\frac{d\sigma_r}{dr} + n\left(1 - q^*\right)\frac{\sigma_r}{r} = n\frac{\sigma_c^*}{r}. \tag{7.53}$$

The solution of the above differential equation is

$$\sigma_r = C\, r^{n\left(q^* - 1\right)} - \frac{\sigma_c^*}{q^* - 1}. \tag{7.54}$$

The integration constant C is obtained from the boundary condition $\sigma_r = p_i$ at $r = a$ as

$$C = \left(p_i + \frac{\sigma_c^*}{q^* - 1}\right)\frac{1}{a^{n(q^*-1)}}, \tag{7.55}$$

where p_i is the internal or support pressure. Thus, the stresses now take the following forms:

$$\sigma_r = \left(p_i + \frac{\sigma_c^*}{q^* - 1} \right) \left(\frac{r}{a} \right)^{n\left(q^* - 1\right)} - \frac{\sigma_c^*}{q^* - 1}, \tag{7.56}$$

$$\sigma_\theta = q^* \left(p_i + \frac{\sigma_c^*}{q^* - 1} \right) \left(\frac{r}{a} \right)^{n\left(q^* - 1\right)} - \frac{\sigma_c^*}{q^* - 1}. \tag{7.57}$$

Solving the differential equation obtained by inserting the relation given by Eq. (7.52b) in Eq. (7.50) yields:

$$\varepsilon_\theta = \frac{A}{r^{f^* + 1}}. \tag{7.58}$$

The integration constant A is determined from the continuity of the tangential strain at the perfect-residual plastic boundary $r = R_{pb}$ as

$$A = \varepsilon_\theta^{pb} R_{pb}^{f^* + 1}, \tag{7.59}$$

ε_θ^{pb} in Eq. (7.59) is the tangential strain at the perfect-residual plastic boundary ($r = R_{pb}$) and it is specifically given by

$$\varepsilon_\theta^{pb} = \eta_{sf} \varepsilon_\theta^{ep} \qquad \eta_{\text{sf}} = \frac{\varepsilon_{sf}}{\varepsilon_e}, \tag{7.60}$$

where η_{sf} is the tangential strain level at the perfect-residual plastic boundary (Figure 7.13); ε_θ^{ep} is the tangential strain at the elastic-perfect plastic boundary as

$$\varepsilon_\theta^{\text{ep}} = \frac{1 + \nu}{nE} \left(p_0 - \sigma_{rp} \right). \tag{7.61}$$

σ_{rp} in Eq. (7.61) is the radial stress at the elastic-perfect plastic boundary. As a result, the tangential strain in the surrounding rock becomes

$$\varepsilon_\theta = \frac{1 + \nu}{nE} \left(p_0 - \sigma_{rp} \right) \eta_{sf} \left(\frac{R_{pb}}{r} \right)^{f^* + 1}. \tag{7.62}$$

2. Perfectly-Plastic Zone ($R_{pb} \leq r \leq R_{pp}$)
   Inserting the yield criterion Eq. (7.51a) into the governing Eq. (7.48) with $\sigma_3 = \sigma_r$ and $\sigma_1 = \sigma_\theta$ gives

$$\frac{d\sigma_r}{dr} + n\,(1-q)\frac{\sigma_r}{r} = n\frac{\sigma_c}{r}. \tag{7.63}$$

The solution of the above differential equation is

$$\sigma_r = Cr^{n(q-1)} - \frac{\sigma_c}{q-1}. \tag{7.64}$$

The integration constant C is obtained from the boundary condition $\sigma_r = \sigma_{rp}$ at $r = R_{pp}$ as

$$C = \left(\sigma_{rp} + \frac{\sigma_c}{q-1}\right)\frac{1}{R_{pp}^{n(q-1)}}. \tag{7.65}$$

Thus, the stresses now take the following forms:

$$\sigma_r = \left(\sigma_{rp} + \frac{\sigma_c}{q-1}\right)\left(\frac{r}{R_{pp}}\right)^{n(q-1)} - \frac{\sigma_c}{q-1}, \tag{7.66}$$

$$\sigma_\theta = q\left(\sigma_{rp} + \frac{\sigma_c}{q-1}\right)\left(\frac{r}{R_{pp}}\right)^{n(q-1)} - \frac{\sigma_c}{q-1}. \tag{7.67}$$

Since the derivation of the tangential strain is similar to the previous case, the final expression takes the following form:

$$\varepsilon_\theta = \frac{1+\nu}{nE}\left(p_0 - \sigma_{rp}\right)\left(\frac{R_{pp}}{r}\right)^{f+1}. \tag{7.68}$$

The relation between the plastic zone radii is also found from the requirement of the continuity of tangential strain at $r = R_{pb}$ and relation Eq. (7.62) as:

$$\frac{R_{pp}}{R_{pb}} = \eta_{sf}^{\frac{1}{f+1}}. \tag{7.69}$$

3. Elastic Zone ($R_{pp} \le r$)
   The derivation of stresses and displacement expressions for the cylindrical opening was previously given in detail with the consideration of initially stressed elastic medium by a far-field hydrostatic in-situ stress (p_0). The final forms of the expressions for radially symmetric openings are of the following forms:

$$\sigma_r = p_0 - \left(p_0 - \sigma_{rp}\right)\left(\frac{R_{pp}}{r}\right)^{n+1}, \tag{7.70}$$

$$\sigma_\theta = p_0 + \frac{1}{n}\left(p_0 - \sigma_{rp}\right)\left(\frac{R_{pp}}{r}\right)^{n+1}, \tag{7.71}$$

$$\varepsilon_\theta = \frac{1+\nu}{nE}\left(p_0 - \sigma_{rp}\right)\left(\frac{R_{pp}}{r}\right)^{n+1}. \tag{7.72}$$

The specific form for σ_{rp} is obtained from the continuity condition of tangential stresses at $r = R_{pp}$ by equality Eqs. (7.68) and (7.72) as:

$$\sigma_{rp} = \frac{p_0 + n\left(p_0 - \sigma_c\right)}{1 + n\,q}. \tag{7.73}$$

ii. Plastic Zones Radius Around Opening

1. Perfectly Plastic-Residual Plastic Zone Boundary Radius (R_{pb})
The perfectly plastic-residual plastic zone boundary radius is found from the requirement of the continuity of radial stresses, i.e. by equality of Eqs. (7.73) and (7.67), at $r = R_{pb}$ as:

$$\frac{R_{pb}}{a} = \left\{\frac{\frac{(1+n)\left[(q-1)+\alpha\right]}{(1+nq)\,(q-1)}\left(\eta_{sf}\right)^{\frac{n(1\text{-}q)}{f+1}} - \frac{\alpha}{q-1} + \frac{\alpha^*}{q^*-1}}{\beta + \frac{\alpha^*}{q^*-1}}\right\}^{\frac{1}{n(q^*-1)}}, \tag{7.74}$$

where β is the support pressure normalized by overburden pressure and is given as:

$$\beta = \frac{p_i}{p_0}, \tag{7.75}$$

and α is also a competency factor as:

$$\alpha = \frac{\sigma_c}{p_0}. \tag{7.76}$$

2. Perfectly Plastic and Elastic Zone Boundary Radius (R_{pp})
The perfectly plastic and elastic zone boundary radius is also found by inserting σ_{rp} given by Eq. (7.73) in the radial stress Eq. (7.68) with $\sigma_r = p_i$ at $r = a$ as:

$$\frac{R_{pp}}{a} = \left\{\frac{(1+n)\left[(q-1)+\alpha\right]}{(1+nq)\left[(q-1)\beta+\alpha\right]}\right\}^{\frac{1}{n(q-1)}}. \tag{7.77}$$

iii. Normalized Opening Wall Strains

1. Elastic State
   Tangential strain at the opening wall can be obtained as:

$$\varepsilon_\theta^a = \frac{1+\nu}{nE}\left(p_0 - p_i\right), \tag{7.78}$$

$$\sigma_\theta^a = \frac{n+1}{n} p_0 - \frac{1}{n} p_i. \tag{7.79}$$

If the opening is strained to its elastic limit, then $\sigma_\theta^a = \sigma_c$ for $p_i = 0$. Thus, we have the elastic strain limit as:

$$\varepsilon_\theta^e = \frac{1+\nu}{E} \cdot \frac{\sigma_c}{n+1}. \tag{7.80}$$

Using the above relation in Eq. (7.78), one obtains the normalized opening wall strain (ξ) as:

$$\xi = \frac{\varepsilon_\theta^a}{\varepsilon_\theta^e} = \frac{n+1}{n}\left(\frac{1-\beta}{\alpha}\right) \leq 1. \tag{7.81}$$

2. Perfectly-Plastic State
   Tangential strain at the opening wall can be obtained as:

$$\varepsilon_\theta^a = \frac{1+\nu}{nE}\left(p_0 - \sigma_{rp}\right)\left(\frac{R_{pp}}{a}\right)^{f+1}, \tag{7.82}$$

elastic limit is given as:

$$\varepsilon_\theta^e = \frac{1+\nu}{nE}\left(p_0 - \sigma_{rp}\right). \tag{7.83}$$

Using the above relation in Eq. (7.82), one obtains the normalized opening wall strain as:

$$\xi = \frac{\varepsilon_\theta^a}{\varepsilon_\theta^e} = \left\{\frac{(1+n)\left[(q-1)+\alpha\right]}{(1+nq)\left[(q-1)\beta+\alpha\right]}\right\}^{\frac{f+1}{n(q-1)}}. \tag{7.84}$$

3. Residual Plastic State
   Tangential strain at the opening wall can be obtained as

$$\varepsilon_\theta^a = \frac{1+\nu}{nE}\left(p_0 - \sigma_{rp}\right)\eta_{sf}\left(\frac{R_{pb}}{a}\right)^{f^*+1}. \tag{7.85}$$

Using Eqs. (7.81) and (7.83), one obtains the normalized opening wall strain as:

$$\xi = \frac{\varepsilon_\theta^a}{\varepsilon_\theta^e} = \eta_{sf}\left\{\frac{\frac{(1+n)\left[(q-1)+\alpha\right]}{(1+nq)\,(q-1)}\left(\eta_{sf}\right)^{\frac{n(1-q)}{f+1}} - \frac{\alpha}{q-1} + \frac{\alpha^*}{q^*-1}}{\beta + \frac{\alpha^*}{q^*-1}}\right\}^{\frac{f^*+1}{n\left(q^*-1\right)}}, \tag{7.86}$$

where

$$\alpha^* = \frac{\sigma_c^*}{p_0}. \tag{7.87}$$

b. Consideration of the Support System
Although shotcrete, rockbolts and steel ribs are principal support members and they are widely used, it is very rare to find any fundamental study on the proper design method for the reinforcement effect of support systems consisting of shotcrete, rockbolts and steel ribs. As a certain displacement of ground takes place before the installation of supports, the resulting internal pressure provided by the support members must be evaluated in terms of relative displacement. Depending upon the state of rock before the installation of supports, there may be several combinations. In this section, it is assumed that rock is in an elastic state at the time of installation of rockbolts, shotcrete and steel ribs (Figure 7.15), and relative displacement at each region is evaluated for this special case as below (refer to Aydan (1989, 2018) for other combinations).

At the time of installation of support members, if rock behaves elastically, the radial deformation of the tunnel is given by

$$u_{in} = \frac{1+\nu}{nE} r\left(p_0 - \sigma_{rp}\right)\left(\frac{a}{r}\right)^{n+1} \tag{7.88}$$

Subtracting this deformation from Eqs. (7.62), (7.68) and (7.72), we obtain the following relations:

1. Flow region $\left(a \le r \le R_{pb}\right)$

$$\Delta u = \frac{1+\nu}{nE} r\left[\left(p_0 - \sigma_{rp}\right)\eta_{sf}\left(\frac{R_{pb}}{r}\right)^{f^*+1} - \left(p_o - p_{in}\right)\left(\frac{a}{r}\right)^{n+1}\right] \tag{7.89}$$

2. Perfectly plastic region $\left(R_{pb} \le r \le R_{pp}\right)$

$$\Delta u = \frac{1+\nu}{nE} r\left[\left(p_0 - \sigma_{rp}\right)\left(\frac{R_{pp}}{r}\right)^{f+1} - \left(p_o - p_{in}\right)\left(\frac{a}{r}\right)^{n+1}\right] \tag{7.90}$$

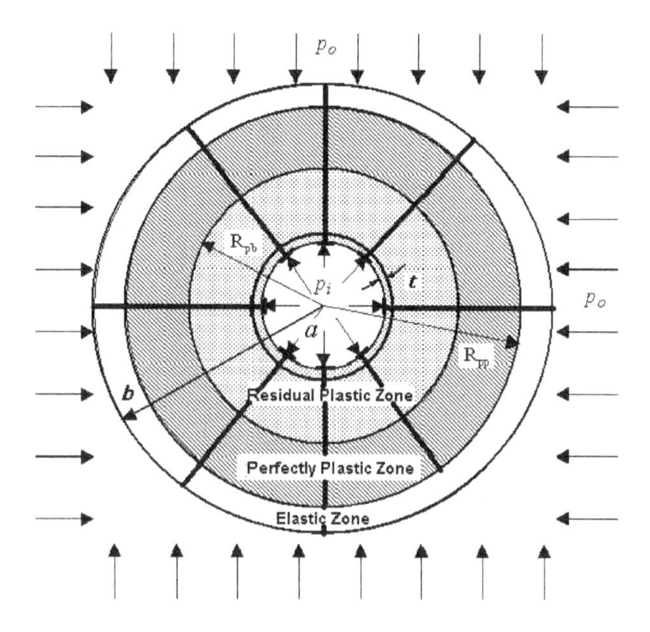

Figure 7.15 Support system for a radially symmetric opening and notations.

3. Elastic region $\left(R_{pp} \leq r\right)$

$$\Delta u = \frac{1+\nu}{nE} r \left[\left(p_0 - \sigma_{rp}\right)\left(\frac{R_{pp}}{r}\right)^{n+1} - \left(p_o - p_{in}\right)\left(\frac{a}{r}\right)^{n+1} \right] \tag{7.91}$$

i. Modeling rockbolts

For a prescribed displacement (Δu_h) at the grout-rock interface, the governing equation of rockbolts takes the following form (Aydan 1989):

$$\frac{d^2 u_{ax}}{d\xi^2} - \alpha^2 \left(u_{ax} - \Delta u_h\right) = 0; \quad \alpha^2 = \frac{2}{r_b^2 \ln\left(r_h/r_b\right)} \frac{G_g}{E_b} \tag{7.92}$$

where $\xi = r$, r_h is the borehole radius, r_b is the bar radius, G_g is the shear modulus of grout and E_b is the elastic modulus of rockbolt. If the relative displacements of rock are introduced in the above non-homogeneous second order differential equation, the solution of the above equation by elementary methods of integration becomes impossible, and the use of a numerical integration becomes necessary. For the sake of simplification, we introduce the following form for differential displacement:

$$f(\xi) = C_o^* e^{-D\xi} \tag{7.93}$$

Constants C_o^* and D are determined from the values of Δu_h at each respective interface of regions. Specifically, they are as follows.

4. Flow region ($a \le r \le R_{pb}$)

$$D_{pb} = \frac{\ln\left(\Delta u_a / \Delta u_{pb}\right)}{\left(R_{pb} - a\right)}; \quad C_o^{*pb} = \Delta u_{pb} e^{D_{pb} R_{pb}} \tag{7.94a}$$

$$\Delta u_a = \frac{1+v}{nE} a\left[\left(p_0 - \sigma_{rp}\right)\eta_{sf}\left(\frac{R_{pb}}{a}\right)^{f^*+1} - \left(p_o - p_{in}^b\right)\right] \tag{7.94b}$$

$$\Delta u_{pb} = \frac{1+v}{nE} R_{pb}\left[\left(p_0 - \sigma_{rp}\right)\eta_{sf} - \left(p_o - p_{in}^b\right)\left(\frac{a}{R_{pb}}\right)^{n+1}\right] \tag{7.94c}$$

5. Perfectly plastic region $\left(R_{pb} \le r \le R_{pp}\right)$

$$D_{pp} = \frac{\ln\left(\Delta u_{pb} / \Delta u_{pp}\right)}{\left(R_{pp} - R_{pb}\right)}; \quad C_o^{*pp} = \Delta u_{pp} e^{D_{pp} R_{pp}} \tag{7.95a}$$

$$\Delta u_{pp} = \frac{1+v}{nE}\left[\left(p_0 - \sigma_{rp}\right) R_{bp} \eta_{sf}\left(\frac{R_{pp}}{R_{pb}}\right)^{f^*+1} - \left(p_o - p_{in}^b\right)\frac{a^{n+1}}{R_{pp}^n}\right] \tag{7.95b}$$

6. Elastic region $\left(R_{pp} \le r\right)$

$$D_{pe} = \frac{\ln\left(\Delta u_b / \Delta u_{pe}\right)}{\left(b - R_{pp}\right)}; \quad C_o^{*pe} = \Delta u_b e^{D_{pe} b} \tag{7.96a}$$

$$\Delta u_{pe} = \frac{1+v}{nE} R_{pe}\left[\left(p_0 - \sigma_{rp}\right)\eta_{sf} - \left(p_o - p_{in}^b\right)\left(\frac{a}{R_{pe}}\right)^{n+1}\right] \tag{7.96b}$$

$$\Delta u_b = \frac{1+v}{nE} b\left[\left(p_0 - \sigma_{rp}\right)\left(\frac{R_{pe}}{b}\right)^{n+1}\frac{R_{pb}}{b}\eta_{sf} - \left(p_o - p_{in}^b\right)\left(\frac{a}{b}\right)^{n+1}\right] \tag{7.96c}$$

The axial displacement and axial stress of a rockbolt take the following form as its solution:

$$u_{ax} = A_1 e^{-a\xi} + A_2 e^{a\xi} + C_o e^{-D\xi} \tag{7.97a}$$

$$\sigma_b = E_b \alpha\left[A_1 e^{-a\xi} - A_2 e^{a\xi} + \frac{D}{\alpha} C_o e^{-D\xi}\right] \tag{7.97b}$$

Integration constants (A_1 and A_2) at each region are determined from the continuity condition and boundary conditions. Accordingly, the internal pressure provided by rockbolts can be obtained from the well-known formula:

$$\Delta p_i^b = \sigma_b(r = a)\frac{A_b}{e_t e_l} \tag{7.98}$$

where A_b, e_t and e_l are cross-section area and spacing of a typical rockbolt.

ii. Modeling shotcrete
The thin wall cylinder or thick wall cylinder approach is commonly used to assess the internal pressure effect of shotcrete in tunneling. For radially symmetric situations, a similar approach can be adopted. One can easily derive the relation for the displacement and outer pressure p_{io}^s acting on the shotcrete with internal pressure $p_{ii}^s = 0$

$$u = \frac{1+\nu_s}{nE_s} p_{io}^s \frac{a_o^{n+2}}{a_o^{n+1} - a_i^{n+1}}\left[\frac{1-2\nu_s}{1-\nu_s+n\nu_s} + \frac{1}{n}\left(\frac{a_i}{a_o}\right)^{n+1}\right] \tag{7.99}$$

The incremental form of the above equation is

$$\Delta u = \frac{1+\nu_s}{nE_s} \Delta p_{io}^s \frac{a_o^{n+2}}{a_o^{n+1} - a_i^{n+1}}\left[\frac{1-2\nu_s}{1-\nu_s+n\nu_s} + \frac{1}{n}\left(\frac{a_i}{a_o}\right)^{n+1}\right] \tag{7.100}$$

and its inverse is

$$\Delta p_{io}^s = K_s \Delta u \tag{7.101}$$

where

$$K_s = \frac{nE_s}{1+\nu_s}\frac{a_o^{n+1} - a_i^{n+1}}{a_o^{n+2}}\left[\frac{(1-\nu_s+n\nu_s)a_o^{n+1}}{n(1-2\nu_s)a_o^{n+1} + (1-\nu_s+n\nu_s)a_i^{n+1}}\right] \tag{7.102}$$

If the thickness of shotcrete is negligible as compared with the radius of opening, then we have the following:

$$K_s = \frac{nE_s}{1+\nu_s}\frac{t}{a_o^2}\frac{(1-\nu_s+n\nu_s)}{(1-\nu_s)} \tag{7.103}$$

iii. Modeling steel ribs
If the steel rib is modeled as a one-dimensional rib, its radial deformation is given by

$$u = \frac{1}{nE_{rb}} p_i^{rb} \frac{a_o^2 e_l}{A_{rb}} \tag{7.104}$$

Its incremental form becomes

$$\Delta u = \frac{1}{nE_{rb}} \Delta p_i^{rb} \frac{a_o^2 e_l}{A_{rb}} \tag{7.105}$$

Or inversely, we have

$$\Delta p_i^{rb} = K_{rb} \Delta u \text{ and } K_{rb} = nE_{rb} \frac{A_{rb}}{a_o^2 e_l} \tag{7.106}$$

The total internal pressure of the support system may be given as

$$p_i^{ss} = \Delta p_i^b + \Delta p_{io}^s + \Delta p_i^{rb} \tag{7.107}$$

If support members yield during the deformation of the surrounding rock, their behaviors are assumed to be elastic-perfectly plastic. This will particularly require further formulation of axial stress evaluation of rockbolts.

c. Consideration of Body Forces in the Plastic Zone
In a general sense, the consideration of body forces violates the radial symmetry of the governing equation. However, there are some proposals in the literature for this purpose (i.e. Fenner 1938; Hoek and Brown 1980; Aydan 1989; Sezaki et al. 1994). Some slight modifications to the developed solutions are described to consider the effect of body forces in residual-plastic zones. As the perfectly plastic region sustains its original strength, the effect of body forces should be negligible in this region. However, the effect of body forces may be important in the residual plastic (flow) region. Using a similar approach proposed by Aydan (1989), the maximum body force on the support system in the residual plastic region may be approximately obtained as follows (Figure 7.15):

$$p_{ib} = \left(1 - \left(\frac{a}{R_{pb}} \right)^{n\left(q^* - 1\right)} \right) \frac{\gamma a \cos\theta}{n\left(q^* - 1\right) - 1} \tag{7.108}$$

where γ is the unit weight of the residual plastic zone. The values of angle (θ) for crown, sidewall and invert are 0, $\pi / 2$ and π, respectively. It should be noted that this assumption is only valid provided that the radial symmetry of the problem is not violated.

d. Consideration of Creep Failure
It is well known that deformation and strength characteristics of rocks depend upon the stress rate or strain rate used in tests (i.e. Bieniawski 1970; Lama and Vutukuri 1978; Aydan et al. 1994; Aydan 2010). Creep tests and relaxation tests are commonly used to determine the time-dependent characteristics of rocks. There is a common conception, that is, the creep of rocks does not occur unless the applied stress is greater than a threshold stress value, which is called the creep threshold

(Aydan et al. 1995a, b; Aydan and Nawrocki 1998). This threshold stress level is generally related to the stress level at which fractures are initiated. The so-called transient creep is likely to be a result of the actual visco-elastic behavior of rock. The secondary creep is, on the other hand, due to the stable crack propagation and the tertiary creep is due to the unstable crack propagation. Therefore, the secondary and tertiary creep are a visco-plastic phenomenon rather than a visco-elastic phenomenon as they involve energy dissipation by fracturing. Figure 7.16 shows the normalized uniaxial creep stress by the short-term uniaxial strength for various rocks as a function of the failure time (time elapsed until the failure). The shrinkage of the uniaxial strength of rock may be represented by the following functional form:

$$\frac{\sigma_c}{\sigma_c^o} = F(t, t_s) \tag{7.109}$$

where t and t_s are time and short-term test durations. σ_c^o is the short-term uniaxial strength of rock-mass. The first authors explored several specific forms of function (F) to evaluate the experimental results. Figure 7.16 compares the experimental results with the following functional form

$$\frac{\sigma_c}{\sigma_c^o} = 1 - b \ln\left(\frac{t}{t_s}\right) \tag{7.110}$$

This simple concept may be also extended to a multi-axial stress state in which the yield surface of rocks under multi-axial loading conditions is modeled as the shrinkage of the yielding surface. An illustration of this concept on the Mohr–Coulomb yield criterion is shown in Figure 7.17.

If the time-dependent characteristics of the uniaxial compressive strength of rocks are known, it may be possible to determine the time-dependent variation of various mechanical properties of rocks from the relationships obtained by Aydan et al. (1993, 1995a, b). Using the approach originally proposed by Ladanyi (1974) together with the time-dependent variation of mechanical properties involved in the equations given in the previous subsection, it is possible to determine the time-dependent deformation of tunnels. Under multi-axial initial stress conditions and

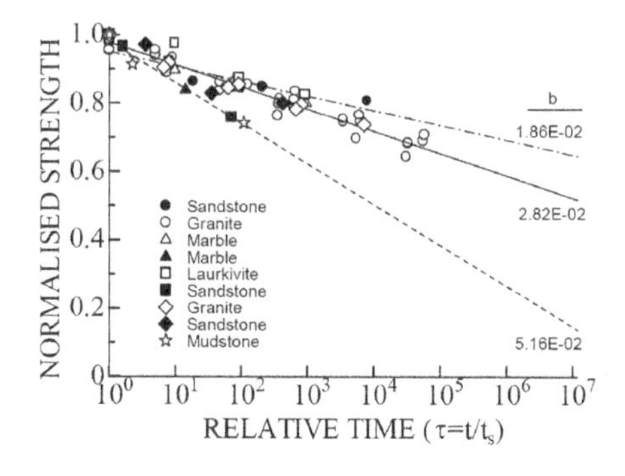

Figure 7.16 Creep strength of various rocks. (From Aydan and Nawrocki 1998.)

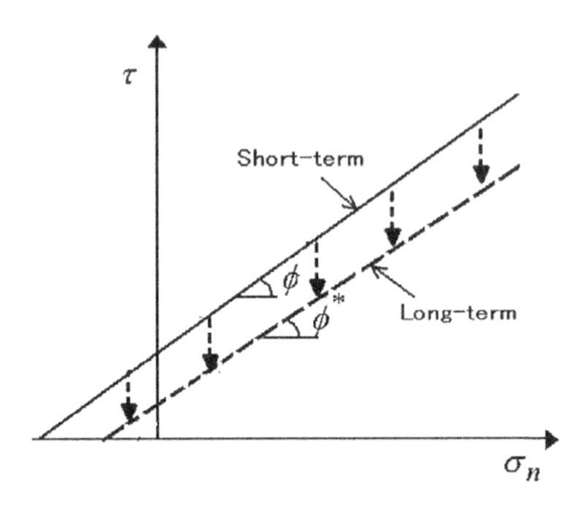

Figure 7.17 Illustration of triaxial long-term strength. (Modified from Aydan et al. 1995b.)

complex tunnel geometry, excavation scheme and boundary conditions, the use of numerical techniques becomes necessary. For numerical analyses, the time-dependent behavior of rocks may be modeled by an approach proposed by Aydan et al. (1995b, 1996).

e. Applications

Elasto-plastic responses of cylindrical and spherical openings were obtained by using the analytical solution, which is described in this study. It is assumed that cylindrical and spherical opening with a diameter of 6 m is excavated at a depth of below 1000 m from the ground surface in rock mass having $\sigma_c = 20$ MPa and $\gamma = 25$ kN/m^3. Empirical equations given by Aydan et al. (1993, 1996) were used for the determination of the properties of rock mass. In these equations, the properties of rock mass were related to the uniaxial compressive strength of rock material. The selected parameters used in the analyses are given in Table 7.1.

Firstly, the ground reaction curves were obtained for the cylindrical and spherical openings. The opening wall strain was related to the normalized ground pressure and normalized plastic radius about the opening (Figure 7.18). At the same internal support pressure, plastic zone radius and wall strain around the cylindrical openings are greater than those for the spherical opening (Figures 7.18 and 7.19). In the present example, the elastic-perfect plastic zone radius and residual-perfect plastic zone radius about cylindrical openings are 5.84 and 5.07 m, respectively. On the other hand, no residual plastic zone was observed, but an elastic-perfect plastic zone radius with 3.19 m was found about the spherical opening.

Table 7.1 The properties of strength and deformation of rock material and rock mass used in the parametric study

σ_c (MPa)	σ_c^* (MPa)	E (GPa)	υ	γ (kN/m^3)	φ (°)	φ^* (°)	η_{sf}	f	f^*	p_0 (MPa)
20	0.05	5.5	0.25	25	42.3	54.8	1.311	2.60	3.96	25

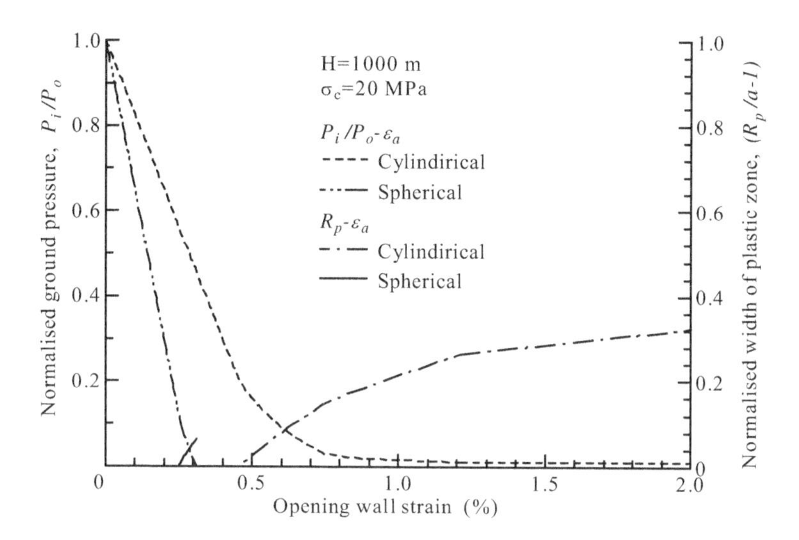

Figure 7.18 Strain, plastic zone radius around cylindrical and spherical openings and ground reaction curves.

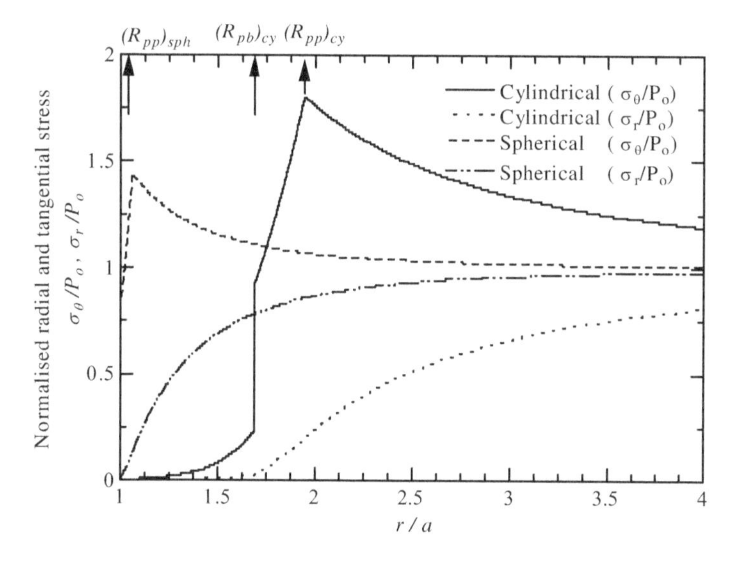

Figure 7.19 Tangential and radial stresses distribution around cylindrical and spherical openings ($P_i = 0$).

Then, the distribution of tangential and radial stresses is obtained around cylindrical and spherical openings for the condition, that is, the internal pressure is zero. Tangential stresses around the cylindrical openings are greater than those for the spherical openings. In addition, the radial and tangential stresses around the spherical opening approach faster to the in-situ stress as a function of distance from the ground surface. As a result, the radius of the plastic zone about the spherical opening is smaller than that of a cylindrical one (Figure 7.19).

The next comparison was concerned with the comparison of the elastic-perfectly plastic-residual plastic model with a strain-softening model. The tunnel was assumed to be 200 m below the ground surface in a rock mass having a uniaxial strength of 2.5 MPa. Empirical equations given by Aydan et al. (1993, 1996) were used for determining other properties of rock mass for analyses. Figures 7.20 and 7.21 show the computed ground reaction curve and stress distributions for two constitutive models. As expected, there is almost no difference regarding ground response curves if the consideration of the softening part of the constitutive model is considered by the proposed scheme. The only difference is associated with the distribution of tangential stress in rock mass as seen in Figure 7.21.

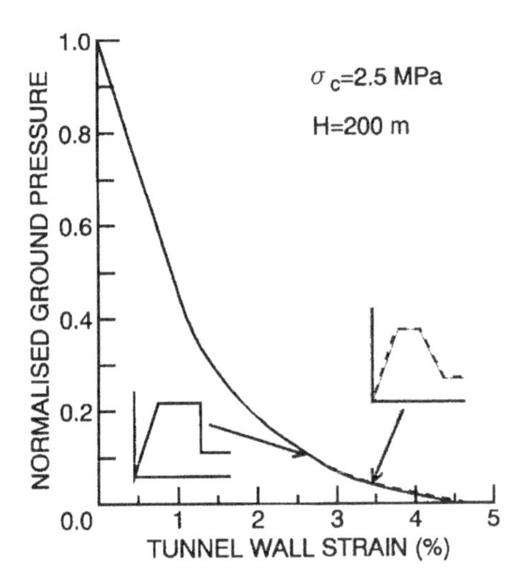

Figure 7.20 Comparison of ground response curves for different constitutive relations.

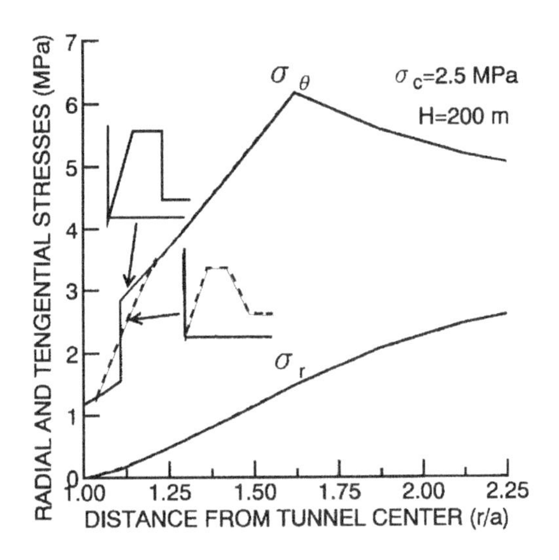

Figure 7.21 Comparison of stress distributions for different constitutive relations.

7.2.6 Foundations-bearing capacity

The stress and strain field induced in the impression experiments is close to the compression of the rock under a rigid indenter (Figure 7.22). Timoshenko and Goodier (1951, 1970) developed the following relation for a circular rigid indentation of elastic half-space problem:

$$\frac{\delta}{D} = \frac{\pi}{4}\frac{1-\nu^2}{E} p \text{ with } p = \frac{4F}{\pi D^2} \tag{7.111}$$

where F, ν and E are applied load, Poisson ratio and elastic modulus of rock. While the displacement distribution is uniform beneath the indenter, the contact pressure induced by the indenter would not be uniform. Jaeger and Cook (1979) discussed the initiation of yielding due to compression and they suggested that the yielding stress level under compression should correspond to one to two times the uniaxial compressive strength of rock and the yielding would occur at a depth of the order of the radius of the indenter.

Aydan et al. (2008) showed that the following relation should exist between uniaxial compression stress–strain rate ($\sigma - \varepsilon$) and applied pressure (p) and nominal strain of indenter with a cylindrical flat-end ($\varepsilon_i = \delta/D$) with the use of a spherical cavity approach:

$$\frac{\sigma}{\varepsilon} = \frac{1+\nu}{4}\frac{p}{\varepsilon_i} \tag{7.112}$$

where ν is Poisson's ratio of rock and D is the diameter of the indenter. It should be noted that radial stress and strain of impression experiments are analogous to stress–strain of the uniaxial stress state. As discussed by Aydan et al. (2008), there may be at least two stress levels for initiating the yielding of rock beneath the indenter. The first yield stress level would correspond to twice the tensile strength level of rock and the other one would correspond to the uniaxial compressive strength level. However, the effect of tensile yielding is generally difficult to differentiate as the deformation moduli before and after yielding in tension remains fairly the same. The overall deformation

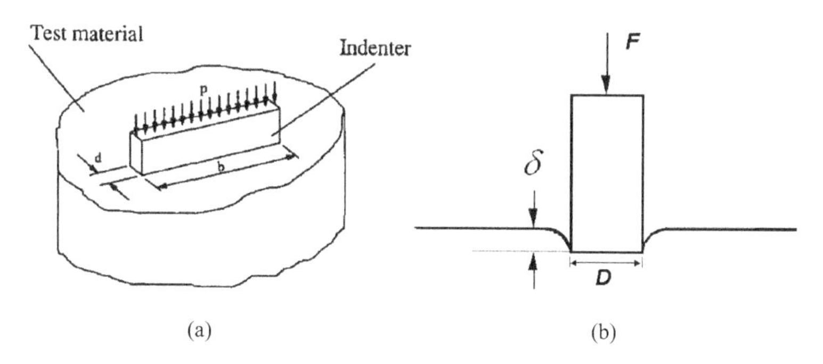

Figure 7.22 Geometrical illustration of various models. (a) The model of Hyde et al. (1996) and (b) the model of Timoshenko-Goodier.

modulus may change after yielding in compression. The experiments also indicate that the ultimate strength value (p_u) cannot be greater than a stress level given by

$$p_u = \frac{2}{1-\sin\varphi}\sigma_c \tag{7.113}$$

The equation above implies that the ultimate strength for a frictionless cohesive medium would be twice its uniaxial strength or four times its cohesion. However, it should be noted that this type of equation implies that considerable yielding should take place beneath the indenters. Aydan et al. (2008) developed the following formulas for three different situations of rock beneath the indenter (Figure 7.23):

Elastic behavior ($p_i \leq 2\sigma_t$)

$$\frac{u_a}{a} = \frac{1+\nu}{2E} p_i \tag{7.114}$$

Radially ruptured (no tension) plastic behavior ($2\sigma_t < p_i \leq \sigma_c$)

$$\frac{u_a}{a} = \frac{1+\nu}{2E} p_t \frac{R_t}{a} + \frac{p_i}{2E}\left(1 - \frac{a}{R_t}\right); \quad \frac{R_t}{a} = \left(\frac{p_i}{p_t}\right)^{1/2}; \quad p_t \leq 2\sigma_t \tag{7.115}$$

Crushed plastic behavior ($p_i \leq 2\sigma_t$)

$$\frac{u_a}{a} = \left[\frac{1+\nu}{2E} p_t \frac{R_t}{a} + \frac{p_c}{2E}\frac{R_c}{a}\left(1 - \frac{R_c}{R_t}\right)\right]\left(\frac{p_i}{p_c}\right)^{\frac{q}{(q-1)}}; \quad \frac{R_c}{a} = \left(\frac{p_i}{p_c}\right)^{\frac{q}{2(q-1)}}; \quad \frac{R_c}{R_t} = \left(\frac{p_t}{p_c}\right)^{1/2} \tag{7.116}$$

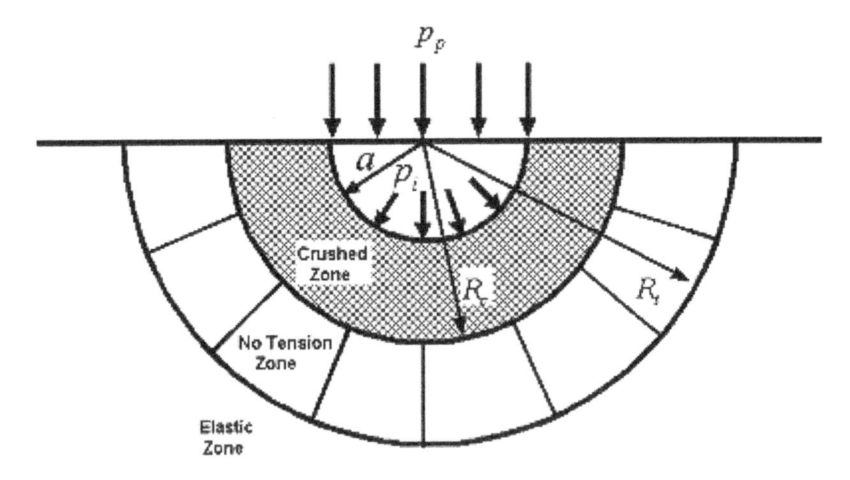

Figure 7.23 An illustration of zones formed beneath the loading plate and notation.

Furthermore, the applied pressure is equal to radial pressure on the walls of the spherical body in view of the equivalence of work done by the pressure of the indenter to that induced by the wall of the spherical body on the surrounding medium as

$$p_p = p_i \tag{7.117}$$

Assuming that the volume of the hemispherical body beneath the indenter remains for a given impression displacement (δ), the outward displacement (u_a) of the hemispherical cavity wall can be easily related to the impression displacement (δ) as follows

$$\delta = 2u_a \tag{7.118}$$

7.2.7 Two-dimensional closed-form solution methods

It is generally difficult to derive closed-form solutions for surface and underground excavations with complex geometry and complex material behavior. Most of the solutions would be limited to one-dimensional in space and time may be incorporated in certain solutions. For deformation-stress analyses, it is rare to find closed-form solutions for surface structures while there are some closed-form solutions for underground openings in two-dimensional elastic space. The most famous solution would be probably that of Kirsch (1898) for a circular hole under a biaxial stress state. Solutions for arbitrary shape openings are also developed and the readers are referred to the textbooks by Timoshenko and Goodier (1970), Savin (1961), Muskhelishvili (1962) and Jaeger and Cook (1979). It becomes more difficult to obtain analytical solutions when surrounding media start to behave in an elasto-plastic manner. The simple yet often used closed-form solutions are for openings with a circular geometry excavated in elasto-plastic media. Several solutions are developed using different yield criteria and post-yielding models (e.g. Talobre 1957; Terzaghi 1946). There are some solutions for underground openings in elasto-plastic rock supported by rockbolts, shotcrete and steel ribs (Hoek and Brown 1980; Aydan et al. 1993). Galin (1946) was the first to obtain closed-form solutions around circular openings enclosed completely by a plastic zone under a biaxial stress state in Tresca-type perfectly plastic materials. Detournay (1983, 1986) attempted to obtain solutions for the same situation with the Mohr–Coulomb yield criterion and discussed several cases using the same solution technique.

There are also a limited number of analytical solutions for the problems of seepage, heat flow and diffusion since most of the solutions would be valid for given boundary and initial conditions.

The analytical solutions for displacement, strain and stress field around cavities exhibiting a non-linear behavior under non-hydrostatic conditions are generally difficult to obtain. However, some analytical solutions were obtained by Kirsch (1898), for circular cavities, by Inglis (1913) for elliptical cavities and by Mindlin (1949) for circular cavities in gravitating media when the surrounding medium behaves elastically. Muskhelishvili (1962) devised a general method based on complex variable functions for arbitrarily shaped cavities.

The stress state around a circular cavity in an elastic medium under bi-axial far-field stresses is first obtained by Kirsch (1898) using Airy's stress function. These solutions are modified to incorporate the effect of uniform internal pressures

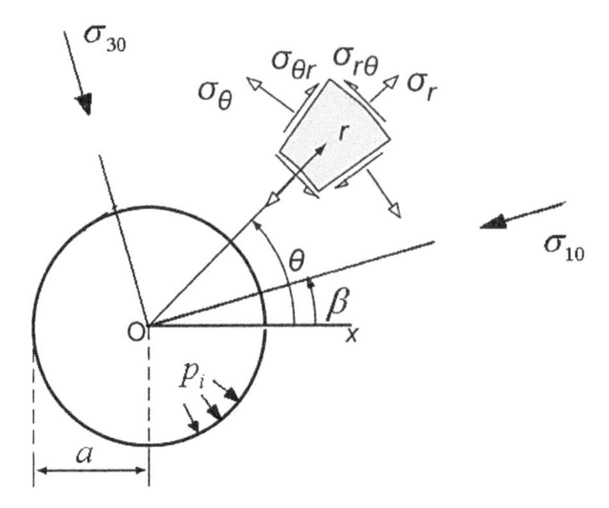

Figure 7.24 Stress tensor components and far-field stresses around a circular hole.

(Jaeger and Cook 1979). In the polar coordinate system, the radial, tangential and shear stresses around the circular cavity can be written in the following forms (Figure 7.24):

$$\sigma_r = \frac{\sigma_{10}+\sigma_{30}}{2}\left(1-\left(\frac{a}{r}\right)^2\right)-\frac{\sigma_{10}-\sigma_{30}}{2}\left(1-4\left(\frac{a}{r}\right)^2+3\left(\frac{a}{r}\right)^4\right)\cos 2(\theta-\beta)+p_i\left(\frac{a}{r}\right)^2$$

$$\sigma_\theta = \frac{\sigma_{10}+\sigma_{30}}{2}\left(1+\left(\frac{a}{r}\right)^2\right)+\frac{\sigma_{10}-\sigma_{30}}{2}\left(1+3\left(\frac{a}{r}\right)^4\right)\cos 2(\theta-\beta)-p_i\left(\frac{a}{r}\right)^2 \quad (7.119)$$

$$\tau_{r\theta} = \frac{\sigma_{10}-\sigma_{30}}{2}\left(1-4\left(\frac{a}{r}\right)^2+3\left(\frac{a}{r}\right)^4\right)\sin 2(\theta-\beta)$$

where σ_{10}, σ_{30} are the far-field principal stresses; a is the radius of hole; r is the radial distance; β is the inclination of σ_{10} far-field stress from the horizontal; θ is the angle of the point from the horizontal and p_i is the internal pressure applied onto the hole perimetry.

The yield criteria available in rock mechanics are:

Mohr–Coulomb

$$\sigma_1 = \sigma_c + q\sigma_3 \quad (7.120a)$$

Drucker and Prager (1952)

$$\alpha I_1 + \sqrt{J_2} = k \quad (7.120b)$$

Hoek and Brown (1980)

$$\sigma_1 = \sigma_3 + \sqrt{m\sigma_c\sigma_3 + s\sigma_c^2} \tag{7.120c}$$

Aydan (1995b)

$$\sigma_1 = \sigma_3 + \left[S_\infty - (S_\infty - \sigma_c)e^{-b_1\sigma_3}\right]e^{-b_2T} \tag{7.120d}$$

where

$$I_1 = \sigma_I + \sigma_{II} + \sigma_{III}; \quad J_2 = \frac{1}{6}(\sigma_I - \sigma_{II})^2 + (\sigma_{II} - \sigma_{III})^2 + (\sigma_{III} - \sigma_I)^2$$

$$\alpha = \frac{2\sin\phi}{\sqrt{3}(3+\sin\phi)}; \; k = \frac{6c\cos\phi}{\sqrt{3}(3+\sin\phi)}; \; c: \text{cohesion}; \phi: \text{friction angle}, \; q = \frac{1+\sin\phi}{1-\sin\phi}$$

σ_∞ is the ultimate deviatoric strength, T is the temperature and m, s, b_1,b_2 are empirical constants. Mohr– Coulomb and Drucker–Prager yield criteria are linear functions of confining or mean stress while the criteria of Hoek–Brown and Aydan are of non-linear type. Furthermore, Aydan's criterion also accounts for the effect of temperature. Figure 7.25 compares several yield criteria for different rocks. When Aydan's criterion is used, the effect of temperature is omitted in Figure 7.25 for the sake of comparison. It should be also noted that the criterion of Hoek and Brown often fails to represent triaxial strength data if it is required to represent tensile and compressive strength contrary to the common belief, that is, the best yield criterion for rocks (Aydan et al. 2012a).

If the yield criterion is chosen to be a function of minimum and maximum principal stresses, they can be given in the following form in terms of stress components given by Eq. (7.119).

$$\sigma_1 = \frac{\sigma_\theta + \sigma_r}{2} + \sqrt{\left(\frac{\sigma_\theta - \sigma_r}{2}\right)^2 + \tau_{r\theta}^2} \tag{7.121a}$$

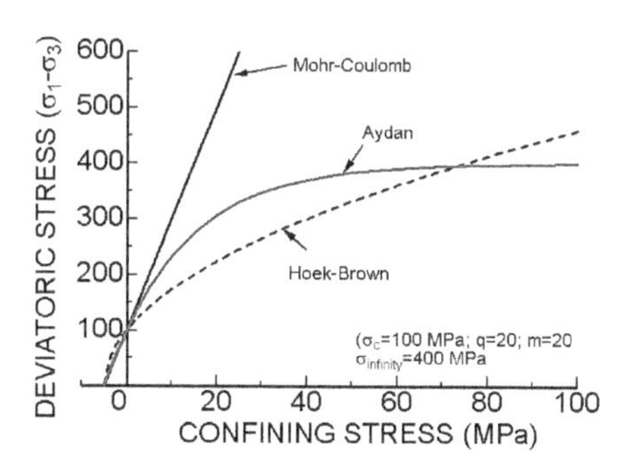

Figure 7.25 Comparison of yield criteria.

$$\sigma_3 = \frac{\sigma_\theta + \sigma_r}{2} - \sqrt{\left(\frac{\sigma_\theta - \sigma_r}{2}\right)^2 + \tau_{r\theta}^2} \tag{7.121b}$$

The damage zone around the blasthole under high internal pressure can be estimated using one of the yield criteria listed above. It should be noted that the yielding is induced by the high internal pressure in the blasthole, which is essentially different from in-situ stress-induced borehole-breakout. In other words, there will always be a damage zone around the blast-hole perimeter when the blasting technique is employed. The blasthole pressure depends upon the characteristics of the surrounding medium, the amount, layout and type of explosive, blasting velocity and the geometry of the blasthole. The blasthole pressure ranges from 100 MPa to 10 GPa (i.e. Jaeger and Cook 1979; Brady and Brown 1985).

First, we assume that the properties of surrounding rock have the values as given in Table 7.2 and the blasthole (internal) pressure has a value of 400 MPa. The rock chosen roughly corresponds to an igneous rock such as granitic rocks. The maximum far-field stress is inclined at an angle of 30° from the horizontal and the lateral stress coefficient has a value of 4. Figure 7.26 shows an example of computation for the given conditions. The largest yield zone is obtained for the Hoek–Brown criterion while the tension cut-off criterion results in a smaller yield zone. The criterion of Aydan estimates a slightly larger yield zone than the Mohr–Coulomb criterion. It is very interesting to note that the yielding propagates in the direction of maximum far-field stress. In other words, the elongation direction of the yield zone would be the best indicator of the maximum far-field stress in the plane of the blast-hole.

Table 7.2 Values of in-situ stress parameters and properties of yield functions

σ_{10} (MPa)	k	σ_c (MPa)	σ_t (MPa)	ϕ (°)	m	σ_∞ (MPa)	b
50.0	4.0	100.0	5.0	60	20	400	31.39

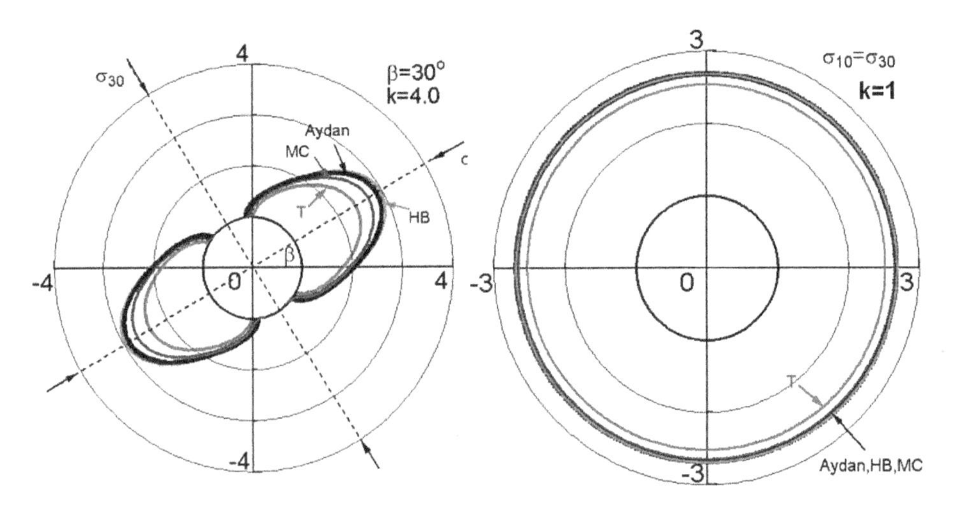

Figure 7.26 Estimated yield zones around the blasthole.

In the next example, the far-field stress state is assumed to be isotropic while keeping the values of parameters the same. Except for the tension cut-off criterion, all yield criteria estimate almost the same size yield zones. It should be however noted that all yield criteria satisfy the same values of tensile and compressive strength. Therefore, the similarity of the size of yield zones should not be surprising.

If the strength of the surrounding rock is anisotropic, the yield functions considering the effect of anisotropy should be used. If the elastic constants of rock are anisotropic, closed-form solutions capable of representing anisotropy should be used instead of those given by Eq. (7.112). It should be also noted that the plastic zone developed using the actual elasto-plastic analyses would be larger than those estimated from the elastic solutions (7.27). If such discrepancies are expected to be larger, it would be better to use the elasto-plastic finite element method. Nevertheless, the basic conceptual model would be the same.

Gerçek (1988, 1996) proposed a semi-numerical technique to obtain the integration constant stress functions based on Muskhelishvili's method. Galin (1946, see Savin 1961 for English version) was the first to develop analytical solutions for circular holes in the Tresca-type material under a non-hydrostatic initial stress state. His solution was extended to Mohr–Coulomb materials by Detournay (1983). He further discussed the problem of a non-enveloping yield zone around the circular hole. Kastner (1961) proposed a method for estimating the approximate yield zone around circular cavities under non-hydrostatic stress conditions using Kirsch's solutions. This method is also employed by Zoback et al. (1980) to estimate the shape of borehole breakouts, which was used to infer the in-situ stress state from borehole breakouts. Gerçek and Geniş (1999) also used the same concept for arbitrarily shaped cavities to estimate the extent of the possible yield zone. Although this method estimates the extent of the yield zone smaller than the actual one as shown by the first author (Aydan 1987) for circular cavities under hydrostatic stress state as shown in Figure 7.27, it yielded the estimated shape of yield zone similar to that by exact solutions. Furthermore, it may also provide some rough guidelines for the anticipated zone for reinforcement by

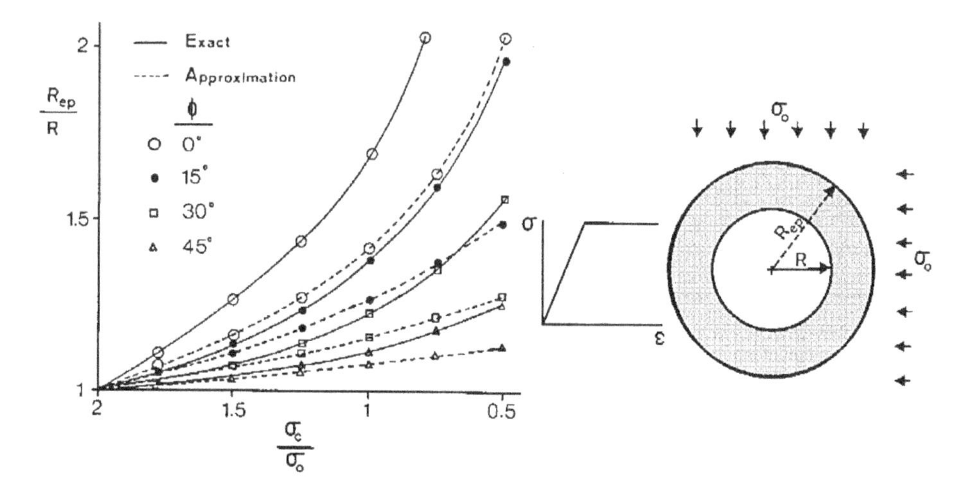

Figure 7.27 Comparison of approximately estimated and exact yield zone. (From Aydan 1987.)

rockbolts. Aydan and Geniş (2010) extended the same concept to estimate overstressed zones about cavities of arbitrary shape based on the stress state computation method proposed by Gerçek (1996) and using strain energy, distortion energy, extension strain and no-tension criteria in addition to the Mohr–Coulomb yield criterion.

Figure 7.28 shows the overstressed zones around a tunnel subjected to hydrostatic initial stress state at different stages of excavations and the contours of maximum principal stress. The most critical stress state is during the excavation of the top heading and the stress state becomes more uniform as the excavation approaches the shape of circular. Furthermore, the extent of the tensile stress zone gradually decreases in size as the excavation progresses.

An interesting yield zone developed around a circular opening excavated in a granodioritic hard rock at a level of 420 m in URL in Winnipeg, Canada. Figure 7.29 shows the prediction of an overstressed zone around the circular opening at URL. Except

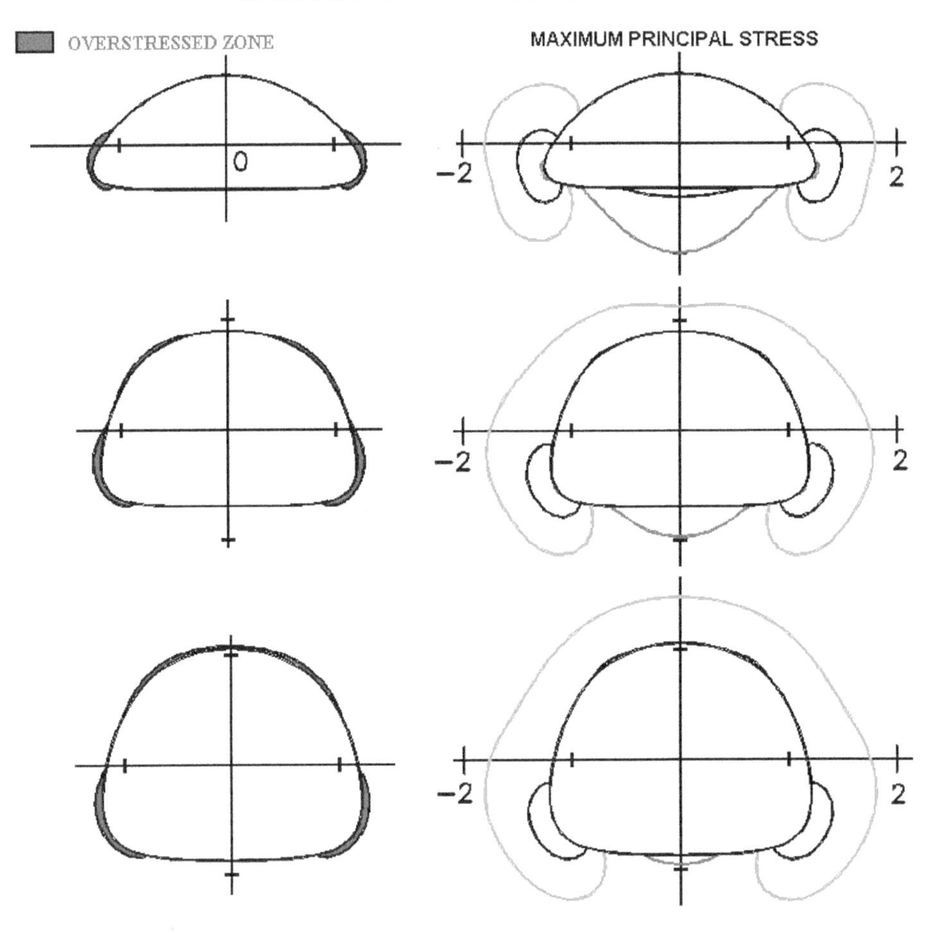

Figure 7.28 The overstressed zones around a tunnel subjected to hydrostatic initial stress state at different stages of excavations and the contours of maximum principal stress.

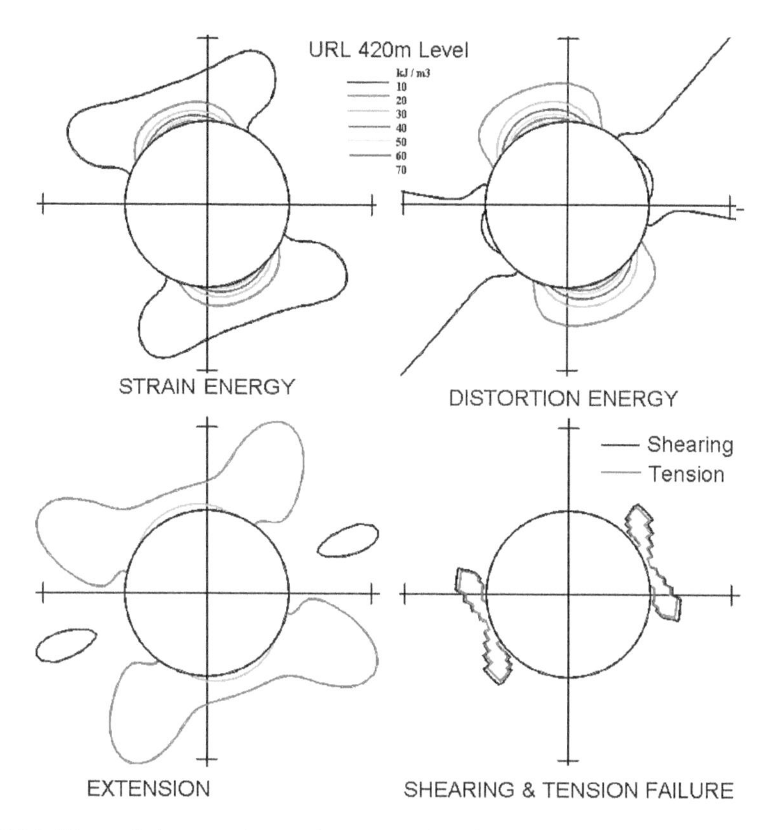

Figure 7.29 View of the opening and estimated yield zones by different methods.

for the shearing and tension yielding model, all methods estimated the most-likely location of the yield zone. In addition to the estimations by the approximate approach, FEM analyses incorporating the yield criteria adopted in the approximate method were performed (Figure 7.29). As noted from the figure, it seems that the distortion energy concept yields results that are close to the observations as shown in Figure 7.29.

7.2.8 Three-dimensional closed-form solutions

The solutions for three-dimensional situations are quite rare except for a very few solutions. Boussinesq (1885) derived solutions for the distribution of stresses in a half-space resulting from surface loads and they are largely used in various applications. Kelvin considered a half-space problem with a point load of P (Figure 7.30). The final expressions are given in the following form:

Displacement components

$$u_r = \frac{Pzr}{4\pi G(1-\nu)R^3};\ u_\theta = 0;\ u_z = \frac{P}{4\pi G(1-\nu)}\left[\frac{2(1-2\nu)}{R} + \frac{1}{R} + \frac{z^2}{R^3}\right] \tag{7.122}$$

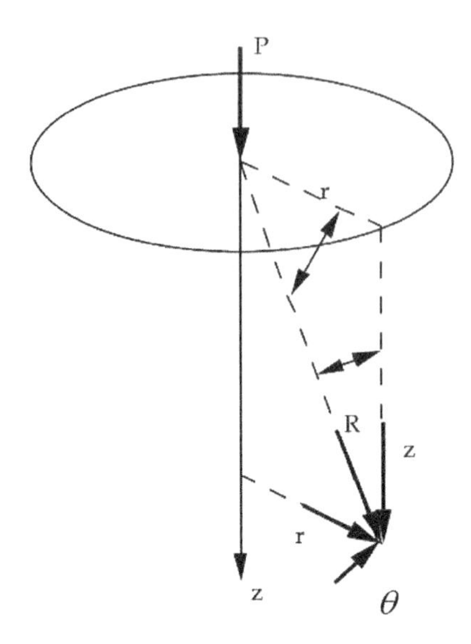

Figure 7.30 Notation for half-space under a point load *P*.

Stress components

$$\sigma_r = \frac{P}{2\pi(1-\nu)}\left[\frac{2(1-2\nu)z}{R^3} - \frac{3r^2z}{R^5}\right]; \sigma_\theta = \frac{P(1-2\nu)z}{2\pi(1-\nu)R^3};$$

$$\sigma_z = \frac{P}{2\pi(1-\nu)}\left[\frac{(1-2\nu)z}{R^3} - \frac{3z^3}{R^5}\right]$$

$$\tau_{r\theta} = \frac{P}{2\pi(1-\nu)}\left[\frac{(1-2\nu)z}{R^3} - \frac{3rz^2}{R^5}\right]; \tau_{r\theta} = \tau_{z\theta} = 0 \tag{7.123}$$

where

$$R = \sqrt{z^2 + r^2}$$

7.3 Closed-form solutions for fluid flow through porous rocks

7.3.1 *Some considerations on the Darcy law for rocks and discontinuities*

The Darcy law (Eq. 5.3) is generally used as a constitutive model for the fluid flow through porous rock and rock discontinuities together with the assumption of laminar flow. A brief description of Darcy's law is presented in this subsection.

Darcy performed a series of experiments on a porous column in 1856. From these experiments, he found out that the volume discharge rate Q is directly proportional to the head drop $h_2 - h_1$ and to the cross-sectional area A, but it is inversely proportional to the length difference $l_2 - l_1$. Calling the proportionality constant K as the hydraulic conductivity, Darcy's law is written as:

$$Q = -KA\frac{h_2 - h_1}{l_2 - l_1} \tag{7.124}$$

The negative sign signifies that groundwater flows in the direction of head loss.

Darcy's law is now widely accepted and used in modeling fluid flow in porous or fractured media. It is elaborated and written in a differential form which is given below for a one-dimensional case as:

$$v = -K\frac{\partial h}{\partial x} \tag{7.125}$$

This law is analogous to Fourier's law in heat flow presented in Chapter 5. Darcy's law is theoretically derived for tube-like pores and slit-like discontinuities in this subsection (Aydan et al. 1997a,b; Aydan & Üçpırtı 1997).

a. Darcy's law for rock with cylindrical pores (Figure 7.31)
   Equilibrium equation for the x-direction is given as

$$\sum F_x = p\pi\left[(r+\Delta r)^2 - r^2\right] - (p+\Delta p)\pi\left[(r+\Delta r)^2 - r^2\right]$$
$$+(\tau+\Delta\tau)2\pi(r+\Delta r)\Delta x - \tau 2\pi r\Delta x = 0 \tag{7.126}$$

   Rearranging the resulting expression and taking the limit and omitting the second-order components yield:

$$\frac{dp}{dx} - \frac{d\tau}{dr} - \frac{\tau}{r} = 0 \tag{7.127}$$

   Assuming that the flow is laminar and a linear relationship holds between shear stress and strain rate $\dot{\gamma}$ as:

$$\tau = \eta\dot{\gamma} \quad \dot{\gamma} = \frac{d\dot{u}}{dr} = \frac{dv}{dr} \quad v = \dot{u} = \frac{du}{dt} \tag{7.128}$$

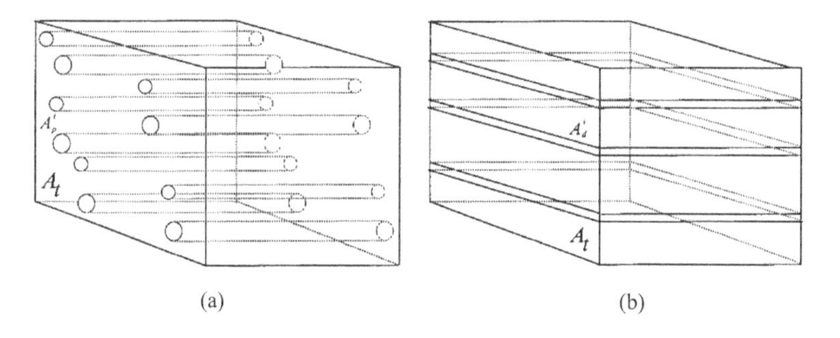

Figure 7.31 Geometrical models for the Darcy law. (a) Rock with cylindrical pores and (b) rock with slit-like discontinuities.

Now, let us insert the above relation into Eq. (7.127), we have the following partial differential equation:

$$\frac{dp}{dx} - \eta \frac{d^2 v}{dr^2} - \frac{\eta}{r}\frac{dv}{dr} = 0 \tag{7.129}$$

Integrating the above partial differential equation for the r-direction yields the following:

$$v = \frac{1}{\eta}\frac{dp}{dx}\frac{r^2}{4} + C_1 \ln r + C_2 \tag{7.130}$$

Introducing the following boundary conditions:

$$v = v_0 \quad \text{at} \quad r = \frac{D}{2}$$

$$\tau = 0 \quad \text{at} \quad r = 0$$

yields the integration constants C_1 and C_2 as:

$$C_1 = 0, \quad C_2 = v_0 - \frac{1}{\eta}\frac{dp}{dx}\frac{D^2}{16}$$

where D is the diameter of the pore. If velocity v_0 is given in the following form

$$v_0 = -\alpha \frac{1}{\eta}\frac{dp}{dx}\frac{D^2}{16} \tag{7.131}$$

integration coefficient C_2 can be obtained as follows:

$$C_2 = -(1+\alpha)\frac{1}{\eta}\frac{dp}{dx}\frac{D^2}{16}$$

The flow rate q passing through the discontinuity for a unit time is:

$$q = \int_0^{2\pi}\int_{r=0}^{y=\frac{D}{2}} v r dr d\theta \tag{7.132}$$

The explicit form of q is obtained as

$$q = -\frac{\pi}{\eta}\frac{D^4}{128}\frac{dp}{dx} \tag{7.133}$$

If the flow rate q is redefined in terms of an average velocity $\bar{v}$ over the pore area as

$$q = -\bar{v}\pi\frac{D^2}{4} \tag{7.134}$$

we have the following expression

$$\bar{v} = -(1+\alpha)\frac{1}{\eta}\frac{D^2}{32}\frac{dp}{dx} \tag{7.135}$$

This relation is known as the Hagen–Poiseuille equation for $\alpha = 0$. In an analogy to the Darcy law, we can rewrite the expression above as

$$\bar{v} = -\frac{k}{\eta}\frac{dp}{dx} \tag{7.136}$$

where

$$k = (1+\alpha)\frac{D^2}{32} \quad \text{or} \quad k = (1+\alpha)\frac{a^2}{8}; \quad a = \frac{D}{2}$$

This is known as the actual permeability of the pores. Let us assume that the ratio (porosity) n of the area of pores over the total area is given by (Figure 7.31a)

$$n = \frac{1}{A_t}\sum_{i=1}^{N}\pi\frac{D_i^2}{4}, \quad \text{or} \quad n = \frac{N\pi\bar{D}^2}{4A_t} \tag{7.137}$$

Then, the apparent permeability k_a is related to the actual permeability as

$$k_a = nk \tag{7.138}$$

b. Darcy law for slit-like discontinuities
For the x-direction, the force equilibrium equation for fluid can be given as follows (Figure 7.31b):

$$\sum F_x = p_{(x)}\Delta y - p_{(x+\Delta x)}\Delta y + \tau_{(y+\Delta y)}\Delta x - \tau_{(y)}\Delta x = 0 \tag{7.139}$$

where p is the pressure and τ is the shear stress. Equation (7.139) takes the following partial differential form by taking Taylor expansions of p and τ as:

$$\frac{dp}{dx} - \frac{d\tau}{dy} = 0 \tag{7.140}$$

Assuming that flow is laminar and the relation between shear stress τ and shear strain rate $\dot{\gamma}$ is linear:

$$\tau = \eta\dot{\gamma}, \quad \dot{\gamma} = \frac{d\dot{u}}{dy} = \frac{dv}{dy}, \quad v = \dot{u} = \frac{du}{dt} \tag{7.141}$$

where η is the viscosity and $\dot{u}$ is the deformation rate. Substituting the above relations into Eq. (7.140) yields the following partial differential equation:

$$\frac{dp}{dx} - \eta \frac{d^2 v}{dy^2} = 0 \tag{7.142}$$

Integrating the equation above for the y-direction yields the following expression for flow velocity v

$$v = \frac{1}{\eta}\frac{dp}{dx}\frac{y^2}{2} + C_1 y + C_2 \tag{7.143}$$

Introducing the following boundary conditions in Eq. (7.143)

$$v = v_o \quad \text{at} \quad y = \frac{h}{2}, \quad \tau = 0 \quad \text{at} \quad y = 0$$

yields the integration constants C_1 and C_2 as:

$$C_1 = 0, \quad C_2 = v_o - \frac{1}{\eta}\frac{dp}{dx}\frac{h^2}{8} \tag{7.144}$$

where h is the aperture of discontinuity. If it is assumed that the following relation exists for v_o

$$v_o = -\alpha \frac{1}{\eta}\frac{dp}{dx}\frac{h^2}{8} \tag{7.145}$$

then, the integration constant C_2 can be written as:

$$C_2 = -(1+\alpha)\frac{1}{\eta}\frac{dp}{dx}\frac{h^2}{8} \tag{7.146}$$

The total flow rate v_t through the discontinuity at a given time is:

$$v_t = 2\int_{y=0}^{y=\frac{h}{2}} v\,dy \tag{7.147}$$

The explicit form of v_t is obtained as:

$$v_t = -(1+\alpha)\frac{1}{\eta}\frac{h^3}{12}\frac{dp}{dx} \tag{7.148}$$

For $\alpha = 0$, the equation above is well known as a "cubic law" equation in groundwater hydrogeology (Snow 1965) and it is introduced to the field of geomechanics by Polubarinova-Kochina in 1962. Let us redefine the flow rate v_t in terms of an average velocity $\bar{v}$ and the discontinuity aperture h as

$$v_t = \bar{v}h \tag{7.149}$$

Inserting this equation into Eq. (7.148) yields the following:

$$\bar{v} = -\frac{1}{\eta}\frac{h^2}{12}\frac{dp}{dx} \tag{7.150}$$

In an analogy to the Darcy law, the above equation may be rewritten as:

$$\bar{v} = -\frac{k_d}{\eta}\frac{dp}{dx} \tag{7.151}$$

where

$$k_d = (1+\alpha)\frac{h^2}{12}$$

k_d in the above equation is called the permeability of discontinuity. If discontinuity porosity n_d is defined as the ratio of the total area $\sum_{i=1}^{N} A_d^i$ of discontinuities to the total area A_t (Figure 7.31b):

$$n_d = \frac{1}{A_t}\sum_{i=1}^{N} A_d^i \tag{7.152}$$

the following relation between apparent permeability k_{d_a} and actual permeability k_d is obtained as:

$$k_{d_a} = n_d k_d \tag{7.153}$$

7.3.2 Permeability tests based on a steady-state flow

7.3.2.1 Pressure difference and flow velocity method

During an experiment, pressures, which are applied at the ends of a test specimen, and flow velocity are measured. If the change of density of fluid with respect to pressure is negligible, permeability can be obtained from the following equation:

$$k = \frac{v_t}{2\pi r_1 H}\frac{\ln(r_2/r_1)\eta}{p_1 - p_2} \tag{7.154}$$

where v_t is the flow rate and H is the specimen length.

If a gas is used as a permeation fluid, the above equation will have the following form:

$$k = \frac{v_t}{\pi H} \frac{\ln(r_2/r_1)}{p_1^2 - p_2^2} p_1 \eta \tag{7.155}$$

7.3.2.2 *Pressure difference method*[1]

For this kind of tests, gas is used as a permeation fluid. If the change of pressure with respect to time is linear, the velocity of gas will also change linearly. The relationship between the mass compressibility coefficient and volumetric compressibility coefficient is given by:

$$c = \rho c^* \tag{7.156}$$

For a compressed gas in a reservoir with a constant volume, the variation of the gas mass for a unit of time can be written as:

$$q = \rho c^* V \frac{dp_1}{dt} \tag{7.157}$$

For a given time, the mass passing through a specimen with a cross-section area A may be given by

$$q = \rho A \bar{v} \tag{7.158}$$

Equating Eqs. (7.157) and (7.158), using (7.155), and rearranging the resulting expression yield the final equation of permeability (Aydan et al. 1996):[2]

$$k = \frac{V\eta}{\pi L} \frac{ln(r_2/r_1)}{\left(p_1^2 - p_2^2\right)} \frac{dp_1}{dt} \tag{7.159}$$

where V is the applicable volume (the sum of the volumes of the supplemental reservoir, the pressure injection tubing and the sample injection hole), η is the gas viscosity, r_2 is the radial distance from the center of the gas injection hole to the periphery of the specimen, r_1 is the radius of the gas injection hole, dp_1/dt is the time rate change of the injection pressure, L is the length of the gas injection hole and p_1 and p_2 are pressures at reservoirs 1 and 2.

1 This test is valid if pressure rate remains constant with time.

2 This relation was also derived by Zeigler (1976). However, how he derived Eq. (7.159) is not known to the author.

7.3.3 *Permeability tests based on a non-steady-state flow (transient flow tests)*

7.3.3.1 The falling head test method for radial flow

The experimental set-up used for this kind of test is shown in Figure 7.32 (Aydan et al. 1997a). As seen from the figure, there is a pipe placed on the top of the cylindrical hole drilled in the middle of the test specimen. The cross-section area of this pipe is denoted as A_h. During the test, the change of pressure and velocity of flow can be measured. The height of water at the outside surface of the test specimen (h_2) is assumed to be constant. When the experiment starts, the flow rate inside the pipe can be given as:

$$q = -\rho g A_h \frac{\partial h_1}{\partial t} \tag{7.160}$$

where h_1 is the height of water inside the pipe. At a given time, the flow rate through a cross-section area of hole (A_p) inside a test specimen is given by

$$v_t = \bar{v} A_p \tag{7.161}$$

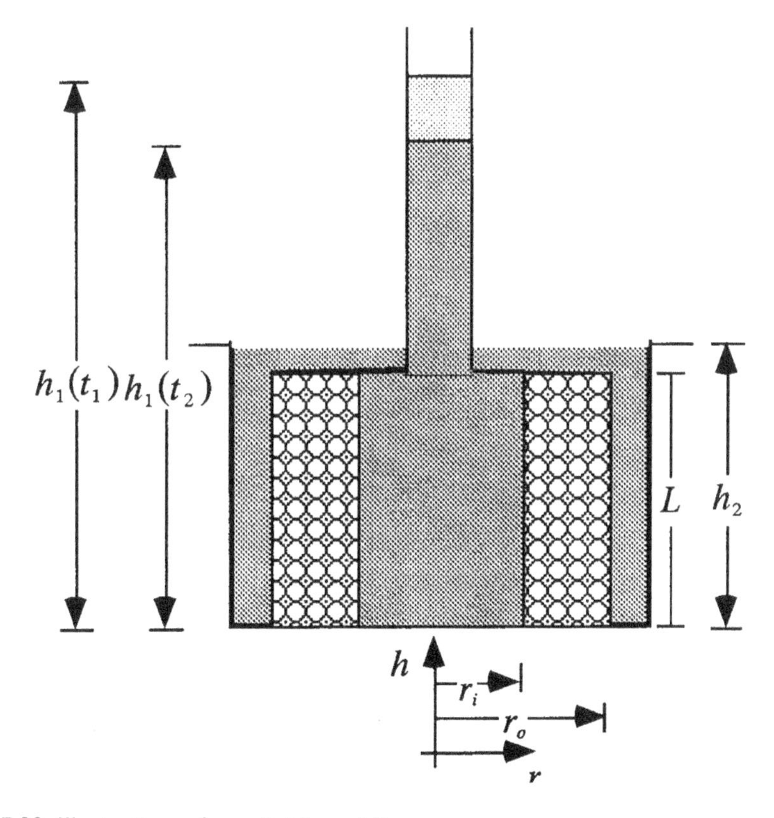

Figure 7.32 Illustration of a radial free-fall test.

It is assumed that the flow rate through the hole perimetry should be equal to the flow rate of the pipe. Then, the pressure gradient in the specimen can be given in the following form:

$$\frac{\partial p}{\partial r} \approx -\frac{\partial}{\partial r}\left(\rho g\left(h_1 - h\right)_2\right) = -\rho g\frac{\partial(h_1 - h_2)}{\partial r} = -\rho g\frac{\left(h_1 - h_2\right)}{r\ln\left(r_o/r_i\right)} \tag{7.162}$$

Substituting Eq. (7.162) together with Eq. (7.154) into Eq. (7.161) and equalizing the resulting equation to Eq. (7.160) yield the following differential equation for the change of water height h_1:

$$\frac{\partial h_1}{h_1 - h_2} = -\frac{k}{\eta}\frac{A_p}{A_h}\frac{1}{r_i\ln\left(r_o/r_i\right)}\partial t \tag{7.163}$$

The solution of the above differential equation is:

$$h_1 = h_2 + Ce^{-\alpha t} \tag{7.164}$$

where

$$\alpha = \frac{k}{r_i\ln\left(r_o/r_i\right)}\frac{A_p}{A_h}\frac{1}{\eta}$$

Introducing the following initial conditions

$$t = 0 \text{ at } \quad h_1 = h_{10}$$

yields the integration constant C as:

$$C = h_{10} - h_2 \tag{7.165}$$

If the integration constant is inserted into Eq.(7.164), the following equation is obtained:

$$-\alpha t = \ln\left(\frac{h_1 - h_2}{h_{10} - h_2}\right) \tag{7.166}$$

Now, if α is substituted into the above equation, the following expression for permeability is obtained

$$k = \eta r_i\ln\left(r_o/r_i\right)\frac{A_h}{A_p}\ln\left(\frac{h_{10} - h_2}{h_1 - h_2}\right)\frac{1}{t} \tag{7.167}$$

7.3.3.2 Transient pulse method for radial flow

Brace et al. (1968) proposed a transient pulse method for longitudinal flow tests. Aydan et al. (1997a) proposed a permeability test for radial flow. Their method is explained in detail herein. This method is fundamentally very similar to Brace's

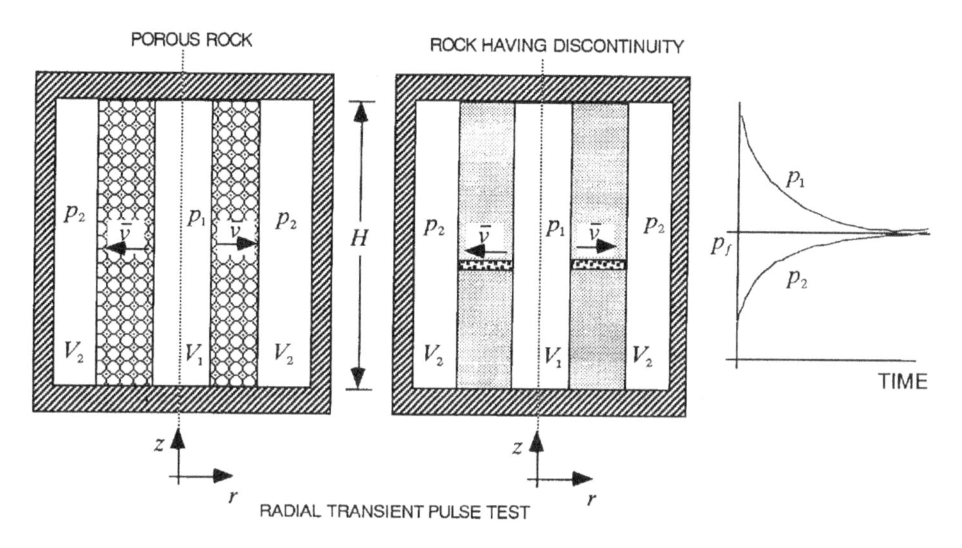

Figure 7.33 Transient pulse radial flow set-up for intact rock and discontinuities.

method (Figure 7.33). The volumetric strain of the fluid inside reservoirs V_1 and V_2 can be written as follows:

$$\varepsilon_V^1 \approx \frac{\Delta V_1}{V_1}, \quad \varepsilon_V^2 \approx \frac{\Delta V_2}{V_2} \tag{7.168}$$

Similarly, for the volumetric strain rate of fluid, the following relations can also be written:

$$\dot{\varepsilon}_V^1 \approx \frac{\Delta \dot{V}_1}{V_1}, \quad \dot{\varepsilon}_V^2 \approx \frac{\Delta \dot{V}_2}{V_2} \tag{7.169}$$

or

$$\Delta \dot{V}_1 = \dot{\varepsilon}_V^1 V_1, \quad \Delta \dot{V}_2 = \dot{\varepsilon}_V^2 V_2 \tag{7.170}$$

If the following relation exists between the volumetric strain of fluid and pressure:

$$\varepsilon_V^1 = -c_f P_1, \quad \varepsilon_V^2 = -c_f P_2 \tag{7.171}$$

and, compressibility coefficient (c_f) is constant, for the volumetric strain rate, the following relation can be also written as

$$\dot{\varepsilon}_V^1 = -c_f \dot{P}_1, \quad \dot{\varepsilon}_V^2 = -c_f \dot{P}_2 \tag{7.172}$$

The flow rate may be given as

$$v_{t1} = \Delta \dot{V}_1, \quad v_{t2} = \Delta \dot{V}_2 \tag{7.173}$$

Using Eqs. (7.171)–(7.173), the flow rate can be rewritten in the following form:

$$v_{t1} = -c_f V_1 \frac{\partial P_1}{\partial t}, \quad v_{t2} = -c_f V_2 \frac{\partial P_2}{\partial t} \tag{7.174}$$

Introducing the following boundary conditions

$$r = r_1 \quad \text{at} \quad P = P_1, \quad r = r_2 \quad \text{at} \quad P = P_2$$

and using Eq. (7.174), the following relation can be obtained for the flow rate

$$v_{t1} = -\frac{kA_{p1}}{\eta}\left(\frac{dP_1}{dr}\right)_{r=r1}, \quad v_{t2} = -\frac{kA_{p2}}{\eta}\left(\frac{dP_2}{dr}\right)_{r=r2} \tag{7.175}$$

where A_{p1} is the surface area of the pressure injection hole and A_{p2} is the area of the pressure release surface. Pressure gradients in the above equations are as follows:

$$\frac{dP_1}{dr} \approx -\frac{1}{r_1}\frac{(P_1 - P_2)}{\ln(r_2/r_1)}, \quad \frac{dP_2}{dr} \approx -\frac{1}{r_2}\frac{(P_2 - P_1)}{\ln(r_2/r_1)} \tag{7.176}$$

Inserting the above equation into Eq. (7.175) and equalizing the resulting equation to Eq. (7.174) yield the following set of equations

$$\frac{\partial P_1}{\partial t} = -\beta \frac{A_{p1}}{V_1 r_1}\frac{(P_1 - P_2)}{\ln(r_2/r_1)} \tag{7.177}$$

$$\frac{\partial P_2}{\partial t} = \beta \frac{A_{p2}}{V_2 r_2}\frac{(P_1 - P_2)}{\ln(r_2/r_1)} \tag{7.178}$$

where

$$\beta = \frac{k}{c_f \eta}$$

Equation (7.177) can be rearranged as follows:

$$P_2 = P_1 + \frac{V_1 r_1 \ln(r_2/r_1)}{\beta A_{p1}}\frac{\partial P_1}{\partial t} \tag{7.179}$$

Taking the time derivative of the above equation, the following expression is obtained:

$$\frac{\partial P_2}{\partial t} = \frac{\partial P_1}{\partial t} + \frac{V_1 r_1 \ln(r_2/r_1)}{\beta A_{p1}}\frac{\partial^2 P_1}{\partial t^2} \tag{7.180}$$

Substituting Eqs. (7.179) and (7.180) into Eq. (7.178) and rearranging the resulting equation yield the following homogeneous differential equation:

$$\frac{\partial^2 P_1}{\partial t^2} + \alpha \frac{\partial P_1}{\partial t} = 0 \tag{7.181}$$

where

$$\alpha = \beta \frac{V_2 r_2 A_{p1} + V_1 r_1 A_{p2}}{\ln(r_2/r_1) V_1 V_2 r_2 r_1}$$

The general solution of this differential equation is

$$P_1 = C_1 + C_2 e^{-\alpha t} \tag{7.182}$$

Introducing the following initial conditions

$$t = 0 \quad \text{at} \quad P_1 = P_i, \quad t = \infty \quad \text{at} \quad P_1 = P_f$$

yields the integration constants C_1 and C_2 as:

$$C_1 = P_f, \quad C_2 = P_i - P_f \tag{7.183}$$

Inserting these integration constants into Eq. (7.182) gives the following equation

$$P_1 = P_f + (P_i - P_f) e^{-\alpha t} \tag{7.184}$$

Taking the time derivative of the above equation

$$\frac{\partial P_1}{\partial t} = -(P_i - P_f) \beta \frac{V_2 r_2 A_{p1} + V_1 r_1 A_{p2}}{V_2 r_2 V_1 r_1 \ln(r_2/r_1)} e^{-\alpha t} \tag{7.185}$$

Substituting Eqs. (7.184) and (7.185) into Eq. (7.179) and rearranging yield the following equation:

$$P_2 = P_f - (P_i - P_f) \frac{V_1 r_1 A_{p2}}{V_2 r_2 A_{p1}} e^{-\alpha t} \tag{7.186}$$

For the following initial condition for

$$t = 0 \quad \text{at} \quad P_2 = P_0$$

Equation (7.186) takes the following form

$$(P_i - P_f) = (P_f - P_0) \frac{V_2 r_2 A_{p1}}{V_1 r_1 A_{p2}} \tag{7.187}$$

The above equation can be rewritten in a different way for P_i–P_0 as follows:

$$(P_i - P_f) = (P_i - P_0) \frac{V_2 r_2 A_{p1}}{V_1 r_1 A_{p2} + V_2 r_2 A_{p1}} \tag{7.188}$$

Inserting this equation into Eq. (7.184) and rearranging yield the following:

$$-\alpha t = \ln\left(\frac{P_1 - P_f}{P_i - P_0}\frac{V_1 r_1 A_{p2} + V_2 r_2 A_{p1}}{V_2 r_2 A_{p1}}\right) \tag{7.189}$$

where

$$\alpha = \frac{k}{c_f \eta}\frac{V_2 r_2 A_{p1} + V_1 r_1 A_{p2}}{V_2 r_2 V_1 r_1 \ln(r_2/r_1)}$$

From the above equations, one gets the following equation to compute permeability

$$k = \eta c_f \frac{V_2 r_2 V_1 r_1 \ln(r_2/r_1)}{V_2 r_2 A_{p1} + V_1 r_1 A_{p2}} \ln\left(\frac{P_1 - P_f}{P_i - P_0}\frac{V_2 r_2 A_{p1}}{V_2 r_2 A_{p1} + V_1 r_1 A_{p2}}\right)\frac{1}{t} \tag{7.190}$$

When gas is used as a permeation fluid, P_1 and P_2 are replaced with $U_1\left(=P_1^2\right)$ and U_2 $\left(=P_2^2\right)$, and permeability can be calculated using the same relation given above. If the volume of reservoir 2 (V_2) is greater than the volume of reservoir 1 (V_1), $(V_2 >> V_1)$ (for instance, the outer side of the specimen is open to air) P_0 and P_f given in the above equation will be equal to atmospheric pressure (P_a). For this particular case, the permeability of a specimen can be obtained from the following equation:

$$k = \eta c_f \frac{V_1 r_1 \ln(r_2/r_1)}{A_{p1}} \ln\left(\frac{P_1 - P_f}{P_i - P_0}\right)\frac{1}{t} \tag{7.191}$$

7.3.3.3 Theory of interface or discontinuity permeability in radial flow tests

For a cylindrical coordinate system, the force equilibrium equation for fluid can be given as follows (Figure 7.33) (Aydan et al. 1997b):

$$\left(\tau_{(z+\Delta z)} - \tau_{(z)}\right) r\Delta r\Delta\theta + \sigma_{r(r)} r\Delta\theta\Delta y - \sigma_{r(r+\Delta r)}(r+\Delta r)\Delta\theta\Delta y + 2\sigma_\theta \Delta r \sin\left(\frac{\Delta\theta}{2}\right) = 0 \tag{7.192}$$

with the use of the following relation

$$\sin\left(\frac{\Delta\theta}{2}\right) \approx \frac{\Delta\theta}{2}$$

The above equation can be rearranged. Then, if the Taylor expansion is used for τ and σ_r in Eq. (7.192), the following partial differential is obtained:

$$\frac{d\sigma_r}{dr} - \frac{\sigma_r - \sigma_\theta}{r} - \frac{d\tau}{dz} = 0 \tag{7.193}$$

For a hydrostatic case $(\sigma_r = \sigma_\theta = p)$, Eq. (7.193) becomes

$$\frac{d\sigma_r}{dr} - \frac{d\tau}{dz} = 0 \tag{7.194}$$

Assuming that flow is laminar and the relation between shear force and shear strain rate $\dot{\gamma}$ is linear:

$$\tau = \eta\dot{\gamma}, \quad \dot{\gamma} = \frac{d\dot{u}}{dz} = \frac{dv}{dz}, \quad \sigma_r = p, \quad v = \dot{u} = \frac{du}{dt} \tag{7.195}$$

Substituting the above relations into Eq. (7.194) yields the following partial differential equation:

$$\frac{dp}{dr} - \eta\frac{d^2v}{dz^2} = 0 \tag{7.196}$$

Integrating the above partial differential equation for the y-direction yields the following expression for flow velocity v

$$v = \frac{1}{\eta}\frac{dp}{dr}\frac{z^2}{2} + C_1 z + C_2 \tag{7.197}$$

Introducing the following boundary conditions in Eq. (7.197)

$$C_1 = 0, \quad C_2 = -\frac{1}{\eta}\frac{dp}{dr}\frac{h^2}{8} \tag{7.198}$$

The total flow rate v_t through the discontinuity at a given time is:

$$v_t = 2\int_0^{2\pi}\int_{z=0}^{z=\frac{h}{2}} vrdzd\theta \tag{7.199}$$

The explicit form of v_t is obtained as:

$$v_t = -\frac{2\pi r}{\eta}\frac{h^3}{12}\frac{dp}{dr} \tag{7.200}$$

Let us redefine the flow rate v_t in terms of an average velocity $\bar{v}$ over the discontinuity aperture area as

$$v_t = 2\pi r\bar{v}h \tag{7.201}$$

and by inserting this equation into Eq. (7.200), the following expression is obtained.

$$\bar{v} = -\frac{1}{\eta}\frac{h^2}{12}\frac{dp}{dr} \tag{7.202}$$

In an analogy to the Darcy law, the above equation may be rewritten as:

$$\bar{v} = -\frac{k_d}{\eta}\frac{dp}{dr} \tag{7.203}$$

where

$$k_d = \frac{h^2}{12} \quad \text{or} \quad h = \sqrt{12k_d}$$

and k_d is called the intrinsic permeability of discontinuity. By inserting Eq. (7.203) into Eq. (7.200), the flow rate v_t can be obtained as:

$$v_t = -2\pi rhk_d \frac{dp}{dr} \tag{7.204}$$

The above equation can be rewritten in the following form:

$$\frac{dr}{r} = -\frac{2\pi hk_d dp}{v_t} \tag{7.205}$$

The integral form of the above equation may be written as:

$$\int_{r_i}^{r_o} \frac{dr}{r} = -\int_{p_i}^{p_o} \frac{2\pi hk_d}{v_t} dp \tag{7.206}$$

If integration is carried out, the permeability of discontinuity (k_d) can be found after some manipulations as follows:

$$k_d = \left[\frac{v_t \ln\left(\frac{r_o}{r_i}\right)}{4\pi\sqrt{3}(p_i - p_o)} \right]^{2/3} \tag{7.207}$$

Figure 7.34 shows pressure responses of Reservoirs 1 and 2 in a transient pulse test on a sandstone sample. Despite some scattering of pressure responses, the variations of reservoir pressures tend to decrease with time and become asymptotic to a stabilizing pressure. The permeability of the sandstone sample was 3.1×10^{-12} m^2.

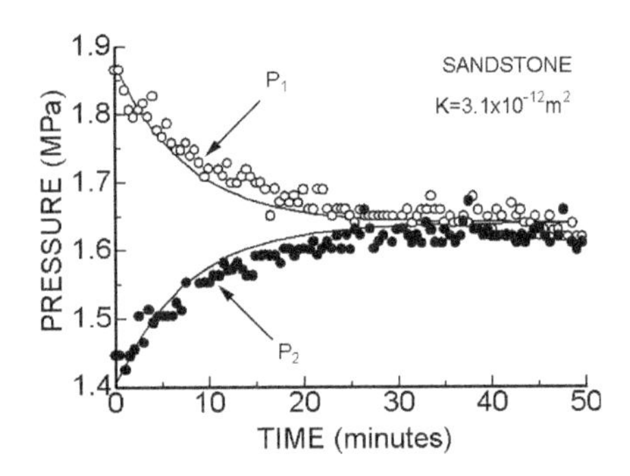

Figure 7.34 Pressure responses of a sandstone sample in a transient pulse test.

7.4 Temperature distribution in the vicinity of geological active faults

As a first case, the geological fault is assumed to be sandwiched between two non-conductive rock slabs and closed-form solutions are derived for temperature rises within the fault due to shearing. Then a more general case is considered such that a seismic energy release takes place within the fault, and adjacent rock is conductive. The solution of the governing equation for this case is solved with the use of the finite element method (Aydan 2017). Several examples were solved by considering some hypothetical energy release functions and their implications are discussed.

A geological fault and its close vicinity may be simplified to a one-dimensional situation as shown in Figure 7.35 by assuming that the mechanical energy release is due purely to shearing with no heat production source. Thus, Eq. (4.30) may be reduced to the following form:

$$\rho c \frac{\partial T}{\partial t} = -\nabla q + \tau \dot{\gamma} \tag{7.208}$$

Let us assume that the heat flux obeys Fourier's law, which is given by

$$q = -k \frac{\partial T}{\partial x} \tag{7.209}$$

Inserting Eq. (7.209) into Eq. (7.208) yields the following equation

$$\rho c \frac{\partial T}{\partial t} = k \frac{\partial^2 T}{\partial x^2} + \tau \dot{\gamma} \tag{7.210}$$

The solution of the above equation will yield the temperature variation with time.

The energy release during earthquakes is a very complex phenomenon. Nevertheless, some simple forms relevant to the overall behavior may be assumed to have some insights into the phenomenon. Two energy release rate functions of the following form are assumed as given below:

$$\dot{E} = \tau \dot{\gamma} = Ate^{-\frac{t}{\theta}} \tag{7.211}$$

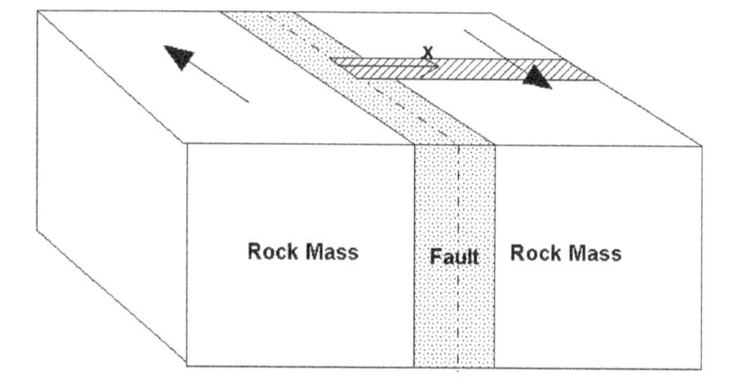

Figure 7.35 Fault model.

$$\dot{E} = \tau\dot{\gamma} = A^{*}e^{-\frac{t}{\theta^{*}}} \tag{7.212}$$

Constants A and A^{*} depend on the shear stress and shear strain rate history with time and fault thickness. Constants θ and θ^{*} are time history constants. For situations illustrated in Figure 7.35, constants A and A^{*} will take the following forms:

$$\text{For Eq. (7.211)} \quad A = \frac{\tau_o u_f}{h\theta^2} \tag{7.213}$$

$$\text{For Eq. (7.212)} \quad A^{*} = \frac{\tau_o u_f}{h\theta^{*}} \tag{7.214}$$

where u_f and h are the final relative displacement and thickness of the fault. τ_o is the shear stress acting on the fault and it is assumed to be constant during the motion. Two specific situations are analyzed, namely:

- Creeping Fault
- Fault with a hill-shaped seismic energy release rate

In the case of creeping fault, the energy release rate is almost constant with time. The geometry of the fault is assumed to be one-dimensional as shown in Figure 7.36. Figures 7.37 and 7.38 show the computed temperature differences at selected locations

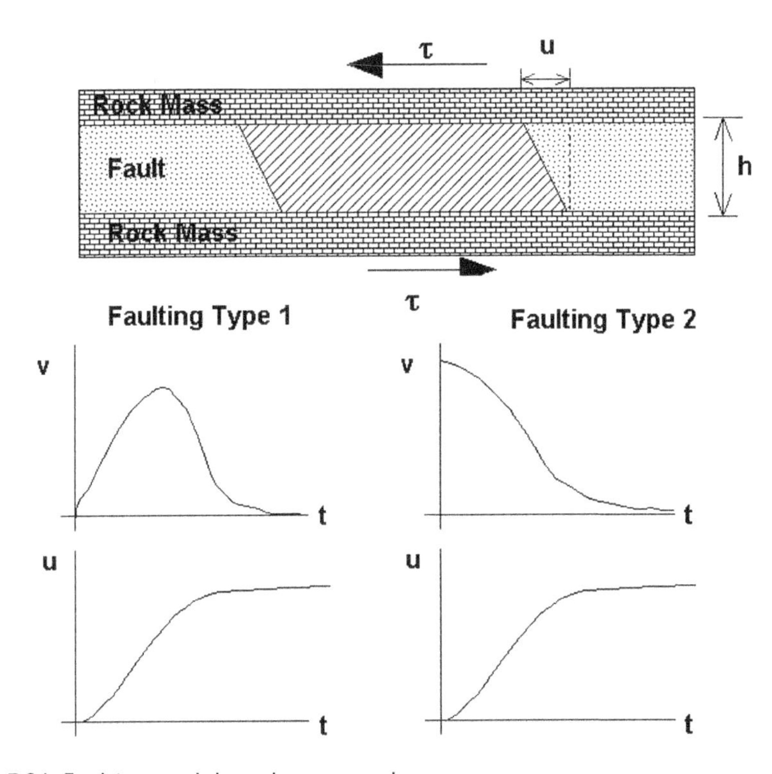

Figure 7.36 Faulting models and energy release types.

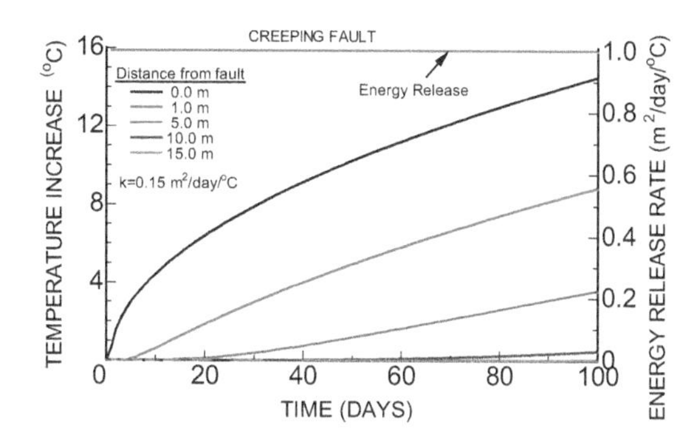

Figure 7.37 Temperature difference variations for a fault sandwiched between conductive rock mass slabs for creeping condition.

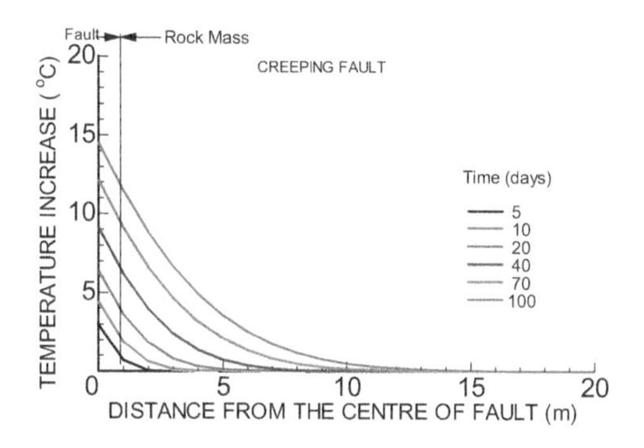

Figure 7.38 Temperature distributions at different time steps for a fault sandwiched between conductive rock mass slabs for creeping condition.

with time and temperature difference distribution throughout the whole domain at selected time steps. In the computations, the energy release rate is assumed to be taking place within the fault zone only. The increase in temperature difference is parabolic and they keep increasing as time goes by. Nevertheless, the temperature difference increases are about 1/10 of those of the fault sandwiched between non-conductive rock mass slabs.

Figures 7.39 and 7.40 show the computed temperature differences at selected locations with time and temperature difference distribution throughout the whole domain at selected time steps for a fault with a hill-like energy release rate. In the computations, the energy release rate is assumed to be taking place within the fault zone only. The increase in temperature difference is parabolic. Temperature difference increases first and then it tends to decay in a manner similar to the assumed seismic energy release rate function.

This situation will be probably quite similar to the actual situation in nature. The temperature difference increases are about 1/10 of those of the fault sandwiched between non-conductive rock mass slabs. These results indicate that the observation of

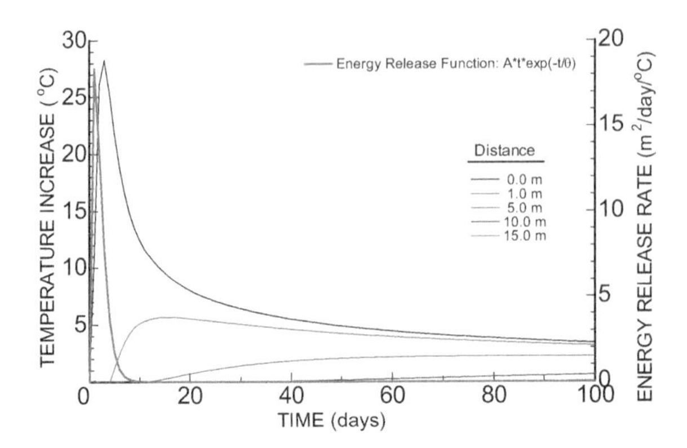

Figure 7.39 Temperature difference variations for a fault sandwiched between conductive rock mass slabs for hill-shaped energy release function.

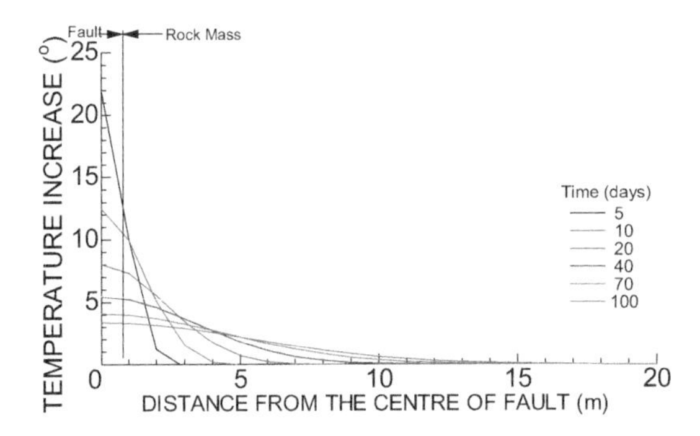

Figure 7.40 Temperature distributions at different time steps for a fault sandwiched between conductive rock mass slabs for hill-shaped energy release function.

ground temperatures may be a very valuable source of information in the predictions of earthquakes. Atmospheric temperature measurements near the ground surface may be quite problematic in interpreting the observations. However, the observation of hot-spring temperature, which reflects the actual ground temperature, may be a very good tool for such measurements without any deep boring.

7.5 Closed-form solutions for diffusion problems

7.5.1 Drying testing procedure

Let us consider a sample with volume V dried in the air with infinite volume as shown in Figure 7.41 (Aydan 2003). Water contained Q in a geo-material sample may be given in the following form

$$Q = \rho_w \theta_w V \tag{7.215}$$

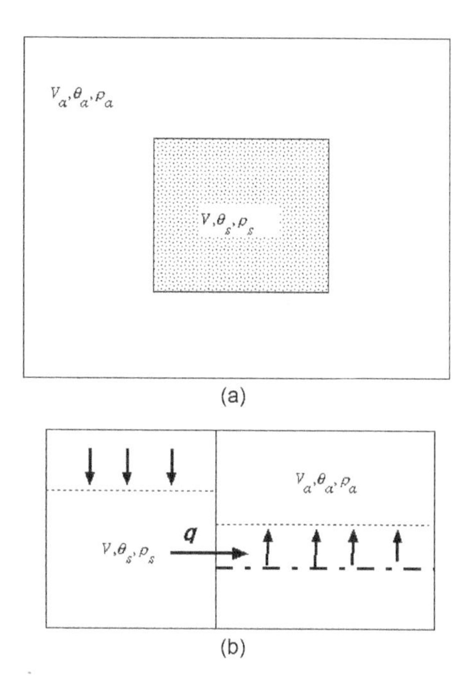

Figure 7.41 (a) Physical and (b) mechanical models for water migration during the drying process.

where p_w, θ_w and V are water density, water content ratio and volume of sample, respectively. If water density and sample volume remain constant, the flux q of water content may be written in the following form

$$q = \frac{dQ}{dt} = -\rho_w V \frac{d\theta_w}{dt} \tag{7.216}$$

Air is known to contain water molecules of 6 g/m^3 when the relative humidity is 100%. When the relative humidity is less than 100%, water is lost from geomaterials to air. If such a situation presents, the water lost from the sample to air may be given in the following form using a concept similar to Newton's cooling law in thermodynamics:

$$q = \rho_w A_s h \Delta\theta = \rho_w A_s h(\theta_w - \theta_a) \tag{7.217}$$

where h and A_s are water loss coefficient and surface area of the sample. Requiring that the water loss rate of the sample should be equal to the water loss into the air based on the mass conservation law, one can easily write the following relation:

$$\rho_w A_s h(\theta_w - \theta_a) = -\rho_w V \frac{d\theta_w}{dt} \tag{7.218}$$

The solution of differential Eq. (7.218) is easily obtained in the following form:

$$\theta_w = \theta_a + Ce^{-\alpha t} \tag{7.219}$$

where

$$\alpha = h\frac{A_s}{V}$$

The integration constant may be obtained from the initial condition, that is,

$$\theta_w = \theta_{w0} \text{ at} \tag{7.220}$$

as follows

$$C = \theta_{w0} - \theta_a \tag{7.221}$$

Thus, the final expression takes the following form:

$$\theta_w = \theta_a + (\theta_{w0} - \theta_a)e^{-\alpha t} \tag{7.222}$$

If the water content migration is considered as a diffusion process, Fick's law in one dimension may be written as follows:

$$q = \rho_w D\frac{\partial \theta_w}{\partial x} \tag{7.223}$$

Requiring water loss rate given by Eq. (7.223) to be equal to that given by Eq. (7.223) yields the following relation

$$D = h\frac{V}{A_s} \tag{7.224}$$

If surface area A_s and volume V of a sample are known, it is easy to determine the water migration diffusion constant D from drying tests easily, provided that the coefficient α and subsequently h are determined from experimental results fitted to Eq. (7.222). If samples behave linearly, water migration characteristics should remain the same during swelling and drying processes. Recent technological developments have made it quite easy to measure the weight of samples and environmental conditions such as temperature and humidity. Figure 7.42 shows an automatic weight and environmental conditions monitoring system developed for such tests. It is also possible to measure the volumetric variations (shrinkage) during the drying process using non-contact type displacement transducers (i.e. laser transducers).

Physical and mechanical properties of materials can be measured using conventional testing methods such as wave velocity measurements, uniaxial compression tests and elastic modulus measurement. Tuff samples used in the tests were from Avanos, Ürgüp and Derinkuyu of the Cappadocia Region in Turkey and Oya in Japan. The samples from the Cappadocia region are gathered from historical and modern underground rock structures. They represent the rocks in which historical and modern underground structures were excavated. These tuff samples bear various clay minerals as given in Table 7.3 (Temel 2002; Aydan and Ulusay 2003). As noted from the table, the clay content is quite high in Avanos tuff and most of the clay minerals are smectite.

In drying experiments, the samples that underwent swelling were dried in a room with an average temperature of 23°C and a relative humidity of 65–70. Figures 7.43–7.45

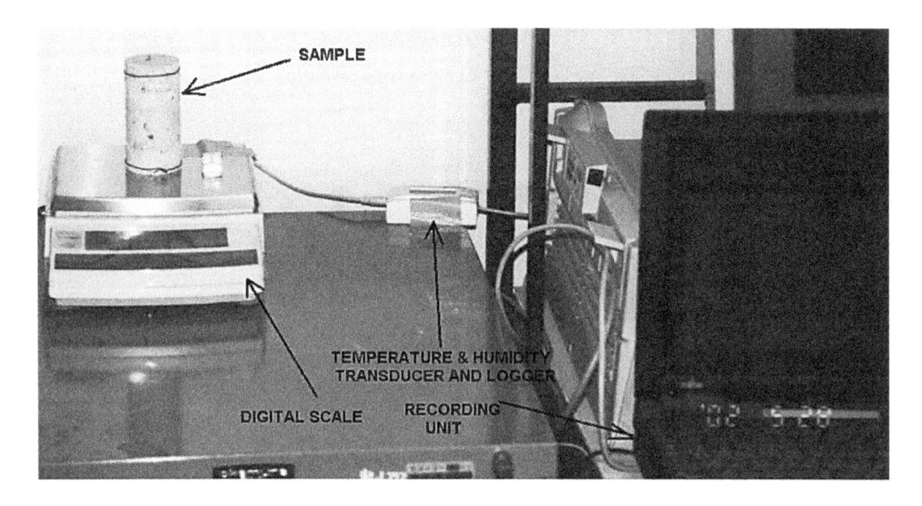

Figure 7.42 The experimental set-up for measuring water content during drying.

Table 7.3 XRD results from the samples of Ürgüp (Kavak tuff) and Avanos

Specimen number	*Clay percentage*	*Clay fraction*		
		Smectite	*Kaolin*	*Illite*
UR-1 (Ürgüp)	74	83	14	3
UR-2 (Ürgüp)	60	67	25	8
AV-1(Avanos)	94	84	13	3
AV-2 (Avanos)	82	95	5	T

T, trace amount.

show the drying test results for some tuff samples from the Cappadocia region in Turkey. As seen from the figures, it takes a longer time for the tuff sample from Avanos compared with Ürgüp and Derinkuyu samples. Derinkuyu sample dries much rapidly than the others. Each sample was subjected to drying twice. Once again it is noted that the drying period increases for Avanos tuff after each run while Derinkuyu tuff tends to dry much rapidly in the second run. From these tests, it may be also possible to determine the diffusion characteristics of each tuff.

The theory derived in the previous section could be applied to the experimental results shown in Figures 9.43–9.45. To obtain the constants of water migration model, Eq. (7.222) may be rewritten as follows:

$$\ln\left(\frac{\theta_w - \theta_a}{\theta_{w0} - \theta_a}\right) = -\alpha t \tag{7.225}$$

The plot of experimental results in the semi-logarithmic space first yields the constant α, from which constant h and diffusion coefficient D can be computed subsequently.

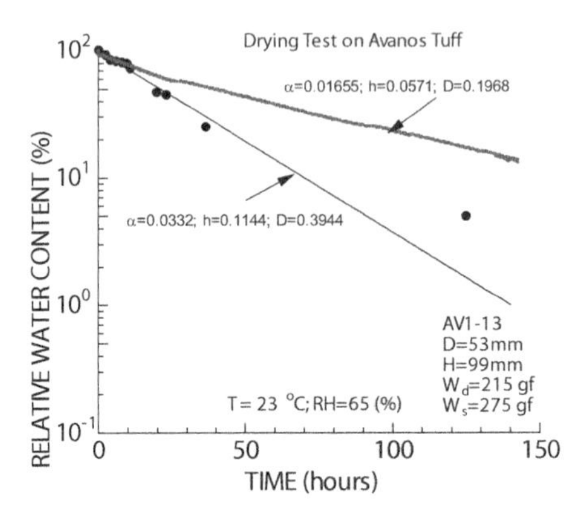

Figure 7.43 Determination of constants for relative water content variation during drying of Avanos tuff.

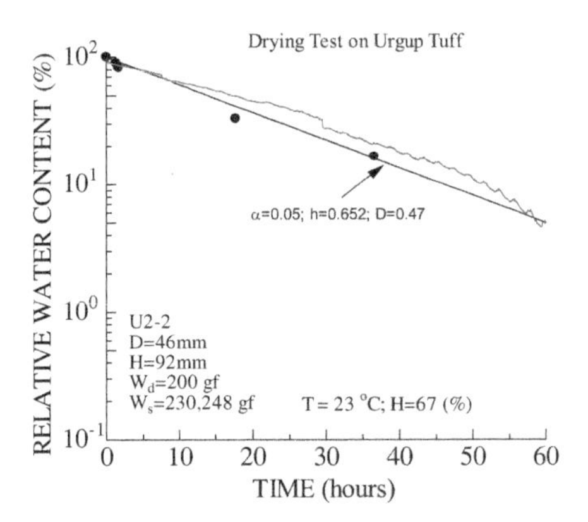

Figure 7.44 Determination of constants for relative water content variation during drying of Ürgüp tuff.

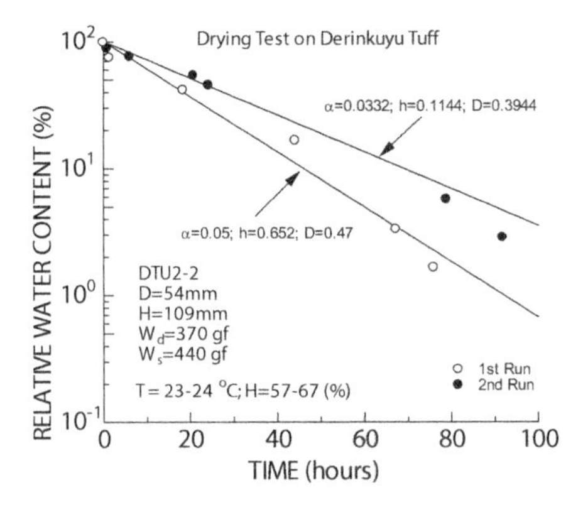

Figure 7.45 Determination of constants for relative water content variation during drying of Derinkuyu tuff.

The results are shown in Figures 9.43–9.45. The units of parameters α, h and D are 1/hr, cm/hr and cm^2/hr, respectively. The computed values of parameters α, h and D are also shown in the same figures.

7.5.2 Saturation testing technique

Initially, dry samples can be subjected to saturation and water migration characteristics may be obtained. The side of the samples can be sealed and subjected to saturation from the bottom. The top surface may be sealed and unsealed as illustrated in Figure 7.46. Samples can be isolated against water migration from sides by sealing while the bottom surface of the samples can be exposed to saturation by immersing in water up to a given depth. There may be two conditions at the top surface, which could be either exposed to air directly or sealed. When the top surface is sealed, the boundary value would be changing with time. The water migration coefficient can be determined from the solution of the following diffusion equation:

$$\frac{\partial \theta_w}{\partial t} = D \frac{\partial^2 \theta_w}{\partial x^2} \tag{7.226}$$

When the top surface is unsealed, the top boundary condition ($x = H$) is

$$\theta_w = \theta_a. \tag{7.227}$$

On the other hand, if the top surface is sealed, the boundary condition is time dependent and it can be estimated from the following condition

$$q_{x=H} = \hat{q}_n(t) \tag{7.228}$$

For some simple boundary conditions, the solution of partial differential Eq. (7.226) can be easily obtained using the technique of separation of variables (i.e. Kreyszig 1983). In the general case, it would be appropriate to solve it using a finite difference technique or finite element method (i.e. Aydan 2003, 2017).

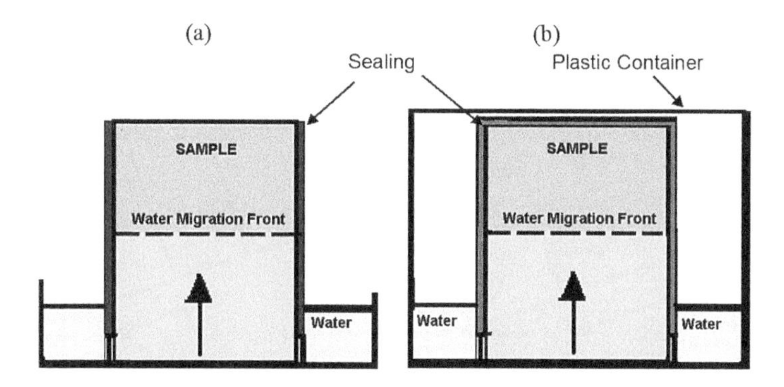

Figure 7.46 Experimental set-ups: (a) top surface unsealed and (b) top-surface sealed.

7.6 Evaluation of creep-like deformation of semi-infinite soft rock layer

The simplified analytical model introduced in this section is based on the theoretical model developed by Aydan (1994a, 1998). A momentum conservation law for an infinitely small element of a ground on a plane with an inclination of α for each respective direction can be written in the following form (Figure 7.47)

x-direction

$$\frac{\partial \tau}{\partial y} = \frac{\partial p}{\partial x} - \rho g \sin\alpha \tag{7.229}$$

y-direction

$$\frac{\partial p}{\partial y} = \rho g \sin\alpha \tag{7.230}$$

where τ, p, ρ and g are shear stress, pressure, density and gravitational acceleration, respectively. The variation of pressure along the x-direction is given by

$$\frac{\partial p}{\partial x} = \rho g \cos\alpha \frac{\partial h}{\partial x} \tag{7.231}$$

If shear stress related to shear strain is linear as given in the following form

$$\tau = G\gamma; \gamma = \frac{\partial u}{\partial y}$$

one can easily obtain the solution given as

$$\tau = \rho g \cos\alpha \left(\tan\alpha - \frac{\partial h}{\partial x} \right)(h - y) \tag{7.232}$$

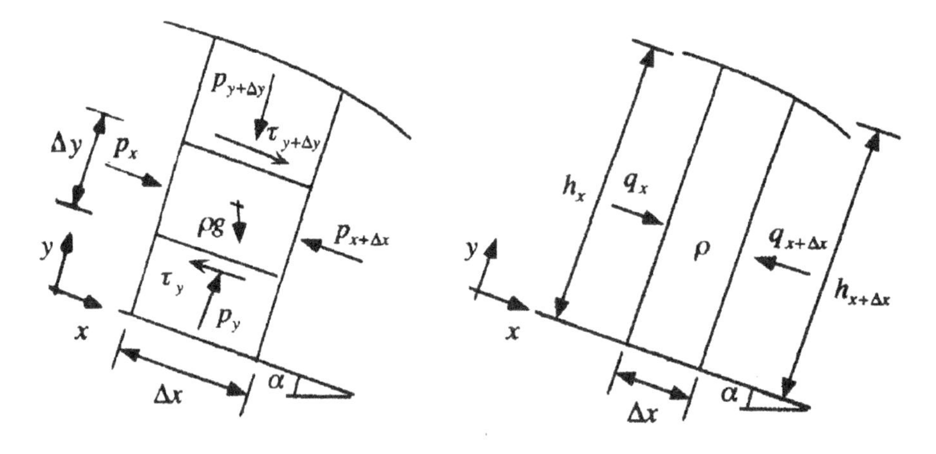

Figure 7.47 Modeling of a layer subjected to shearing. (From Aydan 1994a, 1998.)

If the variation of ground surface height (h) is neglected, the resulting equation for shear stress and displacement takes the following form

$$\tau = \rho g \sin\alpha(h-y);\ u = \frac{\rho g \sin\alpha}{G} y\left(h - \frac{y}{2}\right) \tag{7.233}$$

As well known, the rainfall induces groundwater level fluctuations. However, these fluctuations are not that high as presumed in many limiting equilibrium approaches to analyze the failure of slopes. In other words, the whole body, which is prone to fail, does not become fully saturated. However, the monitoring results indicate that a certain thickness of the layer becomes saturated. In view of experimental results, the deformation modulus would become smaller during the saturation process and recover its original value upon drying. The deformation modulus during saturation may be assumed to be plastic deformation modulus (G_p) and the displacement induced during the saturation period may be viewed as plastic (irrecoverable) deformation (Figure 7.48). With the use of this concept and the analytical model presented above, one can easily derive the following equation for deformation induced by saturation as

$$u_s = \frac{\rho g \sin\alpha}{G_s} y\left(h - \left(t - \frac{y}{2}\right)\right) \tag{7.234}$$

where t is the thickness of the saturated zone in a given cycle of saturation-drying. The plastic deformation would be the difference between displacements induced under saturated and dry states and it will take the following form:

$$u_p = \rho g \sin\alpha y\left(\frac{1}{G_s} - \frac{1}{G_d}\right)\cdot\left(h - \left(t - \frac{y}{2}\right)\right) \tag{7.235}$$

where G_d and G_s are shear modulus for dry and saturated states, respectively. Thus the equivalent shear modulus may be called the plastic deformation modulus (G_p) in this study and can be written as

$$G_p = \frac{G_s G_d}{G_d - G_s} \tag{7.236}$$

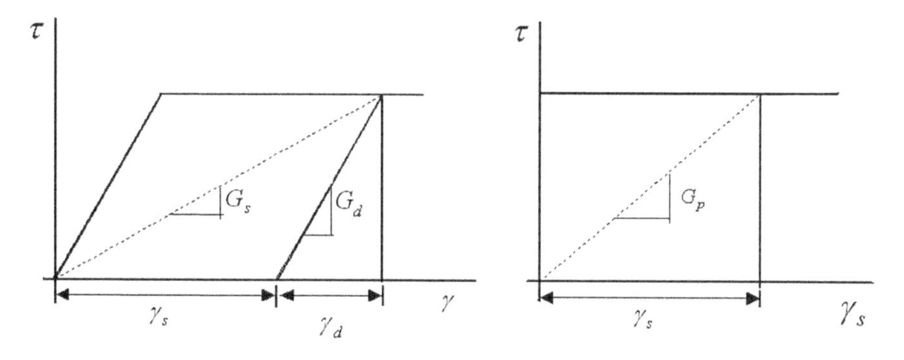

Figure 7.48 Constitutive modeling of cyclic softening-hardening of a marl layer.

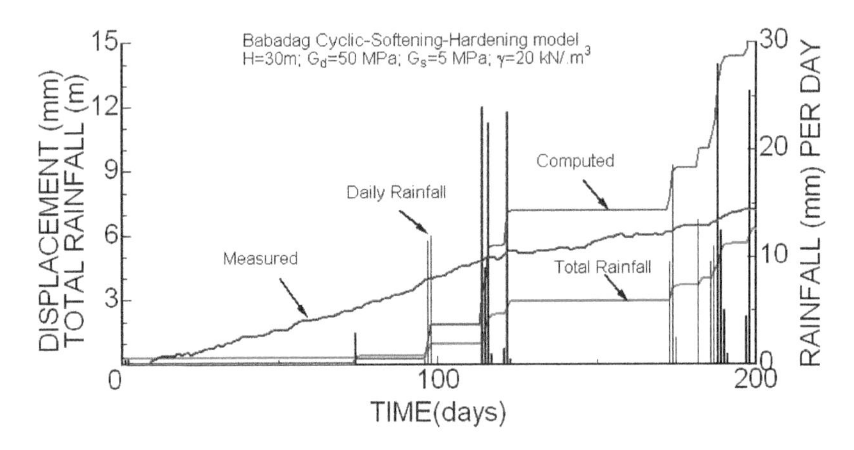

Figure 7.49 Comparison of measured and computed displacements.

The time for saturation and drying of marls is very short (say, in hours). With this observational fact and experimental results, the analysis presented is based on the day unit. Figure 7.49 compares the computed displacement and displacement measured at monitoring station No. 1 in the Gündoğdu district of Babadağ town with the consideration of the thickness of the saturation zone (Kumsar et al. 2016). Despite some differences between computed and measured responses, the analytical model can efficiently explain the overall response of the landslide area of the Gündoğdu district of Babadağ town.

References

Aydan, Ö. 1987. Approximate estimation of plastic zones about underground openings. Nagoya University, Department of Geotechnical Engineering, (unpublished interim report). 7 p.

Aydan, Ö. 1989. The stabilisation of rock engineering structures by rockbolts. Doctorate Thesis, Nagoya University.

Aydan, Ö. 1994a. The dynamic shear response of an infinitely long visco-elastic layer under gravitational loading. *Soil Dynamics and Earthquake Engineering*, Elsevier, 13, 181–186.

Aydan, Ö. 1994b. Thermo-mechanical performance of thick concrete linings cast against frozen rock mass during the hydration of cement. *International Conference on Computational Methods in Structural and Geotechnical Engineering*, Hong Kong, 441–446.

Aydan, Ö. 1995a. Mechanical and numerical modelling of lateral spreading of liquified soil. *The 1st International Conference on Earthquake Geotechnical Engineering*, IS-TOKYO'95, Tokyo, 881–886.

Aydan, Ö. 1995b. The stress state of the earth and the earth's crust due to the gravitational pull. *The 35th US Rock Mechanics Symposium*, Lake Tahoe, 237–243.

Aydan, Ö. 1997. Dynamic uniaxial response of rock specimens with rate-dependent characteristics. *SARES'97*, 322–331.

Aydan, Ö. 1998. A simplified finite element approach for modelling the lateral spreading of liquefied ground. *The 2nd Japan-Turkey Workshop on Earthquake Engineering*, Istanbul.

Aydan, Ö. 2003. The moisture migration characteristics of clay-bearing geo-materials and the variations of their physical and mechanical properties with water content. *2nd Asian Conference on Saturated Soils*, UNSAT-ASIA, 383–388.

Aydan, Ö. 2008. New directions of rock mechanics and rock engineering: geomechanics and geoengineering. *5th Asian Rock Mechanics Symposium (ARMS5)*, Tehran, 3–21.

Aydan, Ö. 2010. An experimental and numerical study on creep characteristics of soft rocks and its use for the long-term response and stability of rock engineering structures. Shimizu, Tokai University, unpublished note, 77 p.

Aydan, Ö. 2017. *Time Dependency in Rock Mechanics and Rock Engineering*, London, CRC Press, Taylor & Francis Group, 241 p.

Aydan, Ö. 2018. *Rock Reinforcement and Rock Support*, London, CRC Press, Taylor & Francis Group, 486 p.

Aydan, Ö. and H. Üçpırtı 1997. The theory of permeability measurement by transient pulse test and experiments. *Journal of the School of Marine Science and Technology*, 43, 45–66.

Aydan, Ö. and P. Nawrocki 1998. Rate-dependent deformability and strength characteristics of rocks. *Proceedings of Symposium on the Geotechnics of Hard Soils-Soft Rock*, Napoli, 1, 403–411.

Aydan, Ö. and R. Ulusay 2003. Geotechnical and geoenvironmental characteristics of man-made underground structures in Cappadocia, Turkey. *Engineering Geology*, 69, 245–272.

Aydan, Ö. and M. Geniş 2010. Rockburst phenomena in underground openings and evaluation of its counter measures. *Journal of Rock Mechanics*, TNGRM (Special Issue, No.17), 1–62.

Aydan, Ö., T. Akagi, T. Ito and T. Kawamoto 1992. Deformation behaviour of tunnels in squeezing rocks and its prediction. *Journal of Japan Society of Civil Engineers, Geotechnical Division*, 448/III-19, 73–82.

Aydan, Ö., T. Akagi and T. Kawamoto 1993. Squeezing potential of rocks around tunnels; Theory and prediction. *Rock Mechanics and Rock Engin*eering, 26(2), 137–163.

Aydan, Ö., T. Akagi, H. Okuda and T. Kawamoto 1994. The cyclic shear behaviour of interfaces of rock anchors and its effect on the long term behaviour of rock anchors. *International Symposium on New Developments in Rock Mechanics and Rock Engineering*, Shenyang, 15–22.

Aydan, Ö., T. Seiki, G.C. Jeong and T. Akagi 1995a. A comparative study on various approaches to model discontinuous rock mass as equivalent continuum. *The 2nd International Conference on Mechanics of Jointed and Fractured Rock*, Wien, 569–574.

Aydan, Ö., T. Akagi, T. Ito, J. Ito and J. Sato 1995b. Prediction of deformation behaviour of a tunnel in squeezing rock with time-dependent chracteristics. *Numerical Models in Geomechanics NUMOG V*, 463–469.

Aydan, Ö., T. Akagi and T. Kawamoto 1996. The squeezing potential of rock around tunnels: theory and prediction with examples taken from Japan. *Rock Mechanics and Rock Engineering*, 29(3), 125–143.

Aydan, Ö., H. Üçpırtı and N. Türk 1997a. Theory of laboratory methods for measuring permeability of rocks and tests. *Kaya Mekaniği Bülteni*, 13, 19–36.

Aydan, Ö., H. Üçpırtı and R. Ulusay 1997b. Theoretical formulation of Darcy's law for fluid flow through porous and/or jointed rock and its validity (in Turkish). *Kaya Mekaniği Bülteni*, 13, 1–18.

Aydan, Ö., S. Watanabe and N. Tokashiki 2008. He inference of mechanical properties of rocks from penetration tests. *5th Asian Rock Mechanics Symposium (ARMS5)*, Tehran, 213–220.

Aydan, Ö., N. Tokashiki and M. Genis 2012. Some considerations on yield (failure) criteria in rock mechanics ARMA 12-640. *46th US Rock Mechanics/Geomechanics Symposium*, Chicago, Paper No. 640, 10 p. (on CD).

Bieniawski, Z.T. 1970. Time-dependent behavior of fractured rock. *Rock Mechanics*, 2, 123–137.

Boussinesq, J. 1885. *Applications des potentiels à l'étude de l'équilibre et mouvement des solides elastiques*, Paris, Gauthier–Villard.

Brace, W.F., J.B. Walsh and W.T. Frangos 1968. Permeability of granite under high pressure. *Journal of Geophysical Research*, 73(6), 2225–2236.

Brady, B.H.G. and E.T. Brown 1985. *Rock Mechanics for Underground Mining*, New York, Boston, London, Moscow, Kluwer Academic Publications, 527 p.
Detournay, E. 1983. Two-dimensional elastoplastic analysis of a deep cylindrical tunnel under non-hydrostatic loading. PhD thesis, University of Minnesota, Minneapolis, MN, 1–133.
Detournay, E. 1986. An approximate statical solution of the elastoplastic interface for the problem of galin with a cohesive-frictional material. *International Journal of Solids Structures*, 22, 1435–1454.
Drucker, D.C. and W. Prager 1952. Soil mechanics and plastic analysis for limit design. *Quarterly of Applied Mathematics*, 10(2), 157–165.
England, A.H. 1971. *Complex Variable Methods in Elasticity*, New York, Dover, Books on Mathematics, 193 p.
Eringen, A.C. 1961. Propagations of elastic waves generated by dynamical loads on a circular cavity. *Journal of Applied Mechanics*, ASME, 28(2), 218–212.
Eringen, A.C. 1980. *Mechanics of Continua*, New York, R. E. Krieger Publishing Company.
Fenner, R. 1938. Researches on the notion of ground stress (in German). *Gluckauf*, 74, 681–695.
Galin, L.A. 1946. Plane elastic-plastic problem: plastic regions around circular holes in plates and beams. *Prikladnaia Matematika i Mechanika*, 10, 365–386.
Gerçek, H. 1988. Calculation of elastic boundary stresses for rectangular underground openings. *Mining Science and Technology*, 7, 173–182.
Gerçek, H. 1996. *Special Elastic Solutions for Underground Openings*. Milestones in Rock Engineering – The Bieniawski Jubilee Collection, Rotterdam, Balkema, 275–290.
Gerçek, H. and M. Genis 1999. Effect of anisotropic in situ stresses on the stability of underground openings. *Proceedings of the 9th International Congress on Rock Mechanics, ISRM*, 1, Balkema, Rotterdam, 367–370.
Green, A.E. and W. Zerna 1986. *Theoretical Elasticity*, London, Oxford at the Clarendon Press.
Hoek, E. and E.T. Brown 1980. *Underground Excavations in Rock*, London, Institute of Mining and Metallurgy.
Hyde, T.H., W. Sun and A.A. Becker 1996. Analysis of the impression creep test method using a rectangular indenter for determining the creep properties in welds. *International Journal of Mechanical Science*, 38, 1089–1102.
Inglis, C.E. 1913. Stresses in a plate due to the presence of cracks and sharp corners, Transactions of Institute of Naval Architecture, 27, 219–230, 338.
Jaeger, J.C. and N.G.W. Cook 1979. *Fundamentals of Rock Mechanics*. 3rd ed., London, Chapman & Hall, pp. 79 and 311.
Kastner, H. 1961. *Statik des Tunnel- and Stollenbaues ("Design of Tunnels")*. 2nd ed., New York, Springer-Verlag.
Kirsch, G. 1898. Die theorie der elastizitat und die bedürfnisse der festigkeitslehre. *Veit Ver Deut Ing*, 42, 797–807.
Kolosov, G.V. 1909. On an application of complex function theory to a plane problem of the mathematical theory of elasticity, Yuriev.
Kreyszig, E. 1983. *Advanced Engineering Mathematics*, New York, John Wiley & Sons.
Kumsar, H., Ö. Aydan, H. Tano, S.B. Çelik and R. Ulusay 2016. An integrated geomechanical investigation, multi-parameter monitoring and analyses of Babadağ-Gündoğdu creep-like landslide. *Rock Mechanics and Rock Engineering*, Special Issue on the Deep-seated landslides, doi: 10.1007/s00603-015-0826-7.
Ladanyi, B. 1974. Use of the long-term strength concept in the determination of ground pressure on tunnel linings. *3rd Congress of the International Society of Rock Mechanics*, Denver, 2B, 1150–1165.
Lama, R.D. and V.S. Vutukuri 1978. *Handbook on Mechanical Properties of Rocks*, Clausthal, Germany, Trans Tech Publications.

Mindlin, R.D. 1949. Compliance of elastic bodies in contact. *Journal of Applied Mechanics*, 16, 259–268.

Muskhelishvili, N.I. 1962. *Some Basic Problems of the Mathematical Theory of Elasticity*, Groningen, Noordhoff.

Polubarinova-Kochina, P.Y.A. 1962. *Theory of Groundwater Movement*, New Jersey, Princeton University Press.

Savin, G.N. 1961. *Stress Concentrations around Holes*, Oxford, Pergamon.

Sezaki, M., Ö. Aydan and T. Kawamoto 1994. A consideration on ground response curve (in Japanese). *Journal of Japan Society of Civil Engineers, Geotechnical Division*, 499/III-28, 77–85.

Snow, D.T. 1965. A parallel plate model of fractured permeable media. Ph.D. dissertation, University of California, Berkeley.

Talobre, J. 1957. *The Mechanics of Rocks* (in French), Paris, Dunod.

Temel, A. 2002. Personal communication. Hacettepe University, Geological Engineering Department, Ankara, Turkey.

Terzaghi, K. 1946. Rock defects and loads on tunnel supports, in R.V. Proctor and T.L. White, eds., *Rock Tunneling with Steel Supports*, Youngstown, Commercial Shearing and Stamping Company, 17–99 pp.

Terzaghi, K. 1960. Stability of steep slopes on hard, unweathered rock. *Geotechnique*, 12, 251–270.

Timoshenko, S.P. and J.N. Goodier 1951. *Theory of Elasticity*. 2nd ed., New York, McGraw-Hill.

Timoshenko, S.P. and J.N. Goodier 1970. *Theory of Elasticity*. 3rd ed., New York, Wiley.

Üçpırtı, H. and Ö. Aydan 1997. An experimental study on the permeability of interface between sealing plug and rock. *The 28th Rock Mechanics Symposium of Japan*, 268–272.

Verruijt, A. 1970. *Theory of Groundwater Flow*, London, UK, MacMillian.

Zachmanoglou, E.C. and D.W. Thoe 1986. *Introduction to Partial Differential Equations with Applications*, New York, Dover Publications Inc.

Zeigler, T.W. 1976. Determination of rock mass permeability. U.S. Army Corps of Engineers Waterways Experiment Station, Vicksburg, Miss. Tech. Rept. S-76-2. 112 p.

Zoback, M.D., H. Tsukahara and S.H. Hickman 1980. Stress measurements at depth in the vicinity of the San Andreas fault: implications for the magnitude of shear stress at depth. *Journal of Geophysical Research*, 85(B11), 6157–6173.

Chapter 8

Methods for approximate solutions

In this chapter, the first part is related to the solutions to the fundamental governing equations using approximate solution methods such as the finite difference method (FDM) and finite element method (FEM) (e.g. Zienkiewicz 1977; Reddy 2006; Segerlind 1976; Verruijt 1982; Owen and Hinton 1980; Crouch and Starfield 1983; Kreyszig 1983; Speigel 1994; Aydan 2018). An illustrative example is chosen in the first part to explain the similarities and dissimilarities of various approximate methods as well as exact solutions. Although formulations for multi-dimensional situations are not presented, they can be easily extended to such situations by just selecting approximate shape functions for multi-dimensional situations as the general forms of equations would remain the same. When the FEM is used to obtain solutions to various partial equations, there may be some repetitions of the derivation and utilization of shape functions. However, each section is intended to be stand-alone in own derivation and applications so that there will be no reference to other sections. In the second part, some numerical procedures available in the field of rock mechanics and rock engineering for rock masses involving discontinuities are presented and several examples are given.

8.1 Comparison of exact and approximate solutions

There are three different approximate methods, namely:

- FDM,
- FEM and
- Boundary element method (BEM).

The characteristics of the closed form and approximate methods are briefly discussed through solving the following ordinary differential equation.

$$\frac{d^2u}{dx^2} - u = 0 \tag{8.1}$$

The boundary conditions are as follows

$u = 0$ at $x = 0$

$u = 1$ at $x = 1$

8.1.1 Exact (closed-form) solution

If the solution of Eq. (8.1) is a series of exponential functions $e^{\lambda x}$, the characteristics equation can be obtained as:

$$\lambda^2 - 1 = 0 \tag{8.2}$$

Hence, the roots (eigenvalues) are

$$\lambda_1 = 1; \quad \lambda_2 = -1 \tag{8.3}$$

Thus, the solution is of the following form:

$$u = C_1 e^{x} + C_2 e^{-x} \tag{8.4}$$

The integration constants are obtained from the boundary conditions as

$$C_1 = 0.4254589 \text{ and } C_2 = -0.4254589$$

8.1.2 Finite difference method

The FDM is the earliest approximate method and it is called a strong form approximate solution as it attempts to solve the differential equation in its original form. It utilizes the Taylor expansion of dependent variable to discretize the governing equation. Let us assume that the domain is discretized into n segments with an equal interval Δx. Equation (8.1) at a node j could be rewritten as

$$\left(\frac{d^2u}{dx^2}\right)_{x=xj} - u_j = 0 \tag{8.5}$$

The Taylor expansions of function u at nodes i, j and k may be written as

$$u_i\left(x_j - \Delta x\right) = u_j - \left(\frac{du}{dx}\right)_{x=xj}\frac{\Delta x}{1!} + \left(\frac{d^2u}{dx^2}\right)_{x=xj}\frac{\Delta x^2}{2!} - 0^3 \tag{8.6}$$

$$u_j\left(x_j\right) = u_j \tag{8.7}$$

$$u_k\left(x_j + \Delta x\right) = u_j + \left(\frac{du}{dx}\right)_{x=xj}\frac{\Delta x}{1!} + \left(\frac{d^2u}{dx^2}\right)_{x=xj}\frac{\Delta x^2}{2!} + 0^3 \tag{8.8}$$

From the above relations, one gets the following relation:

$$\left(\frac{d^2u}{dx^2}\right)_{x=xj} \approx \frac{u_k - 2u_j + u_i}{\Delta x^2} \tag{8.9}$$

Thus, the finite difference form of the above equation takes the following form:

$$\left[\frac{1}{\Delta x^2}u_i - \left(\frac{2}{\Delta x^2}+1\right)u_j + \frac{1}{\Delta x^2}u_k\right] = 0 \tag{8.10}$$

This simultaneous equation system for a domain divided into n segments will result in

$$[K]\{U\} = \{F\} \tag{8.11}$$

Matrix $[K]$ has $n-1$ rows and $n+1$ columns, vector $\{U\}$ has $n+1$ rows and vector $\{F\}$ has $n-1$ rows. However, if the boundary conditions are introduced, it yields the following simultaneous equation system:

$$[K^*]\{U^*\} = \{F^*\} \tag{8.12}$$

The resulting matrix $[K^*]$ has $n-1$ rows and $n-1$ columns. Similarly, the resulting vectors $\{U^*\}$ and $\{F^*\}$ have $n-1$ rows. Therefore, it becomes possible to solve this simultaneous equation system.

Example 1

Let us assume that we have two segments and three nodes. Accordingly, $u_1 = 0$, $u_3 = 1$ and $\Delta x = 0.5$. From Eq. (8.12), we obtain unknown u_2 as

$$u_2 = 0.444444444444 \tag{8.13}$$

Example 2

Let us assume that we have four segments and five nodes. Accordingly, $u_1 = 0$, $u_5 = 1$ and $\Delta x = 0.25$. From Eq. (8.12), we get the following equation system for unknown $\{u^*\}$

$$\begin{bmatrix} -33 & 16 & 0 \\ 16 & -33 & 16 \\ 0 & 16 & -33 \end{bmatrix} \begin{Bmatrix} u_2 \\ u_3 \\ u_4 \end{Bmatrix} = \begin{Bmatrix} 0 \\ 0 \\ -16 \end{Bmatrix} \tag{8.14}$$

The solution of this simultaneous equation system yields the unknowns as:

$$u_2 = 0.215114752376,\ u_3 = 0.443674176776,\ u_4 = 0.69963237225$$

8.1.3 Finite element method

The FEM is a relatively new approximate method, but it is the most widely used method in engineering and science as compared with FDM or other methods. The governing equation is first integrated over the domain and then the resulting integral equation is discretized. Therefore, it is called a *weak form solution*, as there is a possibility that the solution may be different from the actual one.

a. Weak formulation
Taking a dot product of Eq. (8.1) by a trial function δv and integrating it yield the following:

$$\int_0^1 \delta v \cdot \frac{d^2u}{dx^2}dx - \int_0^1 \delta v \cdot u dx = 0 \tag{8.15}$$

Introducing the integral by parts for the first term gives

$$\int_0^1 \frac{d\delta v}{dx} \cdot \frac{du}{dx}dx + \int_0^1 \delta v \cdot u dx = \left[\delta v \cdot \hat{t}\right]_0^1 \tag{8.16}$$

where

$$\hat{t} = \frac{du}{dx}n$$

n in the equation above is the unit normal vector. Let us assume that the trial function v is the same as the function u, which is generally called the *Galerkin approach* in the finite element formulation.

b. Discretization
The domain is discretized into subdomains, which are called finite elements. The function u is approximated by a chosen interpolation function in an element and it is summed up for the whole domain. For this particular problem, let us chose a linear function of the following form:

$$u = ax + b \tag{8.17}$$

Let us assume that the function u at nodes i and j are known. Thus, we can write the following

$$\begin{bmatrix} x_i & 1 \\ x_j & 1 \end{bmatrix} \begin{Bmatrix} a \\ b \end{Bmatrix} = \begin{Bmatrix} u_i \\ u_j \end{Bmatrix} \tag{8.18}$$

Taking the inverse of the above relation, one gets coefficients a and b. Inserting these coefficients into Eq. (8.17) yields the following

$$u = N_i u_i + N_j u_j \tag{8.19}$$

where

$$N_i = \frac{x_j - x}{x_j - x_i}, \quad N_j = \frac{x - x_i}{x_j - x_i}$$

The above equation may be rewritten in a compact form as

$$u = [N]\{U_e\} \quad \text{or} \quad u = \mathbf{N}U_e \tag{8.20}$$

where $[N] = \left[N_i, N_j\right], \{U_e\}^T = \left\{u_i, u_j\right\}$.[1] The derivative of the above equation takes the following form:

$$\frac{du}{dx} = \frac{dN_i}{dx}u_i + \frac{dN_j}{dx}u_j \tag{8.21}$$

The above relation is rewritten in a compact form as

$$\frac{du}{dx} = [B]\{U_e\} \quad \text{or} \quad \frac{du}{dx} = \mathbf{B}U_e \tag{8.22}$$

where $[B] = \left[B_i, B_j\right]$, $B_i = -1/L_e$, $B_j = 1/L_e$, $L_e = x_j - x_i$. The dot product of two vectors is presented in the following form in the FEM:

$$c = \mathbf{a} \cdot b \rightarrow c = \{a\}^T\{b\} \tag{8.23}$$

Equation (8.16), which holds for the whole domain, must also hold for each element as

$$\int_{x_i}^{x_j} \frac{\delta u}{dx} \cdot \frac{du}{dx} dx + \int_{x_i}^{x_j} \delta u \cdot u dx = \left[\delta u \cdot \hat{t}\right]_{x_i}^{x_j} \tag{8.24}$$

Inserting relations given by Eqs. (8.20) and (8.22) into Eq. (8.24) and using the finite element convention for the dot product (Eq. (8.23)) yield the following:

$$\{\delta U_e\}\left(\left[\int_{x_i}^{x_j} [B]^T[B]dx + \int_{x_i}^{x_j} [N]^T[N]dx\right]\{U_e\} - \left[[N]^T\hat{t}\right]_{x_i}^{x_j}\right) = 0 \tag{8.25}$$

The above relation implies the following

$$[K_e]\{U_e\} = \{F_e\} \tag{8.26}$$

where

$$[K_e] = \int_{x_i}^{x_j} [B]^T[B]dx + \int_{x_i}^{x_j} [N]^T[N]dx; \quad \{F_e\} = \left[[N]^T\hat{t}\right]_{x_i}^{x_j}$$

For a typical element, one gets the above relations specifically for a shape function given by Eq. (8.17) as

$$[K_e] = \frac{1}{L}\begin{bmatrix} 1 & -1 \\ -1 & 1 \end{bmatrix} + \frac{L}{6}\begin{bmatrix} 2 & 1 \\ 1 & 2 \end{bmatrix}, \quad \{U_e\} = \begin{Bmatrix} u_i \\ u_j \end{Bmatrix}, \quad \{F_e\} = \begin{Bmatrix} \hat{t}_i \\ \hat{t}_j \end{Bmatrix}$$

1 It should be noted that the horizontally written vector is defined as the transpose of the vector.

The sum-up of the above relation for the whole domain is

$$[K]\{U\}=\{F\} \tag{8.27}$$

where

$$[K]=\sum_{k=1}^{n}[K_e],\quad \{U\}=\sum_{k=1}^{n}\{U_e\},\quad \{F\}=\sum_{k=1}^{n}\{F_e\}.$$

Example 1

Let us assume that we have two elements and three nodes. Accordingly, $u_1=0$, $u_3=0$ and $x_j-x_i=0.5$. From Eq. (8.27), we obtain unknown u_2 as

$$u_2=0.4423076 \tag{8.28}$$

Example 2

Let us assume that we have four elements and five nodes. Accordingly, $u_1=0$, $u_5=0$ and $x_j-x_i=0.25$. The solution of the simultaneous equation system (Eq. (8.27)) yields the following:

$$u_2=0.214787576025,\ u_3=0.443140650725,\ u_4=0.699481489062$$

8.1.4 Comparisons

Solutions obtained from the approximate methods for the example chosen are compared with those by the closed-form solution (CFS). Figure 8.1a and b show comparisons of computations for two element (three-nodes) and four element (five-nodes) discretizations

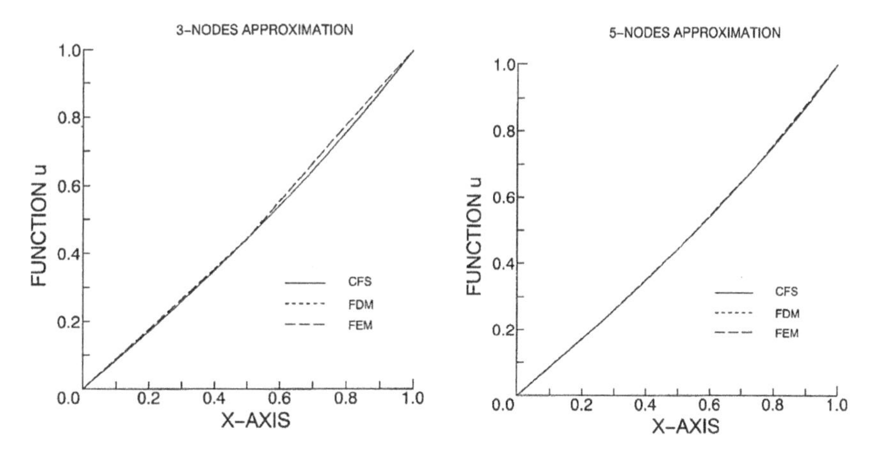

Figure 8.1 Comparison of computations by FDM and FEM with that by CFS.

Table 8.1 Comparison of computed results from CFS, FDM and FEM (two elements and three nodes

Node no.	*Location (x)*	*CFS*	*FDM*	*FEM*
2	0.5	0.443409283	0.44444444	0.4423076

Table 8.2 Comparison of computed results from CFS, FDM and FEM (four elements and five nodes

Node no.	*Location (x)*	*CFS*	*FDM*	*FEM*
2	0.25	0.214941503	0.21511475	0.21478757
3	0.5	0.443409283	0.44367417	0.44314065
4	0.75	0.699723944	0.69963237	0.69948149

of the domain by the FDM and FEM with that by the CFS, respectively. Tables 8.1 and 8.2 compare the specific values of solutions obtained from the CFS, FDM and FEM. As seen from both figures as well as from tables, the approximate solutions almost coincide with the exact ones at nodal points. Increasing the number of nodes result in better solutions and errors caused by discretization decreases. Nevertheless, the small differences between the exact solutions and approximate solutions clearly illustrate the reason why the numerical methods are called "methods for approximate solutions".

8.2 1D hyperbolic problem: equation of motion

As shown in Chapter 3, the equation of momentum for 1D problems can be written as

$$\frac{\partial \sigma}{\partial x} + b = \rho \frac{\partial^2 u}{\partial t^2} \tag{8.29}$$

Let us assume that this equation is subjected to the following boundary and initial conditions:

Boundary conditions

$$\begin{aligned} u(0,t) &= 0 \quad \text{at} \quad x = 0 \\ t(0,t) &= t_0 \quad \text{at} \quad x = L \end{aligned} \tag{8.30}$$

Initial conditions

$$\begin{aligned} u(x,0) &= 0 \quad \text{at} \quad t = 0 \\ \dot{u}(x,0) &= 0 \quad \text{at} \quad t = 0 \end{aligned} \tag{8.31}$$

Let us further assume that the material is of the Kelvin–Voigt type as given below:

$$\sigma = E\varepsilon + C\dot{\varepsilon} \tag{8.32}$$

Displacement – strain and strain rate are given as

$$\varepsilon = \frac{\partial u}{\partial x}; \quad \dot{\varepsilon} = \frac{\partial \dot{u}}{\partial x} \tag{8.33}$$

where

$$\dot{u} = \frac{\partial u}{\partial t}$$

We utilize the FEM to solve the equation system in the following.

8.2.1 Weak form formulation

Taking a variation on the displacement field δu, the integral form of Eq. (8.29) becomes

$$\int_V \delta u \cdot \frac{\partial \sigma}{\partial x} dV + \int_V \delta u \cdot b dV = \int_V \rho \delta u \cdot \frac{\partial^2 u}{\partial t^2} dV \tag{8.34}$$

where $dV = dAdx$.

Applying the integral by parts to the first term on the LHS with respect to x yields the following:

$$\int_A [\delta u \cdot t]_0^L dA + \int_V \delta u \cdot b dV = \int_V \frac{\partial \delta u}{\partial x} \cdot \sigma dV + \int_V \rho \delta u \cdot \frac{\partial^2 u}{\partial t^2} dV \tag{8.35}$$

where $t = \sigma \cdot n$. The above equation is called the weak form of Eq. (8.29). Inserting the constitutive relation given by Eq. (8.32) into Eq. (8.35), we obtain the following:

$$\begin{aligned} \int_A [\delta u \cdot t]_0^L dA + \int_V \delta u \cdot b dV &= \int_V E \frac{\partial \delta u}{\partial x} \cdot \frac{\partial u}{\partial x} dV \\ &+ \int_V C \frac{\partial \delta u}{\partial x} \cdot \frac{\partial u}{\partial x} dV + \int_V \rho \delta u \cdot \frac{\partial^2 u}{\partial t^2} dV \end{aligned} \tag{8.36}$$

8.2.2 Discretization

For this particular problem, a linear function of the following form for the space is chosen:

$$u(t) = ax + b \tag{8.37}$$

Let us assume that the function u at nodes i and j are known. Thus, we can write the following:

$$\begin{bmatrix} x_i & 1 \\ x_j & 1 \end{bmatrix} \begin{Bmatrix} a \\ b \end{Bmatrix} = \begin{Bmatrix} u_i \\ u_j \end{Bmatrix} \tag{8.38}$$

Taking the inverse of the above relation, one gets coefficients a and b. Inserting these coefficients in Eq. (8.37) yields the following:

$$u(t) = N_i u_i(t) + N_j u_j(t) \tag{8.39}$$

where

$$N_i = \frac{x_j - x}{x_j - x_i}, \quad N_j = \frac{x - x_i}{x_j - x_i}$$

The above equation may be rewritten in a compact form as

$$u = [N]\{U_e\} \quad \text{or} \quad u = \mathbf{N} U_e \tag{8.40}$$

where $[N] = [N_i, N_j]$, $\{U_e\}^T = \{u_i, u_j\}$. The derivative of the above equation takes the following form:

$$\frac{du}{dx} = \frac{dN_i}{dx} u_i + \frac{dN_j}{dx} u_j \tag{8.41}$$

The above relation is rewritten in a compact form as

$$\frac{du}{dx} = [B]\{U_e\} \quad \text{or} \quad \frac{du}{dx} = \mathbf{B} U_e \tag{8.42}$$

where $[B] = [B_i, B_j]$, $B_i = -1/L_e$, $B_j = 1/L_e$, $L_e = x_j - x_i$. Equation (8.36), which holds for the whole domain, must also hold for each element as

$$\begin{aligned} &\int_{V_e} \rho \delta u \cdot \frac{\partial^2 u}{\partial t^2} dV + \int_{V_e} C \frac{\partial \delta u}{\partial x} \cdot \frac{\partial u}{\partial x} dV + \int_{V_e} E \frac{\partial \delta u}{\partial x} \cdot \frac{\partial u}{\partial x} dV \\ &= \int_{A_e} [\delta u \cdot t]_{x_i}^{x_j} dA + \int_{V_e} \delta u \cdot b_e dV \end{aligned} \tag{8.43}$$

The discretized form of the above equation becomes

$$\begin{aligned} &\int_{V_e} \rho [N]^T [N] \{\ddot{U}_e\} dV + \int_{V_e} C [B]^T [B] \{\dot{U}_e\} dV + \int_{V_e} E [B]^T [B] \{U_e\} dV \\ &= \int_{A_e} \left[N_k \right]^T t_e \Big]_{x_i}^{x_j} dA + \int_{V_e} [N]^T b_e dV \end{aligned} \tag{8.44}$$

The above equation may be written in a compact form as

$$[M_e]\{\ddot{U}_e\} + [C_e]\{\dot{U}_e\} + [K_e]\{U_e\} = \{F_e\} \tag{8.45}$$

where

$$[M_e] = \int_{V_e} \rho [N]^T [N] dV, \quad [C_e] = \int_{V_e} C [B]^T [B] dV, \quad [K_e] = \int_{V_e} E [B]^T [B] dV,$$

$$\{F_e\} = \int_{A_e} \left[[N]^T t_e \right]_{x_i}^{x_j} dA + \int_{V_e} [N]^T b_e dV$$

For the total domain, we have the following

$$[M]\{\ddot{U}\} + [C]\{\dot{U}\} + [K]\{U\} = \{F\} \tag{8.46}$$

where

$$[M] = \sum_{k=1}^{n} [M_e]_k, [C] = \sum_{k=1}^{n} [C_e]_k, [K] = \sum_{k=1}^{n} [K_e]_k, \{F\} = \sum_{k=1}^{n} \{F_e\}_k, \{U\} = \sum_{k=1}^{n} \{U_e\}_k$$

Equation (8.46) could not be solved as it is. For a time step m, we can rewrite Eq. (8.46) as

$$[M]\{\ddot{U}\}_m + [C]\{\dot{U}\}_m + [K]\{U\}_m = \{F\}_m \tag{8.47}$$

Therefore, we discretized the displacement field $\{U\}$ for a time-domain using the Taylor expansion as it is in the FDM as:

$$\{U\}_{m-1} = \{U\}_m - \frac{\partial \{U\}_m}{\partial t} \frac{\Delta t}{1!} + \frac{\partial^2 \{U_m\}}{\partial t^2} \frac{\Delta t^2}{2!} - 0^3 \tag{8.48}$$

$$\{U\}_m = \{U\}_m \tag{8.49}$$

$$\{U\}_{m+1} = \{U\}_m + \frac{\partial \{U_m\}}{\partial t} \frac{\Delta t}{1!} + \frac{\partial^2 \{U_m\}}{\partial t^2} \frac{\Delta t^2}{2!} + 0^3 \tag{8.50}$$

From the above relations, one easily gets the following equations:

$$\{\dot{U}\}_m = \frac{1}{\Delta t} \left(\{U\}_{m+1} - \{U\}_{m-1} \right) \tag{8.51}$$

$$\{\ddot{U}\}_m = \frac{1}{\Delta t^2} \left(\{U\}_{m+1} - 2\{U\}_m + \{U\}_{m-1} \right) \tag{8.52}$$

Inserting these relations into Eq. (8.47), we get the following

$$\left[M^* \right] \{U\}_{m+1} = \left\{ F^* \right\}_{m+1} \tag{8.53}$$

where

$$\left[M^* \right] = \left[\frac{1}{\Delta t^2} [M] + \frac{1}{\Delta t} [C] \right]$$

$$\left\{ F^* \right\}_{m+1} = \left[\frac{2}{\Delta t^2} [M] - [K] \right] \{U\}_m - \left[\frac{1}{\Delta t^2} [M] - \frac{1}{\Delta t} [C] \right] \{U\}_{m-1} + \{F\}_m$$

8.2.3 Specific example

For a typical two-noded element, the following parameters are obtained:

$$[M_e]=\frac{\rho L_e A_e}{6}\begin{bmatrix} 2 & 1 \\ 1 & 2 \end{bmatrix};\ [C_e]=\frac{CA_e}{L}\begin{bmatrix} 1 & -1 \\ -1 & 1 \end{bmatrix};\ [K_e]=\frac{EA_e}{L}\begin{bmatrix} 1 & -1 \\ -1 & 1 \end{bmatrix};$$

$$\{F_e\}=\begin{Bmatrix} t_i \\ t_j \end{Bmatrix}+\frac{b}{2}\begin{Bmatrix} 1 \\ 1 \end{Bmatrix};\ \{U_e\}=\begin{Bmatrix} u_i \\ u_j \end{Bmatrix}$$

If the space is discretized into two elements, we have the following simultaneous equation system

$$\begin{bmatrix} \left(K_{11}^*\right)^1 & \left(K_{12}^*\right)^1 & 0 \\ \left(K_{21}^*\right)^1 & \left(K_{22}^*\right)^1+\left(K_{11}^*\right)^2 & \left(K_{12}^*\right)^2 \\ 0 & \left(K_{21}^*\right)^2 & \left(K_{22}^*\right)^2 \end{bmatrix}\begin{Bmatrix} U_1 \\ U_2 \\ U_3 \end{Bmatrix}_{m+1}=\begin{Bmatrix} F_1^* \\ F_2^* \\ F_3^* \end{Bmatrix}_{m+1}$$

8.2.4 ID parabolic problem: creep problem

If the inertia term in Eq. (8.29) is negligible and the constitutive law is of the Kelvin–Voigt type, then the finite element form of Eq. (8.29) becomes

$$[C]\{\dot{U}\}+[K]\{U\}=\{F\} \tag{8.54}$$

For a time step m, we get the following equation using the Taylor expansion:

$$[C^*]\{U\}_{m+1}=\{F^*\}_{m+1} \tag{8.55}$$

where

$$\left[C^*\right]=\frac{1}{\Delta t}[C],\ \left\{F^*\right\}_{m+1}=\left[\frac{1}{\Delta t}[C]-[K]\right]\{U\}_m+\{F\}_m$$

8.2.5 ID elliptic problem: static problem

If the inertia term in Eq. (8.29) is negligible and the constitutive law is of Hookean type, then the finite element form of Eq. (8.29) becomes

$$[K]\{U\}=\{F\} \tag{8.56}$$

The solution would correspond to the static solution. If hyperbolic solution is convergent, it would become equivalent to the static solution. As for a parabolic solution, the solution would tend to be equivalent to the static solution as time increases.

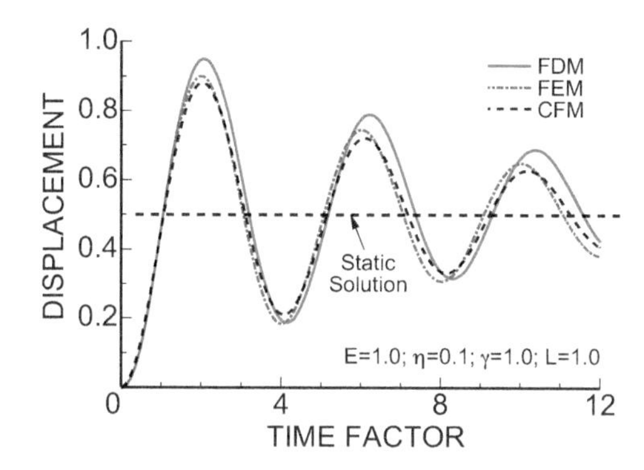

Figure 8.2 Comparison of solutions from different methods.

8.2.6 Computational examples

In the first example, the dynamic response of a layer of infinite length with a thickness of 1 m was analyzed. The body force of the layer was assumed to be applied suddenly and the viscosity coefficient was selected to be 0.1 with other parameters shown in Figure 8.2. Figure 8.2 is also a comparison of displacement responses at the top of the layer obtained from the solutions of exact method (CFM), FDM and FEM. Although the solutions are quite similar to each other, there are some slight differences as explained in Section 8.1. All solutions tend to converge to the static solution as time increases.

In the second example, the dynamic response of a layer of infinite length and of 1 m thickness is also analyzed using the FEM. The body force of the layer was assumed to be applied suddenly and the viscosity coefficient was selected 0.2 and 0.5. Figure 8.3a shows the computed displacement response of some selected points with time. It is interesting to note that fluctuations occur as the viscosity coefficient decreases in magnitude.

In the third example, the same problem was analyzed using hyperbolic, parabolic and elliptic formulations. The results are shown in Figure 8.3b. As seen from the figure, both computed responses of selected points hyperbolic and parabolic converge to those obtained from elliptical (static) formulation. The examples presented herein should give some idea and understanding off what to expect from three different solution techniques irrespective of the dimensions of the domain analyzed.

8.3 Parabolic problems: heat flow, seepage and diffusion

8.3.1 Introduction

The governing equation for heat flow, fluid flow, seepage and difusion problems takes exactly the same form except that the physical meaning of variables is different (e.g. Reddy 2006; Segerlind 1976; Zienkiewicz 1977; Aydan et al. 1985, 1986, 2012). In the following, a finite element formulation of such a governing equation and its discretization are given. Furthermore, some sample computations are presented.

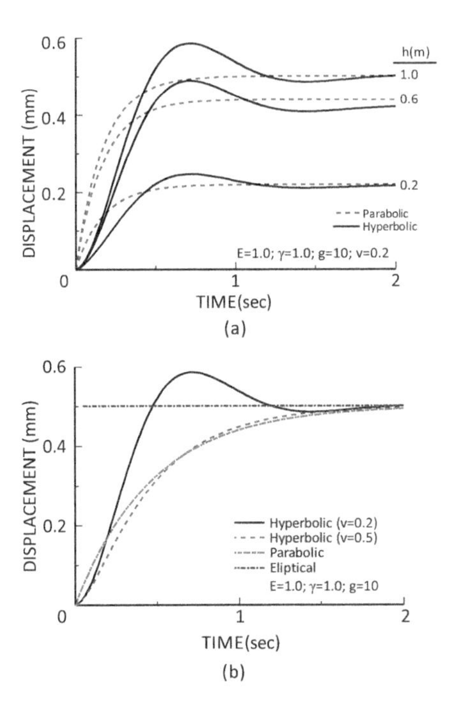

Figure 8.3 Comparison of (a) hyperbolic and parabolic solutions for different depths and (b) hyperbolic, parabolic and elliptical solutions for different viscosities at the top of the layer.

8.3.2 Governing equation

As shown in Chapter 4, the laws of mass conservation and heat flow of non-convective or non-advective type take the following form for 1D problems:

$$-\frac{\partial q}{\partial x}+g=\rho c\frac{\partial T}{\partial t} \tag{8.57}$$

where T can be temperature, water head or mass concentration. Let us assume that this equation is subjected to the following boundary and initial conditions:

Boundary conditions

$$\begin{aligned} &T(0,t)=0 \quad &\text{at} \quad &x=0 \\ &q_n(0,t)=q_0 \quad &\text{at} \quad &x=L \end{aligned} \tag{8.58}$$

Initial conditions

$$\begin{aligned} &T(x,0)=0 \quad &\text{at} \quad &t=0 \\ &\dot{T}(x,0)=0 \quad &\text{at} \quad &t=0 \end{aligned} \tag{8.59}$$

Let us further assume that the material obeys a linear type of constitutive law between flux q and dependent variable T

$$q = -k\frac{\partial T}{\partial x} \tag{8.60}$$

8.3.3 Weak form formulation

Taking a variation on variable δT, the integral form of Eq. (8.57) becomes

$$-\int_V \delta T \cdot \frac{\partial q}{\partial x} dV + \int_V \delta T \cdot g dV = \int_V \rho c \delta T \cdot \frac{\partial T}{\partial t} dV \tag{8.61}$$

where $dV = dAdx$.

Applying the integral by parts to the first term on the LHS with respect to x yields the following:

$$-\int_A [\delta T \cdot q_n]_0^L dA + \int_V \delta T \cdot g dV = -\int_V \frac{\partial \delta T}{\partial x} \cdot q dV + \int_V \rho c \delta T \cdot \frac{\partial T}{\partial t} dV \tag{8.62}$$

where $q_n = q \cdot n$. The above equation is called the weak form of Eq. (8.57). Inserting the constitutive relation given by Eq. (8.60) into Eq. (8.62), we obtain the following:

$$-\int_A [\delta T \cdot q_n]_0^L dA + \int_V \delta T \cdot g dV = \int_V k\frac{\partial \delta T}{\partial x} \cdot \frac{\partial T}{\partial x} dV + \int_V \rho c \delta T \cdot \frac{\partial T}{\partial t} dV \tag{8.63}$$

8.3.4 Discretization

For this particular problem, let us chose a linear function of the following form for the space:

$$T(t) = ax + b \tag{8.64}$$

Let us assume that the function T at nodes i and j are known. Thus, we can write the following:

$$\begin{bmatrix} x_i & 1 \\ x_j & 1 \end{bmatrix} \begin{Bmatrix} a \\ b \end{Bmatrix} = \begin{Bmatrix} T_i \\ T_j \end{Bmatrix} \tag{8.65}$$

Taking the inverse of the above relation, one gets coefficients a and b. Inserting these coefficients in Eq. (8.64) yields the following:

$$T = N_i T_i + N_j T_j \tag{8.66}$$

where

$$N_i = \frac{x_j - x}{x_j - x_i}; \quad N_j = \frac{x - x_i}{x_j - x_i}$$

The above equation may be rewritten in a compact form as

$$T = [N]\{T_e\} \quad \text{or} \quad T = \mathbf{N}T_e \tag{8.67}$$

where $[N] = [N_i, N_j]$, $\{T_e\}^T = \{T_i, T_j\}$. The derivative of the above equation takes the following form:

$$\frac{dT}{dx} = \frac{dN_i}{dx}T_i + \frac{dN_j}{dx}T_j \tag{8.68}$$

The above relation is rewritten in a compact form as

$$\frac{dT}{dx} = [B]\{T_e\} \quad \text{or} \quad \frac{dT}{dx} = \mathbf{B}T_e \tag{8.69}$$

where $[B] = [B_i, B_j]$, $B_i = -1/L_e$, $B_j = 1/L_e$, $L_e = x_j - x_i$. Equation (8.63), which holds for the whole domain, must also hold for each element as

$$\int_{V_e} \rho c \delta T \cdot \frac{\partial T}{\partial t} dV + \int_{V_e} k \frac{\partial \delta T}{\partial x} \cdot \frac{\partial T}{\partial x} dV = \int_{A_e} \left[\delta T \cdot q_n^e \right]_{x_i}^{x_j} dA + \int_{V_e} \delta T \cdot g_e dV \tag{8.70}$$

The discretized form of the above equation becomes

$$\begin{aligned} &\int_{V_e} \rho c [N]^T [N] \{\dot{T}\}_e dV + \int_{V_e} k [B]^T [B] \{U\}_e dV \\ &= \int_{A_e} \left[[N_k]^T q_n^e \right]_{x_i}^{x_j} dA + \int_{V_e} [N]^T g_e dV \end{aligned} \tag{8.71}$$

The above equation may be written in a compact form as

$$[M_e]\{\dot{T}_e\} + [K_e]\{T_e\} = \{F_e\} \tag{8.72}$$

where $[M_e] = \int_{V_e} \rho c [N]^T [N] dV$; $[K_e] = \int_{V_e} k [B]^T [B] dV$;

$\{F_e\} = \int_{A_e} \left[[N]^T q_n^e \right]_{x_i}^{x_j} dA + \int_{V_e} [N]^T g_e dV$

For the total domain, we have the following

$$[M]\{\dot{T}\} + [K]\{T\} = \{F\} \tag{8.73}$$

where

$$[M] = \sum_{k=1}^{n} [M_e]_k, \quad [K] = \sum_{k=1}^{n} [K_e]_k, \quad \{F\} = \sum_{k=1}^{n} \{F_e\}_k, \quad \{T\} = \sum_{k=1}^{n} \{T_e\}_k$$

Equation (8.73) could not be solved as it is. For a time step m, we can rewrite Eq. (8.73) as

$$[M]\{\dot{T}\}_{(m+\theta)} + [K]\{T\}_{(m+\theta)} = \{F\}_{(m+\theta)} \tag{8.74}$$

Therefore, we discretize the dependent variable $\{T\}$ for a time domain for using the Taylor expansion as it is in the FDM as

$$\{T\}_m = \{T\}_{(m+\theta)-\theta} = \{T\}_{(m+\theta)} - \frac{\partial\{T\}_{(m+\theta)}}{\partial t}\frac{\theta\Delta t}{1!} + \frac{\partial^2\{T\}_{(m+\theta)}}{\partial t^2}\frac{\theta^2\Delta t^2}{2!} - 0^3 \quad (8.75)$$

$$\{T\}_{m+1} = \{T\}_{(m+\theta)+(1-\theta)} = \{T\}_{(m+\theta)} + \frac{\partial\{T\}_{(m+\theta)}}{\partial t}\frac{(1-\theta)\Delta t}{1!} + \frac{\partial^2\{T\}_{(m+\theta)}}{\partial t^2}\frac{(1-\theta)^2\Delta t^2}{2!} + 0^3 \quad (8.76)$$

From the above relations, one easily gets the following:

$$\{T\}_{(m+\theta)} = \theta\{T\}_{m+1} + (1-\theta)\{T\}_m, \quad \{\dot{T}\}_{(m+\theta)} = \frac{\{T\}_{m+1} - \{T\}_m}{\Delta t} \quad (8.77)$$

$$\{F\}_{(m+\theta)} = \theta\{F\}_{m+1} + (1-\theta)\{F\}_m, \quad (8.78)$$

Inserting these relations into Eq. (8.74), we get the following:

$$[M^*]\{T\}_{m+1} = \{F^*\}_{m+1} \quad (8.79)$$

where

$$[M^*] = \left[\frac{1}{\Delta t}[M] + \theta[K]\right]$$

$$\left\{F^*\right\}_{m+1} = \left[\frac{1}{\Delta t}[M] - (1-\theta)[K]\right]\{T\}_m + \theta\{F\}_{m+1} + (1-\theta)\{F\}_m$$

8.3.5 Steady-state problem

When time variation of a dependent variable T is negligible, then the problem is called a steady-state problem. This type special case corresponds to elliptical problem. The final finite element form of the discretized governing equation becomes

$$[K]\{T\} = \{F\} \quad (8.80)$$

where

$$[K] = \sum_{k=1}^{n}[K_e]_k, \quad \{F\} = \sum_{k=1}^{n}\{F_e\}_k, \quad \{T\} = \sum_{k=1}^{n}\{T_e\}_k$$

8.3.6 Specific example

For a typical two-noded element, the followings are obtained:

$$[M_e]=\frac{\rho c L_e A_e}{6}\begin{bmatrix} 2 & 1 \\ 1 & 2 \end{bmatrix},\quad [K_e]=\frac{kA_e}{L}\begin{bmatrix} 1 & -1 \\ -1 & 1 \end{bmatrix}$$

If the space is discretized into two elements, we have the following simultaneous equation system.

$$\begin{bmatrix} (K_{11}^*)^1 & (K_{12}^*)^1 & 0 \\ (K_{21}^*)^1 & (K_{22}^*)^1+(K_{11}^*)^2 & (K_{12}^*)^2 \\ 0 & (K_{21}^*)^2 & (K_{22}^*)^2 \end{bmatrix}\begin{Bmatrix} T_1 \\ T_2 \\ T_3 \end{Bmatrix}_{m+1}=\begin{Bmatrix} F_1^* \\ F_2^* \\ F_3^* \end{Bmatrix}_{m+1}$$

8.3.7 Example 1: simulation of a solid body with heat generation

A specific example is given herein by simulating the temperature variation in a solid body for the following three different conditions (Aydan et al. 1985, 1986):

a. Heat generation only,
b. Heat flux input only and
c. Heat generation + heat flux.

The heat generation function is assumed to be of the following form:

$$g = At\exp - t/\tau \tag{8.81}$$

where A and τ are the physical constants determined from heat generation tests of solid, and t is the time. Figure 8.4 compares the results of computations for three different conditions. When the *heat generation only* condition is considered, temperature first increases and then decreases in a manner similar to the heat generation function. When the *heat flux only* condition is considered, temperature increases monotonically and tends to be asymptotic. When the *heat generation + heat flux* condition is considered, temperature first increases and then decreases. Finally, it becomes asymptotic to those computed for *heat flux only condition*.

8.3.8 Example 2: simulation of a diffusion problem

Fundamentally, the governing equations are the same except that the meaning of parameters would be different. This example corresponds to the simulation of one-dimensional diffusion experiments used commonly in practice. In the computations, the concentration was supplied from one end while it is kept at 0 at the other end. Figure 8.5 compares the computed responses by the exact solution (CFM), FDM and FEM. Despite slight differences in computed results, the responses from different methods are quite close to each other.

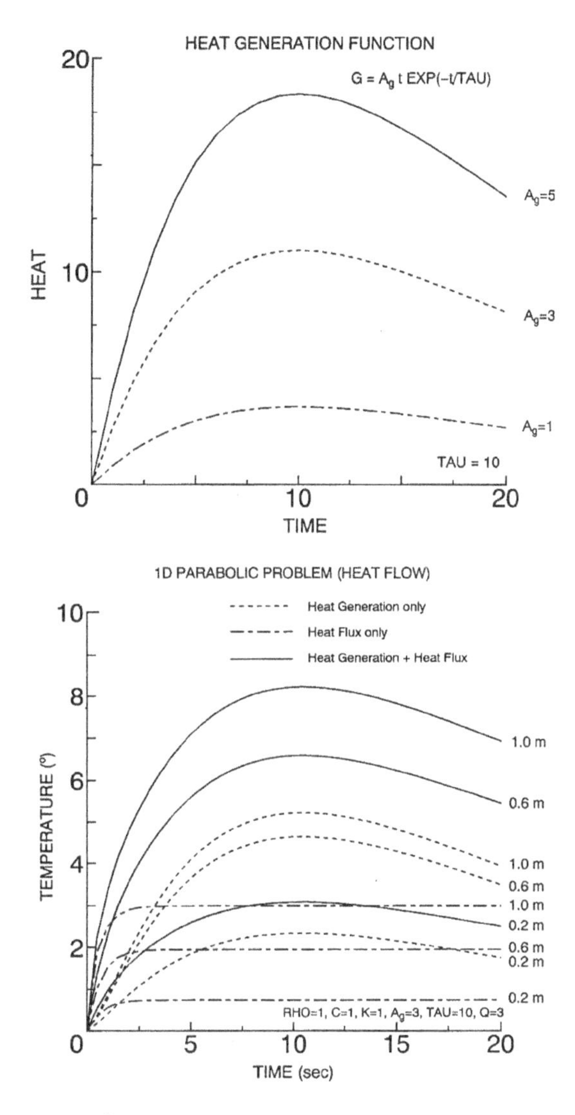

Figure 8.4 Comparison of temperature response of some selected points for various conditions.

8.4 FEM for 1D pseudo-coupled parabolic problems: heat flow and thermal stress; swelling and swelling pressure

8.4.1 Introduction

In the followings, a finite element formulation for a coupled heat flow and stress problem (thermo-mechanics) and its discretization are given (Aydan et al. 1985, 1986, 2012). Although the stress field is coupled with the heat flow, the heat flow field is an uncoupled one. Therefore, such a problem may be called as a pseudo-coupled problem.

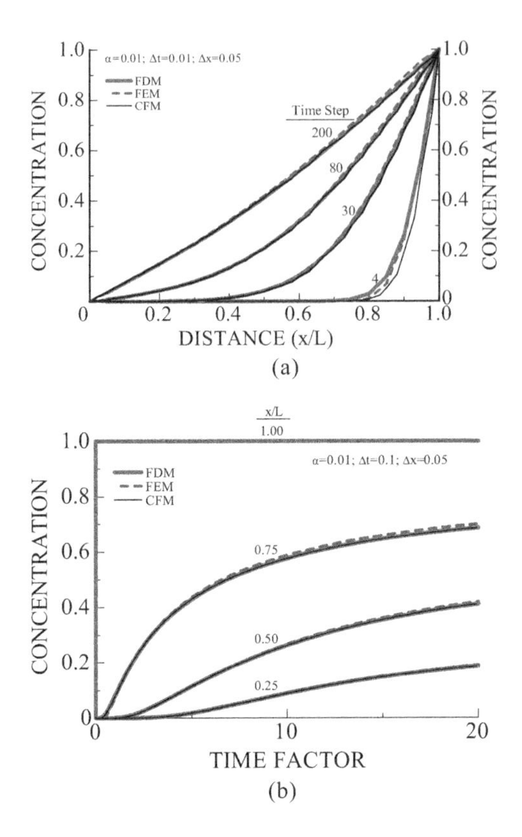

Figure 8.5 Comparison of computed concentration responses by different methods. (a) Variation of concentration with distance at selected time steps and (b) variation of concentration with time at selected locations.

8.4.2 Governing equations

8.4.2.1 Governing equation for a heat flow

As shown in Chapter 3, the laws of mass conservation and heat flow of non-convective or non-advective type take the following form for 1D problems:

$$-\frac{\partial q}{\partial x}+g=\rho c\frac{\partial T}{\partial t} \tag{8.82}$$

where T is the temperature. Let us assume that this equation is subjected to following boundary and initial conditions

Boundary conditions

$$\begin{aligned} &T(0,t)=0 \quad &\text{at} \quad &x=0 \\ &q_n(0,t)=q_0 \quad &\text{at} \quad &x=L \end{aligned} \tag{8.83}$$

Initial conditions

$$\begin{aligned} T(x,0) &= 0 \quad \text{at} \quad t = 0 \\ \dot{T}(x,0) &= 0 \quad \text{at} \quad t = 0 \end{aligned} \tag{8.84}$$

Let us further assume that the material obeys a linear type of constitutive law between flux q and dependent variable T

$$q = -k\frac{\partial T}{\partial x} \tag{8.85}$$

8.4.2.2 Governing equation for stress field

As shown in Chapter 4, the equation of momentum for 1D problems without inertia term can be written as

$$\frac{\partial \sigma}{\partial x} + b = 0 \tag{8.86}$$

Let us assume that this equation is subjected to the following boundary conditions:

$$\begin{aligned} u(0,t) &= 0 \quad \text{at} \quad x = 0 \\ t(0,t) &= t_0 \quad \text{at} \quad x = L \end{aligned} \tag{8.87}$$

Let us further assume that the material is of Hookean type as given below:

$$\sigma = E\varepsilon \tag{8.88}$$

The displacement–strain relation is given as

$$\varepsilon = \frac{\partial u}{\partial x} \tag{8.89}$$

The incremental form of Eq. (8.86) takes the following form if the body force remains constant with time:

$$\frac{\partial \dot{\sigma}}{\partial x} = 0 \tag{8.90}$$

Similarly, the constitutive law and displacement rate and strain rate relations may be rewritten as

$$\dot{\sigma} = E\dot{\varepsilon} \tag{8.91}$$

$$\dot{\varepsilon} = \frac{\partial \dot{u}}{\partial x} \tag{8.92}$$

8.4.3 Coupling of heat and stress fields

If the mechanical energy rate is slow, the effect of heat field on the stress field is coupled through volumetric strain caused by temperature variation. For 1D situations, this is written as

$$\dot{\varepsilon}_t = \dot{\varepsilon} - \dot{\varepsilon}_0 \tag{8.93}$$

where $\dot{\varepsilon}_0 = \alpha\Delta T$, α is the thermal expansion coefficient and ΔT is the temperature variation of a given point per unit time.

8.4.4 Weak form formulation

8.4.4.1 Weak formulation for a heat flow field

Taking a variation on variable δT, the integral form of Eq. (8.82) becomes

$$-\int_V \delta T \cdot \frac{\partial q}{\partial x} dV + \int_V \delta T \cdot g dV = \int_V \rho c \delta T \cdot \frac{\partial T}{\partial t} dV \tag{8.94}$$

where $dV = dAdx$.

Applying the integral by parts to the first term on the LHS with respect to x yields the following:

$$-\int_A [\delta T \cdot q_n]_0^L dA + \int_V \delta T \cdot g dV = -\int_V \frac{\partial \delta T}{\partial x} \cdot q dV + \int_V \rho c \delta T \cdot \frac{\partial T}{\partial t} dV \tag{8.95}$$

where $q_n = q \cdot n$. The above equation is called the weak form of Eq. (8.82). Inserting the constitutive relation given by Eq. (8.85) into Eq. (8.95), we obtain the following:

$$-\int_A [\delta T \cdot q_n]_0^L dA + \int_V \delta T \cdot g dV = \int_V k \frac{\partial \delta T}{\partial x} \cdot \frac{\partial T}{\partial x} dV + \int_V \rho c \delta T \cdot \frac{\partial T}{\partial t} dV \tag{8.96}$$

8.4.4.2 Weak formulation for stress field

Taking a variation on displacement rate field $\delta \dot{u}$, the integral form of Eq. (8.90) becomes

$$\int_V \delta \dot{u} \cdot \frac{\partial \dot{\sigma}}{\partial x} dV = 0 \tag{8.97}$$

where $dV = dAdx$.

Applying the integral by parts to the LHS with respect to x yields the following:

$$\int_A [\delta \dot{u} \cdot \dot{t}]_0^L dA = \int_V \frac{\partial \delta \dot{u}}{\partial x} \cdot \dot{\sigma} dV \tag{8.98}$$

where $\dot{t} = \dot{\sigma} \cdot n$. The above equation is called the weak form of Eq. (8.90). Inserting the constitutive relation given by Eq. (4.10) and Eq. (8.93) into Eq. (8.98), we obtain the following:

$$\int_A [\delta \dot{u} \cdot \dot{t}]_0^L dA + \int_V E\alpha \frac{\partial \delta \dot{u}}{\partial x} \Delta T_e dV = \int_V E \frac{\partial \delta \dot{u}}{\partial x} \cdot \frac{\partial \dot{u}}{\partial x} dV \tag{8.99}$$

8.4.5 Discretization

8.4.5.1 Discretization of heat flow field

For this particular problem, let us choose a linear function of the following form for the space:

$$T(t) = ax + b \tag{8.100}$$

Let us assume that the function T at nodes i and j are known. Thus, we can write the following:

$$\begin{bmatrix} x_i & 1 \\ x_j & 1 \end{bmatrix} \begin{Bmatrix} a \\ b \end{Bmatrix} = \begin{Bmatrix} T_i \\ T_j \end{Bmatrix} \tag{8.101}$$

Taking the inverse of the above relation, one gets coefficients a and b. Inserting these coefficients in Eq. (8.100) yields the following:

$$T = N_i T_i + N_j T_j \tag{8.102}$$

where

$$N_i = \frac{x_j - x}{x_j - x_i}; \quad N_j = \frac{x - x_i}{x_j - x_i}$$

The above equation may be rewritten in a compact form as

$$T = [N]\{T_e\} \quad \text{or} \quad T = \mathbf{N}T_e \tag{8.103}$$

where $[N] = [N_i, N_j]$, $\{T_e\}^T = \{T_i, T_j\}$. The derivative of the above equation takes the following form:

$$\frac{dT}{dx} = \frac{dN_i}{dx} T_i + \frac{dN_j}{dx} T_j \tag{8.104}$$

The above relation is rewritten in a compact form as

$$\frac{dT}{dx} = [B]\{\dot{T}_e\} \quad \text{or} \quad \frac{dT}{dx} = \mathbf{B}\dot{\mathbf{T}}_e \tag{8.105}$$

where $[B] = [B_i, B_j]$, $B_i = -1/L_e$, $B_j = 1/L_e$, $L_e = x_j - x_i$. Equation (8.96), which holds for the whole domain, must also hold for each element as

$$\int_{V_e} \rho c \delta T \cdot \frac{\partial T}{\partial t} dV + \int_{V_e} k \frac{\partial \delta T}{\partial x} \cdot \frac{\partial T}{\partial x} dV = \int_{A_e} \left[\delta T \cdot q_n^e \right]_{x_i}^{x_j} dA + \int_{V_e} \delta T \cdot g_e dV \tag{8.106}$$

The discretized form of the above equation becomes

$$\int_{V_e} \rho c[N]^T [N]\{\dot{T}\}_e dV + \int_{V_e} k[B]^T [B]\{U\}_e dV = \int_{A_e} \left[[N]^T q_n^e\right]_{x_i}^{x_j} dA + \int_{V_e} [N]^T g_e dV \tag{8.107}$$

The above equation may be written in a compact form as

$$[M_e]\{\dot{T}_e\} + [K_e]\{T_e\} = \{F_e\} \tag{8.108}$$

where

$$[M_e] = \int_{V_e} \rho c[N]^T [N] dV; \quad [K_e] = \int_{V_e} k[B]^T [B] dV;$$
$$\{F_e\} = \int_{A_e} \left[[N_k]^T q_n^e\right]_{x_i}^{x_j} dA + \int_{V_e} [N]^T g_e dA$$

For the total domain, we have the following:

$$[M]\{\dot{T}\} + [K]\{T\} = \{F\} \tag{8.109}$$

where

$$[M] = \sum_{k=1}^{n} [M_e]_k; \quad [K] = \sum_{k=1}^{n} [K_e]_k; \quad \{F\} = \sum_{k=1}^{n} \{F_e\}_k; \quad \{T\} = \sum_{k=1}^{n} \{T_e\}_k$$

Equation (8.109) could not be solved as it is. For a time step m, we can rewrite Eq. (8.109) as

$$[M]\{\dot{T}\}_{(m+\theta)} + [K]\{T\}_{(m+\theta)} = \{F\}_{(m+\theta)} \tag{8.110}$$

Therefore, we discretize dependent variable $\{T\}$ for a time domain for using the Taylor expansion as it is in the FDM as:

$$\{T\}_m = \{T\}_{(m+\theta)-\theta} = \{T\}_{(m+\theta)} - \frac{\partial\{T\}_{(m+\theta)}}{\partial t}\frac{\theta\Delta t}{1!} + \frac{\partial^2\{T\}_{(m+\theta)}}{\partial t^2}\frac{\theta^2\Delta t^2}{2!} - 0^3 \tag{8.111}$$

$$\{T\}_{m+1} = \{T\}_{(m+\theta)+(1-\theta)} = \{T\}_{(m+\theta)} + \frac{\partial\{T\}_{(m+\theta)}}{\partial t}\frac{(1-\theta)\Delta t}{1!} + \frac{\partial^2\{T\}_{(m+\theta)}}{\partial t^2}\frac{(1-\theta)^2\Delta t^2}{2!} + 0^3 \tag{8.112}$$

From the above relations, one easily gets the following:

$$\{T\}_{(m+\theta)} = \theta\{T\}_{m+1} + (1-\theta)\{T\}_m, \quad \{\dot{T}\}_{(m+\theta)} = \frac{\{T\}_{m+1} - \{T\}_m}{\Delta t} \tag{8.113}$$

$$\{F\}_{(m+\theta)} = \theta\{F\}_{m+1} + (1-\theta)\{F\}_m \tag{8.114}$$

Inserting these relations into Eq. (8.115), we get the following:

$$[M^*]\{T\}_{m+1} = \{F^*\}_{m+1} \tag{8.115}$$

where

$$\left[M^*\right]=\left[\frac{1}{\Delta t}[M]+\theta[K]\right];\left\{F^*\right\}_{m+1}=\left[\frac{1}{\Delta t}[M]-(1-\theta)[K]\right]\{T\}_m+\theta\{F\}_{m+1}+(1-\theta)\{F\}_m$$

8.4.5.2 Discretization of stress field

For this particular problem, let us choose a linear function of the following form for the space:

$$\dot{u} = ax + b \tag{8.116}$$

Let us assume that the function $\dot{u}$ at nodes i and j are known. Thus, we can write the following:

$$\begin{bmatrix} x_i & 1 \\ x_j & 1 \end{bmatrix}\begin{Bmatrix} a \\ b \end{Bmatrix}=\begin{Bmatrix} \dot{u}_i \\ \dot{u}_j \end{Bmatrix} \tag{8.117}$$

Taking the inverse of the above relation, one gets coefficients a and b. Inserting these coefficients into Eq. (8.116) yields the following:

$$\dot{u} = N_i\dot{u}_i + N_j\dot{u}_j \tag{8.118}$$

where

$$N_i = \frac{x_j - x}{x_j - x_i}, \quad N_j = \frac{x - x_i}{x_j - x_i}$$

The above equation may be rewritten in a compact form as

$$\dot{u} = [N]\{\dot{U}_e\} \quad \text{or} \quad \dot{u} = \mathbf{N}\dot{\mathbf{U}}_e \tag{8.119}$$

where $[N]=\left[N_i,N_j\right]$, $\{\dot{U}_e\}^T=\{\dot{u}_i,\dot{u}_j\}$ The derivative of the above equation takes the following form:

$$\frac{d\dot{u}}{dx} = \frac{dN_i}{dx}\dot{u}_i + \frac{dN_j}{dx}\dot{u}_j \tag{8.120}$$

The above relation is rewritten in a compact form as

$$\frac{d\dot{u}}{dx} = [B]\{\dot{U}_e\} \quad \text{or} \quad \frac{d\dot{u}}{dx} = \mathbf{B}\dot{\mathbf{U}}_e \tag{8.121}$$

where $[B]=\left[B_i,B_j\right]$, $B_i=-1/L_e$, $B_j=1/L_e$, $L_e=x_j-x_i$. Equation (8.99), which holds for the whole domain, must also hold for each element as

$$\int_{V_e} E\frac{\partial\delta\dot{u}}{\partial x}\cdot\frac{\partial\dot{u}}{\partial x}dV = \int_{V_e} E\alpha\frac{\partial\delta\dot{u}}{\partial x}\Delta T_e dV + \int_{A_e}\left[\delta\dot{u}\cdot\dot{t}\right]_{x_i}^{x_j} dA \tag{8.122}$$

The discretized form of the above equation becomes

$$\int_{V_e} E[B]^T[B]\{\dot{U}\}_e dV = \int_{V_e} E\alpha[B]^T\Delta T_e dV + \int_{A_e}\left[[N]^T\dot{t}_e\right]_{x_i}^{x_j} dA \tag{8.123}$$

The above equation may be written in a compact form as

$$[K_e]\{\dot{U}_e\} = \{\dot{F}_e\} \tag{8.124}$$

where

$$[K_e] = \int_{V_e} E[B]^T[B]dV;\quad \{\dot{F}_e\} = \int_{A_e}\left[[N_k]^T\dot{t}_e\right]_{x_i}^{x_j} dA + \int_{V_e} E\alpha[B]^T\Delta T_e dV$$

For the total domain, we have the following

$$[K]\{\dot{U}\} = \{\dot{F}\} \tag{8.125}$$

where

$$[K] = \sum_{k=1}^{n}[K_e]_k,\quad \{\dot{F}\} = \sum_{k=1}^{n}\{\dot{F}\}_{ek},\quad \{\dot{U}\} = \sum_{k=1}^{n}\{\dot{U}_e\}_k$$

8.4.6 Specific example

For a typical two-noded element for heat flow field, the following are obtained:

$$[M_e] = \frac{\rho c L_e A_e}{6}\begin{bmatrix} 2 & 1 \\ 1 & 2 \end{bmatrix},\quad [K_e] = \frac{kA_e}{L}\begin{bmatrix} 1 & -1 \\ -1 & 1 \end{bmatrix},$$

$$\{F_e\} = \begin{Bmatrix} q_i \\ q_j \end{Bmatrix} + \frac{g_e}{2}\begin{Bmatrix} 1 \\ 1 \end{Bmatrix};\quad \{T_e\} = \begin{Bmatrix} T_i \\ T_j \end{Bmatrix}$$

Similarly, for a typical two-noded element for stress field, the following are obtained:

$$[K_e] = \frac{EA_e}{L}\begin{bmatrix} 1 & -1 \\ -1 & 1 \end{bmatrix},\{\dot{F}_e\} = \begin{Bmatrix} \dot{t}_i \\ \dot{t}_j \end{Bmatrix} + \frac{E\alpha A_e\Delta T_e}{L}\begin{Bmatrix} -1 \\ 1 \end{Bmatrix},\quad \{\dot{U}_e\} = \begin{Bmatrix} \dot{u}_i \\ \dot{u}_j \end{Bmatrix}$$

8.4.7 Example: simulation of heat generation and associated thermal stress

A specific example is given herein by simulating the temperature variation in a solid with a length of 1 m and associated stress field for the following conditions:

a. Heat generation + heat flux

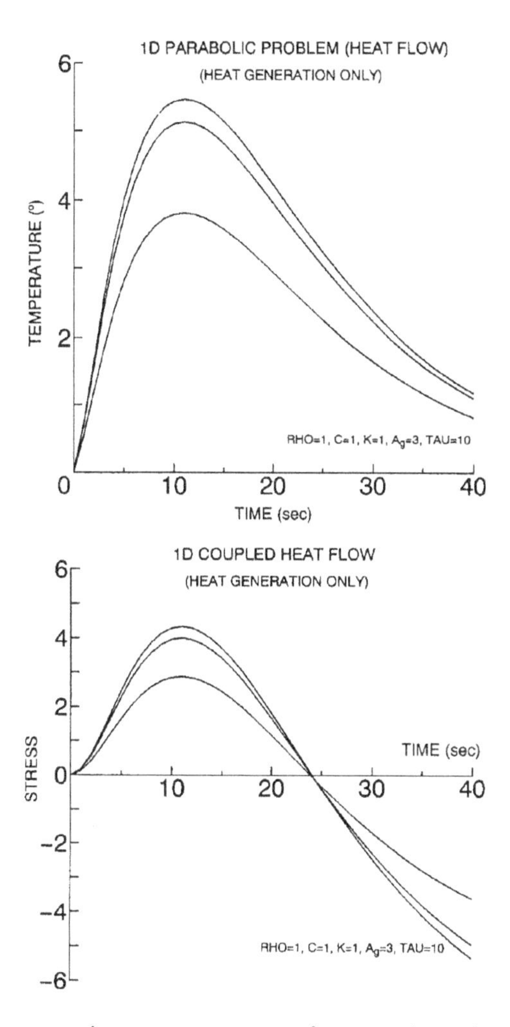

Figure 8.6 Temperature and stress responses of some selected points.

The heat generation function is assumed to be of the following form:

$$g = A_g t e^{-t/\tau_g} \tag{8.126}$$

where A_g and τ_g are physical constants determined from heat generation tests, and t is time. Furthermore, the variation elastic modulus of hardening solid with time is assumed to be of the following form:

$$E(t) = A_e\left(1 - e^{-t/\tau_e}\right) \tag{8.127}$$

where A_e and τ_e are physical constants determined from uniaxial tests of hardening solid and t is time. Figure 8.6 shows the variation of temperature and associated stress of some selected points.

8.5 Hydro-mechanical coupling: seepage and effective stress problem

8.5.1 Introduction

In the following, a finite element formulation for a coupled seepage and stress problem and its discretization are given (Aydan 2017). This is a fully coupled problem and it is generally called as a *consolidation problem* in the geotechnical engineering field.

8.5.2 Governing equations

8.5.2.1 Governing equation for the seepage field

As shown in Chapter 4, the volumetric variation of porous media takes the following form for 1D problems:

$$\frac{\partial \varepsilon}{\partial t} = -\frac{\partial v}{\partial x} \tag{8.128}$$

where v is the relative velocity. Let us assume that this equation is subjected to the following boundary and initial conditions:

Boundary conditions

$$\begin{aligned} v(0,t) &= 0 \quad \text{at} \quad x = 0 \\ q_n(0,t) &= q_0 \quad \text{at} \quad x = L \end{aligned} \tag{8.129}$$

Initial conditions

$$\begin{aligned} p(x,0) &= 0 \quad \text{at} \quad t = 0 \\ \dot{p}(x,0) &= 0 \quad \text{at} \quad t = 0 \end{aligned} \tag{8.130}$$

Let us further assume that the seepage obeys a linear type of constitutive law (Darcy's law) between velocity v and dependent variable pressure p

$$v = -k\frac{\partial p}{\partial x} \tag{8.131}$$

8.5.2.2 Governing equation for stress field

As shown in Chapter 3, the equation of momentum for 1D problems without the inertia term can be written as

$$\frac{\partial \sigma}{\partial x} + b = 0 \tag{8.132}$$

Let us assume that this equation is subjected to the following boundary conditions:

$$\begin{aligned} u(0,t) &= 0 \quad \text{at} \quad x = 0 \\ t(0,t) &= t_0 \quad \text{at} \quad x = L \end{aligned} \tag{8.133}$$

The displacement–strain relation is given as

$$\varepsilon = \frac{\partial u}{\partial x} \tag{8.134}$$

Let us assume that the total stress σ may be related to the effective stress law of Terzaghi through the following relation

$$\sigma = \sigma' - p \tag{8.135}$$

The incremental form of Eq. (8.132) together with the effective stress law takes the following form if the body force remains constant with time:

$$\frac{\partial \dot{\sigma}'}{\partial x} - \frac{\partial \dot{p}}{\partial x} = 0 \tag{8.136}$$

Let us further assume that the material is of Hookean type as given below:

$$\sigma' = E\varepsilon \tag{8.137}$$

Similarly, the constitutive law and displacement rate and strain rate relations may be rewritten as

$$\dot{\sigma}' = E\dot{\varepsilon} \tag{8.138}$$

$$\dot{\varepsilon} = \frac{\partial \dot{u}}{\partial x} \tag{8.139}$$

8.5.3 Weak form formulation

8.5.3.1 Weak form formulation for seepage field

Taking a variation on variable δp, the integral form of Eq. (8.128) becomes

$$\int_V \delta p \cdot \frac{\partial \varepsilon}{\partial t} dV = -\int_V \delta p \cdot \frac{\partial v}{\partial x} dV \tag{8.140}$$

where $dV = dAdx$.

Applying the integral by parts to the RHS with respect to x yields the following:

$$\int_V \delta p \cdot \frac{\partial \varepsilon}{\partial t} dV - \int_V \frac{\partial \delta p}{\partial x} \cdot v dV = -\int_A \left[\delta p \cdot q_n\right]_0^L dA \tag{8.141}$$

where $q_n = v \cdot n$. The above equation is called the weak form of Eq. (8.128). Inserting the constitutive relation given by Eq. (8.131) into Eq. (8.141), we obtain the following:

$$\int_V \delta p \cdot \frac{\partial p}{\partial t} dV + \int_V k \frac{\partial \delta p}{\partial x} \cdot \frac{\partial p}{\partial x} dV = -\int_A \left[\delta T \cdot q_n\right]_0^L dA \tag{8.142}$$

8.5.3.2 Weak form formulation for stress field

Taking a variation on displacement rate field $\delta\dot{u}$, the integral form of Eq. (8.136) becomes

$$\int_V \delta\dot{u}\cdot\frac{\partial(\dot{\sigma}'-\dot{p})}{\partial x}dV=0 \tag{8.143}$$

where $dV=dAdx$.

Applying the integral by parts to the LHS with respect to x yields the following:

$$\int_A \left[\delta\dot{u}\cdot\dot{t}\right]_0^L dA=\int_V \frac{\partial\delta\dot{u}}{\partial x}\cdot(\dot{\sigma}'-\dot{p})dV \tag{8.144}$$

where $\dot{t}=(\dot{\sigma}'-\dot{p})\cdot n$. The above equation is called the weak form of Eq. (8.136). Inserting the constitutive relation given by Eqs. (8.138) and (8.139) into Eq. (8.144), we obtain

$$\int_A \left[\delta\dot{u}\cdot\dot{t}\right]_0^L dA=\int_V E\frac{\partial\delta\dot{u}}{\partial x}\cdot\frac{\partial\dot{u}}{\partial x}dV-\int_V E\frac{\partial\delta\dot{u}}{\partial x}\cdot\dot{p}dV \tag{8.145}$$

8.5.4 Discretization

8.5.4.1 Discretization for physical space

a. Interpolation (shape) function for pressure field

For this particular problem, let us choose a linear function of the following form for the space in a local coordinate system whose origin is at a nodal point i as

$$p(t)=a\xi+b;\quad \xi=x-x_i;\quad d\xi=dx \tag{8.146}$$

Let us assume that the function p at nodes i and k is known. Thus, we can write the following

$$\begin{bmatrix} 0 & 1 \\ L & 1 \end{bmatrix}\begin{Bmatrix} a \\ b \end{Bmatrix}=\begin{Bmatrix} P_i \\ P_k \end{Bmatrix} \tag{8.147}$$

Taking the inverse of the above relation, one gets coefficients a and b. Inserting these coefficients in Eq. (8.146) yields the following:

$$p=N_iP_i+N_kP_k \tag{8.148}$$

where

$$N_i=1-\frac{\xi}{L};\quad N_k=\frac{\xi}{L}$$

The above equation may be rewritten in a compact form as

$$p=[N]_p\{P_e\}\quad \text{or}\quad p=\mathbf{N}_p\mathbf{P}_e \tag{8.149}$$

where $[N]_p = [N_i, N_k]$, $\{P_e\}^T = \{P_i, P_k\}$. The derivative of the above equation takes the following form:

$$\frac{dp}{dx} = \frac{dp}{d\xi} = \frac{dN_i}{d\xi}P_i + \frac{dN_k}{d\xi}P_k \tag{8.150}$$

The above relation is rewritten in a compact form as

$$\frac{dp}{dx} = \frac{dp}{d\xi} = [B]_p\{P_e\} \quad \text{or} \quad \frac{dp}{dx} = \frac{dp}{d\xi} = \mathbf{B}_p\mathbf{P}_e \tag{8.151}$$

where $[B] = [B_i, B_k]$, $B_i = -1/L$, $B_k = 1/L$, $L = x_k - x_i$.

b. Interpolation (shape) function for displacement field
For this particular problem, we have to choose a quadratic function of the following form for the space discretization of displacement field if the shape function for pressure field is linear in a local coordinate system whose origin is at a nodal point i as

$$u(t) = a + b\xi + c\xi^2; \quad \xi = x - x_i; \quad d\xi = dx \tag{8.152}$$

Let us assume that the function u at nodes i, j and k are known. Thus, we can write the following:

$$\begin{bmatrix} 1 & 0 & 0 \\ 1 & \frac{L}{2} & \frac{L^2}{4} \\ 1 & L & L^2 \end{bmatrix} \begin{Bmatrix} a \\ b \\ c \end{Bmatrix} = \begin{Bmatrix} U_i \\ U_j \\ U_k \end{Bmatrix} \tag{8.153}$$

Taking the inverse of the above relation, one gets coefficients a,b and c. Inserting these coefficients in Eq. (8.152) yields the following:

$$u = N_i U_i + N_j U_j + N_k U_k \tag{8.154}$$

where

$$N_i = \left(1 - \frac{2\xi}{L}\right)\left(1 - \frac{\xi}{L}\right); \quad N_j = \frac{4\xi}{L}\left(1 - \frac{\xi}{L}\right); \quad N_k = -\frac{\xi}{L}\left(1 - \frac{2\xi}{L}\right)$$

The above equation may be rewritten in a compact form as

$$u = [N]_u\{U_e\} \quad \text{or} \quad u = \mathbf{N}_u\mathbf{U}_e \tag{8.155}$$

where $[N]_u = [N_i, N_j, N_k]$, $\{U_e\}^T = \{U_i, U_j, U_k\}$. The derivative of the above equation takes the following form:

$$\frac{du}{dx} = \frac{du}{d\xi} = \frac{dN_i}{d\xi}U_i + \frac{dN_j}{d\xi}U_j + \frac{dN_k}{d\xi}P_k \tag{8.156}$$

The above relation is rewritten in a compact form as

$$\frac{du}{dx} = \frac{du}{d\xi} = [B]_u \{U_e\} \quad \text{or} \quad \frac{du}{dx} = \frac{du}{d\xi} = \mathbf{B}_u \mathbf{U}_e \tag{8.157}$$

where $[B] = \left[B_i, B_j, B_k \right]$,

$$B_i = \frac{1}{L^2}(4\xi - 3L); \quad B_j = \frac{4}{L^2}(L - 2\xi); \quad B_k = \frac{1}{L^2}(4\xi - L); \quad L = x_k - x_i$$

Equation (8.142) for seepage field, which holds for the whole domain, must also hold for each element as

$$\int_{V_e} \delta p \cdot \frac{\partial \varepsilon}{\partial t} dV + \int_{V_e} k \frac{\partial \delta p}{\partial x} \cdot \frac{\partial p}{\partial x} dV = -\int_{A_e} \left[\delta p \cdot q_n^e \right]_0^L dA \tag{8.158}$$

The discretized form of the above equation becomes

$$\int_{V_e} [N]_p^T [B]_u dV \left\{ \dot{U}_e \right\} + \int_{V_e} k [B]_p^T [B]_p dV \{P\}_e dV = -\int_{A_e} \left[[\bar{N}]_p^T q_n^e \right]_0^L dA \tag{8.159}$$

The above equation may be written in a compact form as

$$[C_e]_{pu} \left\{ \dot{U}_e \right\} + [K_e]_{pp} \{P_e\} = \{Q_e\} \tag{8.160}$$

where

$$[C_e]_{pu} = \int_{V_e} [N]_p^T [B]_u dV, \quad [K_e]_{pp} = \int_{V_e} k [B]_p^T [B]_p dV, \quad \{Q_e\} = -\int_{A_e} \left[[\bar{N}]_p^T q_n^e \right]_0^L dA$$

For the total domain, we have the following:

$$[C]_{pu} \left\{ \dot{U} \right\} + [K]_{pp} \{P\} = \{Q\} \tag{8.161}$$

where

$$[C]_{pu} = \sum_{k=1}^{n} [C_e]_{pu}^k, \quad [K]_{uu} = \sum_{k=1}^{n} [K_e]_{pp}^k, \quad \{Q\} = \sum_{k=1}^{n} \{Q_e\}_k, \quad \{T\} = \sum_{k=1}^{n} \{T_e\}_k$$

Similarly, Eq. (8.145) for stress field, which holds for the whole domain, must also hold for each element as

$$\int_{V_e} E \frac{\partial \delta \dot{u}}{\partial x} \cdot \frac{\partial \dot{u}}{\partial x} dV + \int_{V_e} E \frac{\partial \delta \dot{u}}{\partial x} \cdot \dot{p} dV = \int_{A_e} \left[\delta \dot{u} \cdot \dot{t} \right]_0^L dA \tag{8.162}$$

The discretized form of the above equation becomes

$$\int_{V_e} E [B]_u^T [B]_u dv \left\{ \dot{U} \right\}_e \int_{V_e} [B]_u^T [N]_p dv \left\{ \dot{P} \right\}_e = \int_{A_e} \left[\left[\bar{N} \right]^T \dot{t}_e \right]_0^L dA \tag{8.163}$$

The above equation may be written in a compact form as

$$[K_e]_{uu}\{\dot{U}_e\}-[C_e]_{up}\{\dot{P}_e\}=\{\dot{F}_e\} \tag{8.164}$$

where

$$[K_e]_{uu}=\int_{V_e} E[B]_u^T B_u dV, \quad \{\dot{F}_e\}=\int_{A_e}\left[[\bar{N}]^T \dot{t}_e\right]_0^L dA$$

For the total domain, we have the following:

$$[K]_{uu}\{\dot{U}\}-[C]_{up}\{\dot{P}\}=\{\dot{F}\} \tag{8.165}$$

where

$$[K]_{uu}=\sum_{k=1}^{n}[K_e]_{uu}^k, \quad [C]_{up}=\sum_{k=1}^{n}[C_e]_{up}^k, \quad \{\dot{F}\}=\sum_{k=1}^{n}\{\dot{F}_e\}_k,$$
$$\{\dot{U}\}=\sum_{k=1}^{n}\{\dot{U}_e\}_k; \quad \{\dot{P}\}=\sum_{k=1}^{n}\{\dot{P}_e\}_k$$

The above Eqs. (8.161) and (8.165) can be written in a compact form as

$$\begin{bmatrix} \mathbf{K}_{uu} & -\mathbf{C}_{up} \\ \mathbf{C}_{pu} & \mathbf{0} \end{bmatrix}\begin{Bmatrix} \dot{\mathbf{U}} \\ \dot{\mathbf{P}} \end{Bmatrix}+\begin{bmatrix} \mathbf{0} & \mathbf{0} \\ \mathbf{0} & \mathbf{K}_{pp} \end{bmatrix}\begin{Bmatrix} \mathbf{U} \\ \mathbf{P} \end{Bmatrix}=\begin{Bmatrix} \dot{\mathbf{F}} \\ \mathbf{Q} \end{Bmatrix} \tag{8.166}$$

The above equation may be rewritten in a more compact form as

$$[M]\{\dot{T}\}+[H]\{T\}=\{Y\} \tag{8.167}$$

where

$$[M]=\begin{bmatrix} \mathbf{K}_{uu} & -\mathbf{C}_{up} \\ \mathbf{C}_{pu} & \mathbf{0} \end{bmatrix}; \quad [H]=\begin{bmatrix} \mathbf{0} & \mathbf{0} \\ \mathbf{0} & \mathbf{K}_{pp} \end{bmatrix}; \quad \{\dot{T}\}=\begin{Bmatrix} \dot{\mathbf{U}} \\ \dot{\mathbf{P}} \end{Bmatrix};$$
$$\{T\}=\begin{Bmatrix} \mathbf{U} \\ \mathbf{P} \end{Bmatrix}; \quad \{Y\}=\begin{Bmatrix} \dot{\mathbf{F}} \\ \mathbf{Q} \end{Bmatrix}$$

8.5.4.2 Discretization for a time domain

Equation (8.167) could not be solved as it is. For a time step m, we can rewrite Eq.(8.167) as

$$[M]\{\dot{T}\}_{(m+\theta)}+[H]\{T\}_{(m+\theta)}=\{Y\}_{(m+\theta)} \tag{8.168}$$

Therefore, we discretize dependent variable $\{T\}$ for a time domain by using the Taylor expansion as it is in the FDM as

$$\{T\}_m = \{T\}_{(m+\theta)-\theta} = \{T\}_{(m+\theta)} - \frac{\partial\{T\}_{(m+\theta)}}{\partial t}\frac{\theta\Delta t}{1!} + 0^2 \tag{8.169}$$

$$\{T\}_{m+1} = \{T\}_{(m+\theta)+(1-\theta)} = \{T\}_{(m+\theta)} + \frac{\partial\{T\}_{(m+\theta)}}{\partial t}\frac{(1-\theta)\Delta t}{1!} + 0^2 \tag{8.170}$$

From the above relations, one easily gets the following:

$$\{T\}_{(m+\theta)} = \theta\{T\}_{m+1} + (1-\theta)\{T\}_m; \quad \{\dot{T}\}_{(m+\theta)} = \frac{\{T\}_{m+1} - \{T\}_m}{\Delta t} \tag{8.171}$$

$$\{Y\}_{(m+\theta)} = \theta\{Y\}_{m+1} + (1-\theta)\{Y\}_m \tag{8.172}$$

Inserting these relations into Eq. (8.168), we get

$$\left[M^*\right]\{T\}_{m+1} = \left\{Y^*\right\}_{m+1} \tag{8.173}$$

where

$$\left[M^*\right] = \left[\frac{1}{\Delta t}[M] + \theta[H]\right]$$

$$\left\{Y^*\right\}_{m+1} = \left[\frac{1}{\Delta t}[M] - (1-\theta)[H]\right]\{T\}_m + \theta\{Y\}_{m+1} + (1-\theta)\{Y\}_m$$

8.5.5 Specific example

For a typical element, the following are obtained:

$$[M_e] = A_e\begin{bmatrix} \frac{7E}{3L} & -\frac{8E}{3L} & \frac{E}{3L} & \frac{5}{6} & \frac{1}{6} \\ -\frac{8E}{3L} & \frac{16E}{3L} & -\frac{8E}{3L} & -\frac{2}{3} & \frac{2}{3} \\ \frac{E}{3L} & -\frac{8E}{3L} & \frac{7E}{3L} & -\frac{1}{6} & -\frac{5}{6} \\ -\frac{5}{6} & \frac{2}{3} & \frac{1}{6} & 0 & 0 \\ -\frac{1}{6} & -\frac{2}{3} & \frac{5}{6} & 0 & 0 \end{bmatrix}, \quad [H_e] = \frac{kA_e}{L}\begin{bmatrix} 0 & 0 & 0 & 0 & 0 \\ 0 & 0 & 0 & 0 & 0 \\ 0 & 0 & 0 & 0 & 0 \\ 0 & 0 & 0 & 1 & -1 \\ 0 & 0 & 0 & -1 & 1 \end{bmatrix},$$

$$\{Y_e\} = \begin{Bmatrix} \dot{F}_i \\ \dot{F}_j \\ \dot{F}_k \\ Q_i \\ Q_k \end{Bmatrix}; \quad \{\dot{T}_e\} = \begin{Bmatrix} \dot{T}_i \\ \dot{T}_j \\ \dot{T}_k \\ \dot{P}_i \\ \dot{P}_k \end{Bmatrix}; \quad \{T_e\} = \begin{Bmatrix} T_i \\ T_j \\ T_k \\ P_i \\ P_k \end{Bmatrix}$$

8.5.6 Example: simulation of settlement under sudden loading

A specific example is given herein by simulating the settlement and pore pressure variation in ground by considering the order of approximation function for pressure and displacement field. Figures 8.7 and 8.8 show the variation of settlement and pore pressure at some selected nodes. The results for both situations are almost exactly the same.

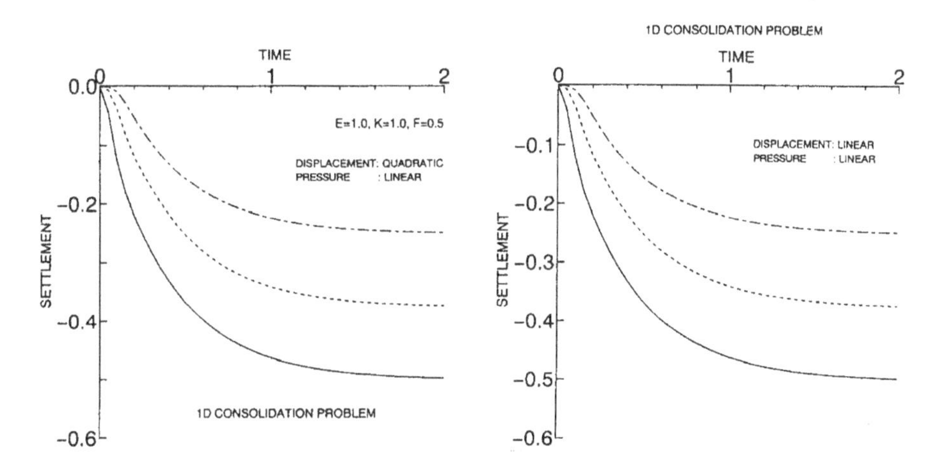

Figure 8.7 Settlement of ground under rapid load for different shape functions.

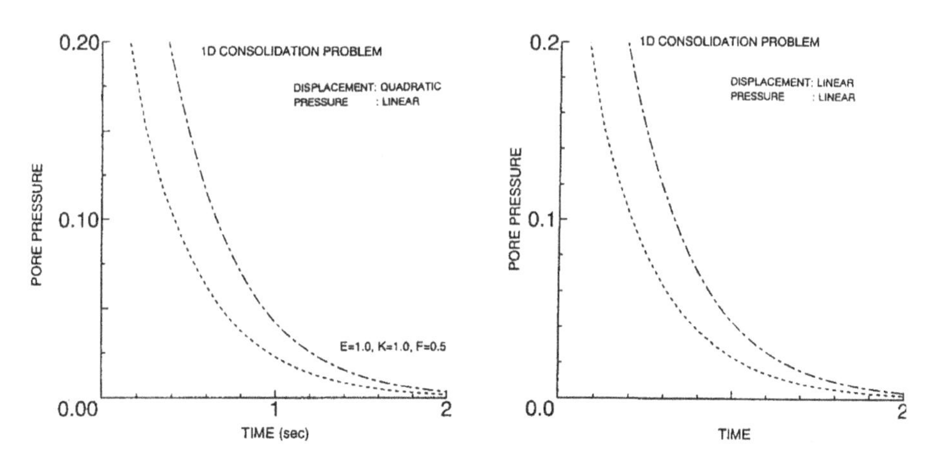

Figure 8.8 Pore pressure of ground under rapid load for different shape functions.

8.6 Biot problem: coupled dynamic response of porous media

8.6.1 Introduction

In the following, a finite element formulation for a coupled seepage and stress problem for dynamic responses saturated porous media and its discretization are given (Aydan 2017). This is a fully coupled problem and it is generally called the *Biot's problem* in geomechanics.

8.6.2 Governing equations

8.6.2.1 Governing equation for a fluid phase

As shown in Chapter 4, the equation of motion for a fluid phase of porous media takes the following form for 1D problems:

$$\frac{\partial \sigma_f}{\partial x} + \rho_f g = \rho_f \frac{\partial^2 u_s}{\partial t^2} + \frac{\rho_f}{n}\frac{\partial^2 w}{\partial t^2} + \frac{1}{K}\frac{\partial w}{\partial t} \tag{8.174}$$

where w is the relative displacement. Let us assume that this equation is subjected to the following boundary and initial conditions:

Boundary conditions

$$\begin{aligned} &w(0,t) = 0 \quad &\text{at} \quad &x = 0 \\ &t_n^f(0,t) = t_0^f \quad &\text{at} \quad &x = L \end{aligned} \tag{8.175}$$

Initial conditions

$$\begin{aligned} &w(x,0) = 0 \quad &\text{at} \quad &t = 0 \\ &\dot{w}(x,0) = 0 \quad &\text{at} \quad &t = 0 \end{aligned} \tag{8.176}$$

Let us further assume that the fluid obeys a linear type of constitutive law (Biot's law) in terms of fluid and skeleton strains and dependent variable fluid pressure p

$$-p = \sigma_f = M\left(\alpha\varepsilon_s + \varepsilon_f\right) \tag{8.177}$$

8.6.2.2 Governing equation for a total stress system

As shown in Chapter 4, the equation of motion for 1D problems takes the following form:

$$\frac{\partial \sigma}{\partial x} + \rho g = \rho \frac{\partial^2 u_s}{\partial t^2} + \rho_f \frac{\partial^2 w}{\partial t^2} \tag{8.178}$$

Let us assume that this equation is subjected to the following boundary conditions:

$$\begin{aligned} &u_s(0,t) = 0 \quad \text{at} \quad x = 0 \\ &t^s(0,t) = t_0^s \quad \text{at} \quad x = L \end{aligned} \tag{8.179}$$

The displacement–strain relation is given as

$$\varepsilon_s = \frac{\partial u_s}{\partial x}, \quad \varepsilon_f = \frac{\partial w}{\partial x} \tag{8.180}$$

Let us assume that the total stress σ may be related to the effective stress law of Biot through the following relation:

$$\sigma = \sigma' - \alpha p \tag{8.181}$$

Let us further assume that the skeleton obeys the following constitutive law:

$$\sigma = (2\mu + \lambda)\varepsilon_s - \alpha p = \left(2\mu + \lambda + \alpha^2 M\right)\varepsilon_s + \alpha M \varepsilon_f \tag{8.182}$$

8.6.3 Weak form formulation

8.6.3.1 Weak form formulation for a fluid phase

Taking a variation on variable δw, the integral form of Eq. (8.174) becomes

$$\begin{aligned} &\int_V \delta w \cdot \rho_f \frac{\partial^2 u_s}{\partial t^2} dV + \int_V \delta w \cdot \frac{\rho_f}{n} \frac{\partial^2 w}{\partial t^2} dV + \int_V \delta w \cdot \frac{1}{K} \frac{\partial w}{\partial t} dV \\ &= -\int_V \delta w \cdot \frac{\partial p}{\partial x} dV + \int_V \delta w \cdot \rho_f g dV \end{aligned} \tag{8.183}$$

where

$$dV = dAdx.$$

Applying the integral by parts to the RHS with respect to x yields the following:

$$\begin{aligned} &\int_V \delta w \cdot \rho_f \frac{\partial^2 u_s}{\partial t^2} dV + \int_V \delta w \cdot \frac{\rho_f}{n} \frac{\partial^2 w}{\partial t^2} dV + \int_V \delta w \cdot \frac{1}{K} \frac{\partial w}{\partial t} dV \\ &- \int_V \frac{\partial \delta w}{\partial x} \cdot p dV = -\int_A \left[\delta w \cdot t_n^f\right]_0^L + \int_V \delta w \cdot \rho_f g dV \end{aligned} \tag{8.184}$$

where $t_n^f = p \cdot n$. The above equation is called the weak form of Eq. (8.174). Inserting the constitutive relation given by Eq. (8.77) into Eq. (8.184), we obtain the following:

$$\int_V \delta w \cdot \rho_f \frac{\partial^2 u_s}{\partial t^2} dV + \int_V \delta w \cdot \frac{\rho_f}{n} \frac{\partial^2 w}{\partial t^2} dV + \int_V \delta w \cdot \frac{1}{K} \frac{\partial w}{\partial t} dV + \int_V \alpha M \frac{\partial \delta w}{\partial x} \cdot \frac{\partial u_s}{\partial x} dV$$
$$+ \int_V M \frac{\partial \delta w}{\partial x} \cdot \frac{\partial w}{\partial x} dV = -\int_A \left[\delta w \cdot t_n^f\right]_0^L dA + \int_V \delta w \cdot \rho_f g dV \quad (8.185)$$

8.6.3.2 *Weak form formulation for a total stress field*

Taking a variation on displacement rate field δu_s, the integral form of Eq. (8.178) becomes

$$\int_V \delta u_s \cdot \frac{\partial \sigma}{\partial x} dV + \int_V \delta u_s \cdot \rho g dV = \int_V \delta u_s \cdot \rho \frac{\partial^2 u_s}{\partial t^2} dV + \int_V \delta u_s \cdot \rho_f \frac{\partial^2 w}{\partial t^2} dV \quad (8.186)$$

where $dV = dAdx$.

Applying the integral by parts to the LHS with respect to x yields the following:

$$\int_A [\delta u_s \cdot t]_0^L dA + \int_V \delta u_s \cdot \rho g dV = \int_V \frac{\partial \delta u_s}{\partial x} \cdot \sigma dV + \int_V \delta u_s \cdot \rho \frac{\partial^2 u_s}{\partial t^2} dV$$
$$+ \int_V \delta u_s \cdot \rho_f \frac{\partial^2 w}{\partial t^2} dV \quad (8.187)$$

where $t^s = \sigma_s \cdot n$. The above equation is called the weak form of Eq. (8.178). Inserting the constitutive relation given by Eq. (8.182) into Eq. (8.187), we obtain the following

$$\int_A [\delta u_s \cdot t]_0^L dA + \int_V \delta u_s \cdot \rho g dV = \left(2\mu + \lambda + \alpha^2 M\right) \int_V \frac{\partial \delta u_s}{\partial x} \cdot \frac{\partial u_s}{\partial x} dV + \alpha M$$
$$\int_V \frac{\partial \delta u_s}{\partial x} \cdot \frac{\partial w}{\partial x} dV + \int_V \delta u_s \cdot \rho \frac{\partial^2 u_s}{\partial t^2} dV + \int_V \delta u_s \cdot \rho_f \frac{\partial^2 w}{\partial t^2} dV \quad (8.188)$$

8.6.4 Discretization

8.6.4.1 *Discretization for physical space*

a. Interpolation (shape) function for a relative displacement field
For this particular problem, let us choose a linear function of the following form for the space in a local coordinate system whose origin is at a nodal point i as

$$w(t) = a\xi + b; \quad \xi = x - x_i; \quad d\xi = dx \quad (8.189)$$

Let us assume that the function w at nodes i and j are known. Thus, we can write the following:

$$\begin{bmatrix} 0 & 1 \\ L & 1 \end{bmatrix} \begin{Bmatrix} a \\ b \end{Bmatrix} = \begin{Bmatrix} w_i \\ w_j \end{Bmatrix} \quad (8.190)$$

Taking the inverse of the above relation, one gets coefficients a and b. Inserting these coefficients in Eq. (8.189) yields the following:

$$w = N_i W_i + N_j W_j \tag{8.191}$$

where

$$N_i = 1 - \frac{\xi}{L}; \quad N_j = \frac{\xi}{L}$$

The above equation may be rewritten in a compact form as

$$w = [N]\{W_e\} \quad \text{or} \quad w = \mathbf{N}\mathbf{W}_e \tag{8.192}$$

where $[N]=[N_i, N_j], \{W_e\}^T = \{W_i, W_j\}$. The derivative of the above equation takes the following form:

$$\frac{dw}{dx} = \frac{dw}{d\xi} = \frac{dN_i}{d\xi}W_i + \frac{dN_j}{d\xi}W_j \tag{8.193}$$

The above relation is rewritten in a compact form as

$$\frac{dw}{dx} = \frac{dw}{d\xi} = [B]p\{W_e\} \quad \text{or} \quad \frac{dw}{dx} = \frac{dw}{d\xi} = \mathbf{B}\mathbf{W}_e \tag{8.194}$$

where $[B]=\left[B_i, B_j\right]$, $B_i = -1/L$, $B_j = 1/L$, $L = x_j - x_i$.

b. Interpolation (shape) function for a skeleton displacement field
As for the displacement field of skeleton, we chose also a linear interpolation function. The resulting expressions will be similar so that it will not be presented herein.

Equation (8.185) for a fluid phase, which holds for the whole domain, must also hold for each element as

$$\int_{V_e} \delta w \cdot \rho_f \frac{\partial^2 u_s}{\partial t^2} dV + \int_{V_e} \delta w \cdot \frac{\rho_f}{n}\frac{\partial^2 w}{\partial t^2} dV + \int_{V_e} \delta w \cdot \frac{1}{K}\frac{\partial w}{\partial t} dV + \int_{V_e} \alpha M \frac{\partial \delta w}{\partial x} \cdot \frac{\partial u_s}{\partial x} dV$$
$$+ \int_{V_e} M \frac{\partial \delta w}{\partial x} \cdot \frac{\partial w}{\partial x} dV = -\int_{A_e} \left[\delta w \cdot t_n^f\right]_0^L dA + \int_{V_e} \delta w \cdot \rho_f g dV \tag{8.195}$$

The discretized form of the above equation becomes

$$\int_{V_e} \rho_f [N]^T [N]\{\ddot{U}_e\} dV + \int_{V_e} \frac{\rho_f}{n} [N]^T [N]\{\ddot{W}_e\} dV$$
$$+ \int_{V_e} \frac{1}{K} [N]^T [N]\{\dot{W}_e\} dV + \int_{V_e} \alpha M [B]^T [B]\{U\} dV$$
$$+ \int_{V_e} M [B]^T [B]\{W\} dV = -\int_{A_e} \left[[\bar{N}]^T t_n^e\right]_0^L dA + \int_{V_e} \rho_f g [N]^T dV \tag{8.196}$$

The above equation may be written in a compact form as

$$[M_e]_1^s\{\ddot{U}_e\}+[M_e]_1^f\{\ddot{W}_e\}+[C_e]_1^f\{\dot{W}_e\}+[K_e]_1\{U_e\}+[K]_1^f\{U_e\}+=\{F_e\}_1 \tag{8.197}$$

where

$$[M_e]_1^s=\int_{V_e}\rho[N]^T[N]dV;\quad [M_e]_1^f=\int_{V_e}\frac{\rho_f}{n}[N]^T[N]dV;\quad [C_e]_1^f=\int_{V_e}\frac{1}{K}[N]^T[N]dV;$$

$$[K_e]_1^s=\alpha M\int_{V_e}[B]^T[B]dV;\quad [K_e]_1^f=M\int_{V_e}[B]^T[B]dV$$

$$\{F_e\}_1=-\int_{A_e}\left[[\bar{N}]^T t_n^e\right]_0^L dA+\int_{V_e}\rho_f g[N]^T dV$$

For the total domain, we have the following:

$$[M]_1^s\{\ddot{U}\}+[M]_1^f\{\ddot{W}\}+[C]_1^f\{\dot{W}\}+[K]_1^s\{U\}+[K]_1^f\{W\}+=\{F\}_1 \tag{8.198}$$

where

$$[M]_1^s=\sum_{k=1}^n[M_e]_1^s,\quad [M]_1^f=\sum_{k=1}^n[M_e]_1^f,\quad [C]_1^f=\sum_{k=1}^n[C_e]_1^f,$$

$$[K]_1^s=\sum_{k=1}^n[K_e]_1^s,\quad [K]_1^f=\sum_{k=1}^n[K_e]_1^f,\quad \{F\}_1=\sum_{k=1}^n\{F_e\}_1$$

Similarly Eq. (8.188) for a stress field, which holds for the whole domain, must also hold for each element as

$$\begin{aligned}\int_{A_e}[\delta u_s\cdot t]_0^L dA+\int_{V_e}\delta u_s\cdot\rho g dV&=\left(2\mu+\lambda+\alpha^2M\right)\int_{V_e}\frac{\partial\delta u_s}{\partial x}\cdot\frac{\partial u_s}{\partial x}dV\\&+\alpha M\int_{V_e}\frac{\partial\delta u_s}{\partial x}\cdot\frac{\partial w}{\partial x}dV+\int_{V_e}\delta u_s\cdot\rho\frac{\partial^2u_s}{\partial t^2}dV+\int_{V_e}\delta u_s\cdot\rho_f\frac{\partial^2 w}{\partial t^2}dV\end{aligned} \tag{8.199}$$

The discretized form of the above equation becomes

$$\begin{aligned}&\int_{V_e}\rho[N]^T[N]dv\{\ddot{U}\}_e\int_{V_e}\rho_f[N]^T[N]dv\{\ddot{W}\}_e+\int_{V_e}\left(2\mu+\lambda+\alpha^2M\right)[B]^T[B]\{U\}dV\\&+\int_{V_e}\alpha M[B]^T[B]\{W\}dV=\int_{A_e}\left[[\bar{N}]^T t_e\right]_0^L dA+\int_{V_e}\rho g[N]^T dV\end{aligned} \tag{8.200}$$

The above equation may be written in a compact form as

$$[M_e]_2^s\{\ddot{U}\}+[M_e]_2^f\{\ddot{W}\}+[K_e]_2^s\{U\}+[K_e]_2^f\{W\}+=\{F\}_2 \tag{8.201}$$

where

$$[M_e]_2^s = \int_{V_e} \rho[N]^T[N]dV, \quad [M_e]_2^f = \int_{V_e} \rho_f[N]^T[N]dV,$$

$$[K_e]_2^s = \left(2\mu + \lambda + \alpha^2 M\right)\int_{V_e} [B]^T[B]dV, \quad [K_e]_2^f = \alpha M\int_{V_e} [B]^T[B]dV,$$

$$\{F_e\}_2 = -\int_{A_e} \left[[\bar{N}]^T t_n^e\right]_0^L dA + \int_{V_e} \rho g[N]^T dV$$

For the total domain, we have the following

$$[M]_2^s\{\ddot{U}\} + [M]_2^f\{\ddot{W}\} + [K]_2^s\{U\} + [K]_2^f\{W\} + = \{F\}_2 \tag{8.202}$$

where

$$[M]_2^s = \sum_{k=1}^{n}[M_e]_2^s, \quad [M]_2^f = \sum_{k=1}^{n}[M_e]_2^f, \quad [K]_2^s = \sum_{k=1}^{n}[K_e]_2^s,$$

$$[K]_2^f = \sum_{k=1}^{n}[K_e]_2^f, \quad \{F\}_2 = \sum_{k=1}^{n}\{F_e\}_2$$

Above Eqs. (8.198) and (8.202) can be written in a compact form as

$$\begin{bmatrix} \mathbf{M}_1^s & \mathbf{M}_1^f \\ \mathbf{M}_2^s & \mathbf{M}_2^f \end{bmatrix}\begin{Bmatrix} \ddot{\mathbf{U}} \\ \ddot{\mathbf{W}} \end{Bmatrix} + \begin{bmatrix} \mathbf{0} & \mathbf{C}_1^f \\ \mathbf{0} & \mathbf{0} \end{bmatrix}\begin{Bmatrix} \dot{\mathbf{U}} \\ \dot{\mathbf{W}} \end{Bmatrix}$$

$$+\begin{bmatrix} \mathbf{K}_1^s & \mathbf{K}_1^f \\ \mathbf{K}_2^s & \mathbf{K}_2^f \end{bmatrix}\begin{Bmatrix} \mathbf{U} \\ \mathbf{W} \end{Bmatrix} = \begin{Bmatrix} \dot{\mathbf{F}}_1 \\ \mathbf{F}_2 \end{Bmatrix} \tag{8.203}$$

The above equation may be rewritten in a more compact form as

$$[M]\{\ddot{T}\} + [C]\{\dot{T}\} + [H]\{T\} = \{Y\} \tag{8.204}$$

where

$$[M] = \begin{bmatrix} \mathbf{M}_1^s & \mathbf{M}_1^f \\ \mathbf{M}_2^s & \mathbf{M}_2^f \end{bmatrix}; \quad [C] = \begin{bmatrix} \mathbf{0} & \mathbf{C}_1^f \\ \mathbf{0} & \mathbf{0} \end{bmatrix}; \quad [H] = \begin{bmatrix} \mathbf{K}_1^s & \mathbf{K}_1^f \\ \mathbf{K}_2^s & \mathbf{K}_2^f \end{bmatrix}$$

$$\{\ddot{T}\} = \begin{Bmatrix} \ddot{\mathbf{U}} \\ \ddot{\mathbf{W}} \end{Bmatrix}; \quad \{\dot{T}\} = \begin{Bmatrix} \dot{\mathbf{U}} \\ \dot{\mathbf{W}} \end{Bmatrix}; \quad \{T\} = \begin{Bmatrix} \mathbf{U} \\ \mathbf{W} \end{Bmatrix}; \quad \{Y\} = \begin{Bmatrix} \mathbf{F}_1 \\ \mathbf{F}_2 \end{Bmatrix}$$

8.6.4.2 Discretization for a time domain

Equation (8.204) could not be solved as it is. For a time step m, we can rewrite Eq. (8.204) as

$$[M]\{\ddot{T}\}_m + [C]\{\dot{T}\}_m + [H]\{T\}_m = \{Y\}_m \tag{8.205}$$

Therefore, we discretize field $\{T\}$ for a time-domain using the Taylor expansion as it is in the FDM as

$$\{T\}_{m-1} = \{T\}_m - \frac{\partial\{T\}_m}{\partial t}\frac{\Delta t}{1!} + \frac{\partial^2\{T\}_m}{\partial t^2}\frac{\Delta t^2}{2!} - 0^3 \tag{8.206}$$

$$\{T\}_m = \{T\}_m \tag{8.207}$$

$$\{T\}_{m+1} = \{T\}_m + \frac{\partial\{T\}_m}{\partial t}\frac{\Delta t}{1!} + \frac{\partial^2\{T\}_m}{\partial t^2}\frac{\Delta t^2}{2!} + 0^3 \tag{8.208}$$

From the above relations, one easily obtains

$$\{\dot{T}\}_m = \frac{1}{\Delta t}\left(\{T\}_{m+1} - \{T\}_{m-1}\right) \tag{8.209}$$

$$\{\ddot{T}\}_m = \frac{1}{\Delta t^2}\left(\{T\}_{m+1} - 2\{T\}_m + \{T\}_{m-1}\right) \tag{8.210}$$

Inserting these relations into Eq. (8.205), we get the following:

$$[M^*]\{T\}_{m+1} = \{Y^*\}_{m+1} \tag{8.211}$$

where

$$[M^*] = \left[\frac{1}{\Delta t^2}[M] + \frac{1}{\Delta t}[C]\right]$$

$$\{Y^*\}_{m+1} = \left[\frac{2}{\Delta t^2}[M] - [H]\right]\{T\}_m - \left[\frac{1}{\Delta t^2}[M] - \frac{1}{\Delta t}[C]\right]\{T\}_{m-1} + \{Y\}_m$$

8.6.5 Specific example

For a typical element, the following is obtained:

$$[M_e] = A_e\begin{bmatrix} \frac{\rho L}{3} & \frac{\rho L}{6} & \frac{\rho_f L}{3} & \frac{\rho_f L}{6} \\ \frac{\rho L}{6} & \frac{\rho L}{3} & \frac{\rho_f L}{6} & \frac{\rho_f L}{3} \\ \frac{\rho_f L}{3} & \frac{\rho_f L}{6} & \frac{\rho_f L}{3n} & \frac{\rho_f L}{6n} \\ \frac{\rho_f L}{6} & \frac{\rho_f L}{3} & \frac{\rho_f L}{6n} & \frac{\rho_f L}{3n} \end{bmatrix}, \quad [C_e] = \frac{A_e L}{6K}\begin{bmatrix} 0 & 0 & 2 & 1 \\ 0 & 0 & 1 & 2 \\ 0 & 0 & 0 & 0 \\ 0 & 0 & 0 & 0 \end{bmatrix},$$

$$[H_e] = A_e \begin{bmatrix} \dfrac{\alpha M}{L} & -\dfrac{\alpha M}{L} & \dfrac{M}{L} & -\dfrac{M}{L} \\ -\dfrac{\alpha M}{L} & \dfrac{\alpha M}{L} & -\dfrac{M}{L} & \dfrac{M}{L} \\ \dfrac{2\mu+\lambda+\alpha^2 M}{L} & -\dfrac{2\mu+\lambda+\alpha^2 M}{L} & \dfrac{\alpha M}{L} & -\dfrac{\alpha M}{L} \\ -\dfrac{2\mu+\lambda+\alpha^2 M}{L} & \dfrac{2\mu+\lambda+\alpha^2 M}{L} & -\dfrac{\alpha M}{L} & \dfrac{\alpha M}{L} \end{bmatrix},$$

$$\{Y_e\} = \begin{Bmatrix} \dot{F}_{1i} \\ F_j^1 \\ F_i^1 \\ F_i^2 \\ F_j^2 \end{Bmatrix} \quad 2mm\{\ddot{T}_e\} = \begin{Bmatrix} \ddot{U}_i \\ \ddot{U}_j \\ \ddot{W}_i \\ \ddot{W}_j \end{Bmatrix} \{\dot{T}_e\} = \begin{Bmatrix} \dot{U}_i \\ \dot{U}_j \\ \dot{W}_i \\ \dot{W}_j \end{Bmatrix} \{T_e\} = \begin{Bmatrix} U_i \\ U_j \\ W_i \\ W_j \end{Bmatrix}$$

8.6.6 *Example: simulation of dynamic response of saturated porous media*

A specific example is given herein by simulating the response of an half-space under sinusoidal cyclic loading. No drainage is allowed at the bottom and the fluid phase is free of traction at the ground surface. Figure 8.9 shows the variation of displacement, velocity and acceleration responses of some selected nodes.

8.7 Introduction of boundary conditions in a simultaneous equation system

8.7.1 *Formulation*

The simultaneous equation system of the FEM results in a $[m \times m]$ square matrix always as given below:

$$[K]\{U\} = \{F\} \tag{8.212}$$

It should be noticed that matrix $[K]$ is a square symmetric matrix and its determinant $|K|$ is zero. Thus, its inverse is not possible due to the singularity problem. The solution becomes possible if and only if the boundary conditions are introduced. The introduction of boundary conditions associated with vectors $\{U\}$ and $\{F\}$ can be partitioned into two unknown and known parts together with the duality concept as given below

$$\{U\} = \begin{Bmatrix} \{U\}_u \\ \{U\}_n \end{Bmatrix}, \text{ and } \{F\} = \begin{Bmatrix} \{F\}_n \\ \{F\}_u \end{Bmatrix} \tag{8.213}$$

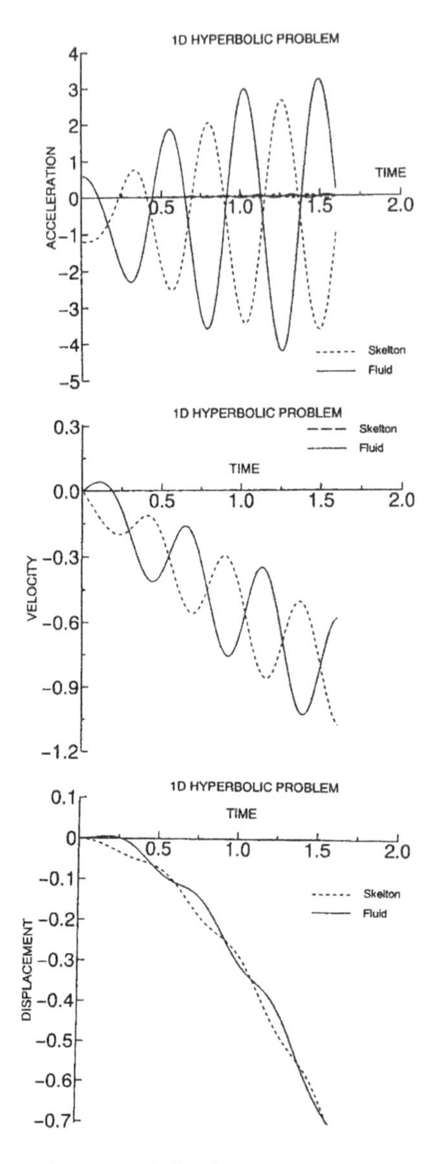

Figure 8.9 Acceleration, velocity and displacement responses.

Accordingly, Eq. (8.212) may be rewritten as

$$\begin{bmatrix} [K]_{uu} & [K]_{un} \\ [K]_{nu} & [K]_{nn} \end{bmatrix} \begin{Bmatrix} \{U\}_u \\ \{U\}_n \end{Bmatrix} = \begin{Bmatrix} \{F\}_n \\ \{F\}_u \end{Bmatrix} \tag{8.214}$$

As the solution of $\{U\}_u$ is required, one can write the following equation:

$$[K]_{uu}\{U\}_u + [K]_{un}\{U\}_n = \{F\}_n \tag{8.215}$$

Rearanging (8.215), we have the following relations:

$$[K]_{uu}\{U\}_u = \{F\}_n - [K]_{un}\{U\}_n \tag{8.216a}$$

or

$$\{U\}_u = [K]_{uu}^{-1}\left(\{F\}_n - [K]_{un}\{U\}_n\right) \tag{8.216b}$$

If the unknown part of vector $\{F\}$ is required, Eq. (8.214) results in the following relation

$$[K]_{nu}\{U\}_u + [K]_{nn}\{U\}_n = \{F\}_u \tag{8.217}$$

One can easily obtain $\{F\}_u$ from Eq. (8.217) by inserting the unknown vector $\{U\}_u$ obtained from Eq. (8.216) in the previous stage.

8.7.2 Actual implementation and solution of Eq. (8.216b)

Rearranging Eq. (8.212) in the form of Eq. (8.213) and solving using the relation given by (8.216b) require some extra computational time. Instead of rearrangement in the form of Eq. (8.213), the following procedure is implemented. For each boundary condition U_k, the force vector is modified by changing i from 1 to m except row i as

$$F_k^* = F_k - K_{ki}U_k$$

Then 1 is assigned to K_{kk}, and the other components of the row and column matrix $[K]$ are assigned to 0. The value of vector F_k^* is assigned to U_k. The actual implementation of this procedure in the FEM program coded in True BASIC programming language is given below:

```
*************************************************************************
!
! ****** Displacement Boundary condition is implemented ***************
!
FOR IB=1 TO NBS
  IBNI=IBS(IB)
  FOR I=1 TO NODE
   GFS(I)=GFS(I)-GKS(IBNI,I)* UDIS(IB)
   IF I<>IBNI THEN
    GKS(IBNI,I)=0.
    GKS(I,IBNI)=0.
   ELSE
    GKS(IBNI,IBNI)=1.0
    GFS(IBNI)=UDIS(IB)
   END IF
  NEXT I
NEXT IB
!
*************************************************************************
```

8.8 Rayleigh damping and its implementation

The final forms of the discretized form of equation of motion (8.47) irrespective of the method of solution (FDM, FEM and BEM) and continuum or discontinuum, depending upon the character of the governing equation, may be written in the following form:

$$[M]\{\ddot{\varphi}\}+[C]\{\dot{\varphi}\}+[K]\{\varphi\}=\{F\} \tag{8.218}$$

The specific forms of matrices $[M]$, $[C]$, $[K]$ and vector $\{F\}$ in the equation above would only differ depending upon the method of solution chosen and dimensions of physical space. Viscosity matrix $[C]$ is associated with rate-dependency of geomaterials. However, in many dynamic solution schemes, viscosity matrix $[C]$ is expressed in the following form using the Rayleigh damping approach as

$$[C]=\alpha[M]+\beta[K] \tag{8.219}$$

where α and β are called the proportionality constants. This approach becomes very convenient in large-scale problems if the central finite difference technique and mass lumping are utilized. However, it should be also noted that it is very difficult to determine these parameters from experiments although it is computationally very convenient. Again, in non-linear problems, the deformation moduli of rocks are reduced in relation to the straining using an approach commonly used in soil dynamics. In such approaches, the reduction of moduli is determined from cyclic tests. Nevertheless, it should be noted that the validity of such an approach for rock and rock masses is quite questionable and it is not recommended for dynamic problems in geomechanics (Aydan 2017).

8.9 Non-linear problems

If the material behavior involves non-linearity, the equation system above must be solved iteratively with the implementation of required conditions associated with the constitutive law chosen. The iteration techniques may be broadly classified as the initial, secant or tangential stiffness method (e.g. Owen and Hinton 1980). In geomechanics, the secant method and the initial stiffness method are recommended to deal with the softening behavior. Particularly, the initial stiffness method would be numerically quite stable although the convergence might be slow. It should be also noted that when the overall equilibrium could not be achieved, it would correspond to the collapse of the structure (Figure 8.10).

8.10 Multi-dimensional situations

The formulations described for 1D situations in previous sections are fundamentally the same for multi-dimensional situations in a symbolic form as given below (e.g. Zienkiewicz 1977; Reddy 2006; Segerlind 1976):

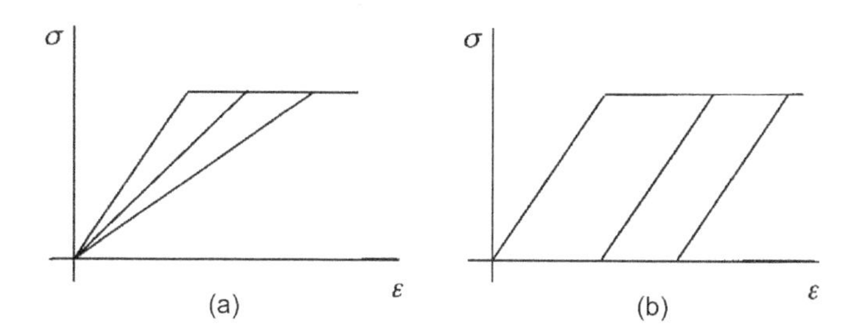

Figure 8.10 Illustration of numerical techniques to deal with non-linearity. (a) Secant method and (b) initial stiffness method.

Hyperbolic

$$[M]\{\ddot{\Phi}\}+[C]\{\dot{\Phi}\}+[K]\{\Phi\}=\{\psi\} \tag{8.220a}$$

Parabolic

$$[C]\{\dot{\Phi}\}+[K]\{\Phi\}=\{\psi\} \tag{8.220b}$$

Elliptical

$$[K]\{\Phi\}=\{\psi\} \tag{8.220c}$$

The basic differences would be the specific forms of shape functions to obtain matrices $[M],[C],[K]$ and vector $\{\psi\}$ depending upon the dimension of the problem. As the physical space would involve two or three independent variables in a multi-dimensional space, the integration procedures would also become quite cumbersome. Under such circumstances, numerical integration techniques would be necessary.

8.10.1 Shape functions

For illustrative purposes, the simplest interpolation function for shape functions in the FEM may be given in the following form for 1D, 2D and 3D spaces for a single variable φ as

1D:

$$\varphi=a+bx \tag{8.221}$$

2D:

$$\varphi=a+bx+cy \tag{8.222}$$

3D:

$$\varphi=a+bx+cy+dz \tag{8.223}$$

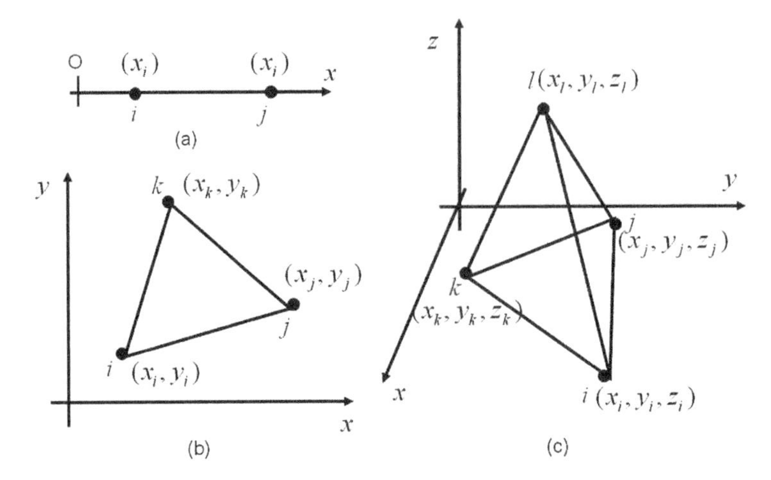

Figure 8.11 Illustration of linear elements in (a) 1D, (b) 2D and (c) 3D.

Based on the fundamental concept of deriving shape functions described in the previous sections, one can easily show that variable φ can be expressed in terms of its values at the nodes of a typical element as given below (Figure 8.11):

1D (linear element)

$$\varphi = N_i\varphi_i + N_j\varphi_j \tag{8.224}$$

where

$$N_i = \frac{x_j - x}{x_j - x_i}, \quad N_j = \frac{x - x_i}{x_j - x_i}$$

2D-triangular element

$$\varphi = N_i\varphi_i + N_j\varphi_j + N_k\varphi_k \tag{8.225}$$

where

$$N_i = E_{11} + E_{21}x + E_{31}y, \quad N_j = E_{12} + E_{22}x + E_{32}y, \quad N_k = E_{13} + E_{23}x + E_{33}y$$

$$E_{11} = \frac{1}{\Delta}\left(x_j y_k - x_k y_j\right), \quad E_{12} = \frac{1}{\Delta}\left(x_i y_k - x_k y_i\right), \quad E_{13} = \frac{1}{\Delta}\left(x_j y_i - x_i y_j\right)$$

$$E_{21} = \frac{1}{\Delta}\left(y_j - y_k\right), \quad E_{22} = \frac{1}{\Delta}\left(y_k - y_i\right), \quad E_{23} = \frac{1}{\Delta}\left(y_i - y_j\right)$$

$$E_{31} = \frac{1}{\Delta}\left(x_k - x_j\right), \quad E_{32} = \frac{1}{\Delta}\left(x_i - x_k\right), \quad E_{33} = \frac{1}{\Delta}\left(x_j - x_i\right)$$

$$\Delta = \left(x_j y_k - x_k y_j\right) + \left(x_k y_i - x_i y_k\right) + \left(x_i y_j - x_j y_i\right)$$

3D-tetrahedron

$$\varphi = N_i\varphi_i + N_j\varphi_j + N_k\varphi_k + N_l\varphi_l \tag{8.226}$$

where

$$N_i = a_i + b_i x + c_i y + d_i z,\quad N_j = a_j + b_j x + c_j y + d_j z,\quad N_k = a_k + b_k x + c_k y + d_k z,$$
$$N_l = a_l + b_l x + c_l y + d_l z$$

$$a_i = \frac{1}{J}\begin{vmatrix} x_j & y_j & z_j \\ x_k & y_k & z_k \\ x_l & y_l & z_l \end{vmatrix},\quad b_i = -\frac{1}{J}\begin{vmatrix} 1 & y_j & z_j \\ 1 & y_k & z_k \\ 1 & y_l & z_l \end{vmatrix},\quad c_i = \frac{1}{J}\begin{vmatrix} 1 & x_j & z_j \\ 1 & x_k & z_k \\ 1 & x_l & z_l \end{vmatrix},$$
$$d_i = -\frac{1}{J}\begin{vmatrix} 1 & x_j & y_j \\ 1 & x_k & y_k \\ 1 & x_l & y_l \end{vmatrix}$$

$$a_j = -\frac{1}{J}\begin{vmatrix} x_i & y_i & z_i \\ x_k & y_k & z_k \\ x_l & y_l & z_l \end{vmatrix},\quad b_j = \frac{1}{J}\begin{vmatrix} 1 & y_i & z_i \\ 1 & y_k & z_k \\ 1 & y_l & z_l \end{vmatrix},\quad c_j = -\frac{1}{J}\begin{vmatrix} 1 & x_i & z_i \\ 1 & x_k & z_k \\ 1 & x_l & z_l \end{vmatrix},$$
$$d_j = \frac{1}{J}\begin{vmatrix} 1 & x_i & y_i \\ 1 & x_k & y_k \\ 1 & x_l & y_l \end{vmatrix}$$

$$a_k = \frac{1}{J}\begin{vmatrix} x_i & y_i & z_i \\ x_j & y_j & z_j \\ x_l & y_l & z_l \end{vmatrix},\quad b_k = -\frac{1}{J}\begin{vmatrix} 1 & y_i & z_i \\ 1 & y_j & z_j \\ 1 & y_l & z_l \end{vmatrix},\quad c_k = \frac{1}{J}\begin{vmatrix} 1 & x_i & z_i \\ 1 & x_j & z_j \\ 1 & x_l & z_l \end{vmatrix},$$
$$d_k = -\frac{1}{J}\begin{vmatrix} 1 & x_i & y_i \\ 1 & x_j & y_j \\ 1 & x_l & y_l \end{vmatrix}$$

$$a_l = -\frac{1}{J}\begin{vmatrix} x_i & y_i & z_i \\ x_j & y_j & z_j \\ x_k & y_k & z_k \end{vmatrix},\quad b_l = \frac{1}{J}\begin{vmatrix} 1 & y_i & z_i \\ 1 & y_j & z_j \\ 1 & y_k & z_k \end{vmatrix},\quad c_l = -\frac{1}{J}\begin{vmatrix} 1 & x_i & z_i \\ 1 & x_j & z_j \\ 1 & x_k & z_k \end{vmatrix},$$
$$d_l = \frac{1}{J}\begin{vmatrix} 1 & x_i & y_i \\ 1 & x_j & y_j \\ 1 & x_k & y_k \end{vmatrix}$$

$$J = \begin{vmatrix} 1 & x_i & y_i & z_i \\ 1 & x_j & y_j & z_j \\ 1 & x_k & y_k & z_k \\ 1 & x_l & y_l & z_l \end{vmatrix}$$

It should be noted that the summation of shape functions of a given element must have the value of 1 at all times irrespective of the dimension of the physical domain. Furthermore, the shape function at its own node must have a value of 1 and 0 at all other nodes.

To increase the accuracy of approximation as well as node number reduction of discretization, the higher-order functions are used and the local (natural) coordinate systems $(0,\xi)$, $(0,\xi,\eta)$ and (O,ξ,η,ς) are preferred over the global coordinate systems in 1D, 2D and 3D, respectively (Figures 8.12–8.14). For this purpose, an iso-parametric element concept is generally introduced and the shape functions for variable/variables and coordinates of nodes are chosen to be the same. The iso-parametric shape functions of linear type utilizing a local (natural) coordinate system are given for 1D, 2D and 3D situations in the following forms.

1D linear element

$$\varphi = N_i\varphi_i + N_j\varphi_j \tag{8.227}$$

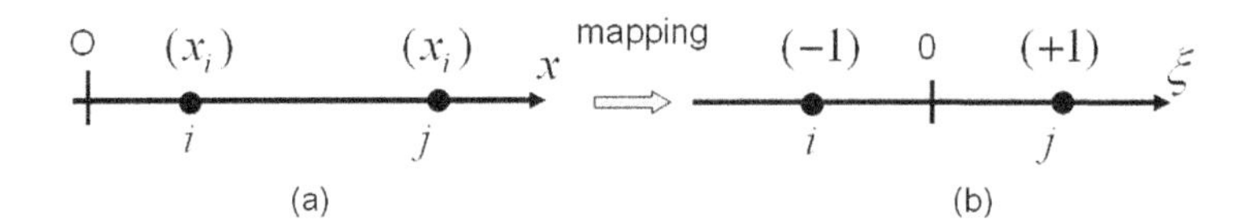

Figure 8.12 Illustration of an iso-parametric element in 1D and mapping. (a) Global coordinates and (b) local (natural) coordinates.

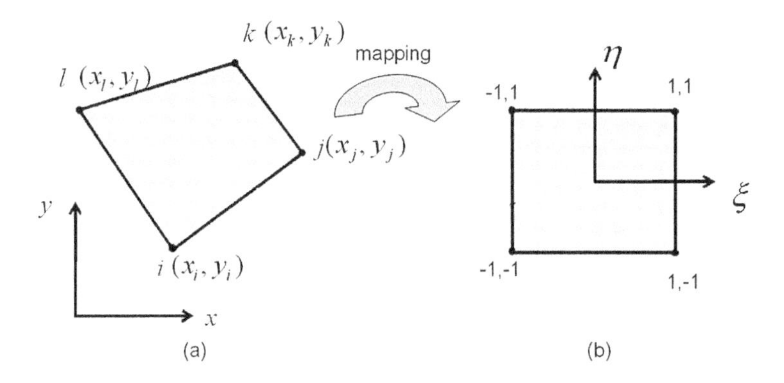

Figure 8.13 Illustration of an iso-parametric element in 2D and mapping. (a) Global coordinates and (b) local (natural) coordinates.

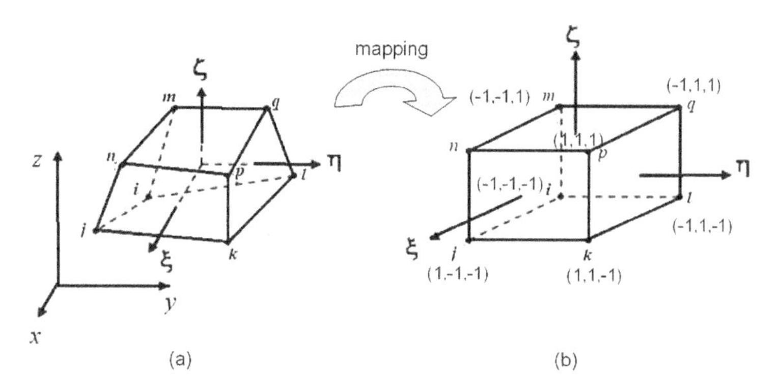

Figure 8.14 Illustration of an iso-parametric element in 3D and mapping. (a) Global coordinates and (b) local (natural) coordinates.

where

$$N_i = \frac{1}{2}(1-\xi); \quad N_j = \frac{1}{2}(1+\xi)$$

2D quadrilateral element

$$\varphi = N_i\varphi_i + N_j\varphi_j + N_k\varphi_k + N_l\varphi_l \tag{8.228}$$

where

$$N_i^e = (1-\xi)(1-\eta)/4, \quad N_j^e = (1+\xi)(1-\eta)/4$$
$$N_k^e = (1+\xi)(1+\eta)/4, \quad N_l^e = (1-\xi)(1+\eta)/4$$

3D brick (hexahedral) element

$$\varphi = N_i\varphi_i + N_j\varphi_j + N_k\varphi_k + N_l\varphi_l + N_m\varphi_m + N_n\varphi_n + N_p\varphi_p + N_q\varphi_q \tag{8.229}$$

where

$$N_i = (1-\xi)(1-\eta)(1-\zeta)/8, \quad N_j = (1+\xi)(1-\eta)(1-\zeta)/8,$$
$$N_k = (1+\xi)(1+\eta)(1-\zeta)/8, \quad N_l = (1-\xi)(1+\eta)(1-\zeta)/8$$

$$N_m = (1-\xi)(1-\eta)(1+\zeta)/8, \quad N_n = (1+\xi)(1-\eta)(1+\zeta)/8,$$
$$N_p = (1+\xi)(1+\eta)(1+\zeta)/8, \quad N_q = (1-\xi)(1+\eta)(1+\zeta)/8$$

8.10.2 Numerical integration

The shape functions of iso-parametric elements would be integrated through numerical integration techniques (e.g. Zienkiewicz 1977; Reddy 2006; Segerlind 1976). For this purpose, the Gauss–Legendre numerical integration technique is generally used.

The numerical integrations in 1D, 2D and 3D in local (natural) coordinate systems are carried out according to the following procedures:

1D

$$I_{1D} = \int_{-1}^{1} f(\xi)d\xi = \sum_{i=1}^{N} f(\xi_i)w_i \tag{8.230}$$

2D

$$I_{2D} = \int_{-1}^{1}\int_{-1}^{1} f(\xi,\eta)d\xi\, d\eta = \sum_{i=1}^{N}\sum_{j=1}^{N} f(\xi_i,\eta_j)w_i w_j \tag{8.231}$$

3D

$$I_{3D} = \int_{-1}^{1}\int_{-1}^{1}\int_{-1}^{1} f(\xi,\eta)d\xi\, d\eta = \sum_{i=1}^{N}\sum_{j=1}^{N}\sum_{k=1}^{N} f(\xi_i,\eta_j,\zeta_k)w_i w_j w_k \tag{8.232}$$

where w_i is the value of weight at a given Gauss integration point and N is the number of integration Gauss points.

8.11 Special numerical methods for media having discontinuities

The existence of discontinuities in rock mass has special importance on the stability of rock engineering structures, directional seepage, diffusion or heat transport and its treatment in any analysis requires a special attention. Various types of FEMs with contact, joint or interface elements, discrete finite element method (DFEM), discrete element method (DEM), displacement discontinuity analysis (DDA) and displacement discontinuity method (DDM) are developed so far. Besides the solution of equation of motion, they are used for seepage, heat transport or diffusion problems as well as coupled processes. Based on a recent review on these methods by Kawamoto and Aydan (1999), the fundamental features of the available methods are described in this section.

8.11.1 No-tension finite element method

The no-tension FEM is proposed by Valliappan in 1968 (Zienkiewicz et al. 1968). The essence of this method lies with the assumption of no tensile strength for rock mass in view of the existence of geological discontinuities. In the finite element implementation, the tensile strength of rock mass is assumed to be nil. It behaves elastically when all principal stresses are compressive. However, some yield criteria can be utilized even when all principal stresses are compressive (e.g. Aydan and Kawamoto 2001). The excess stresses are redistributed to the elastically behaving rock mass using a similar procedure adopted in the FEM with the consideration of the elasto-plastic behavior (Figure 8.15).

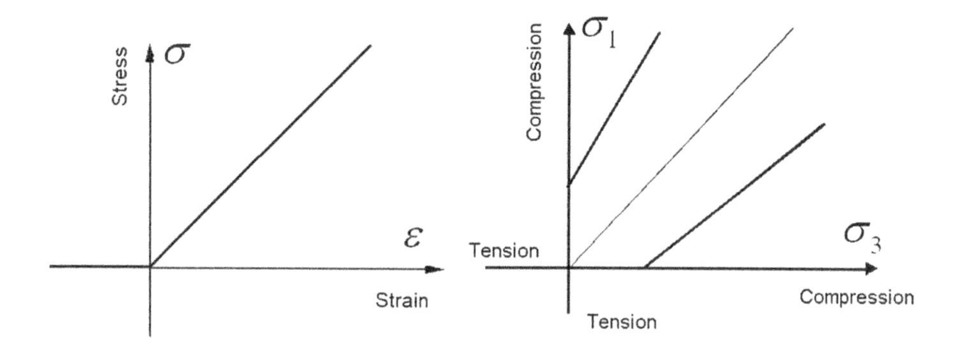

Figure 8.15 An illustration of the no-tension model.

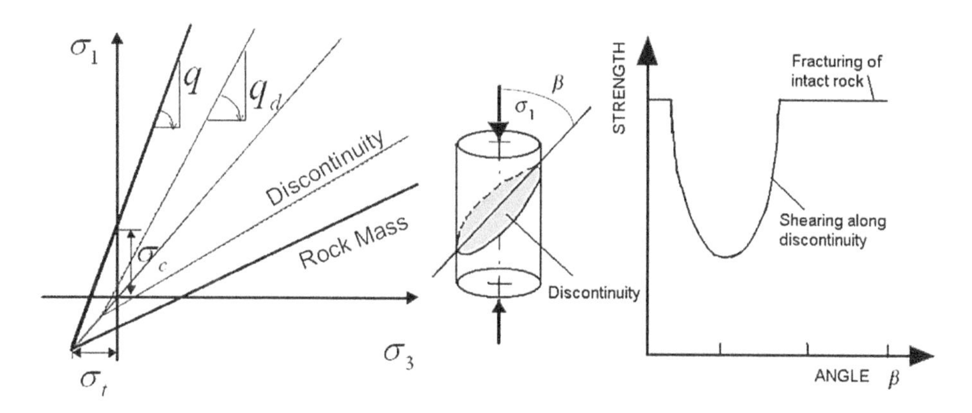

Figure 8.16 An illustration of the pseudo-discontinuum model.

8.11.2 Pseudo discontinuum finite element method

This method was first proposed by Baudendistel et al. in 1970. In this method, the effect of discontinuities in the FEM is considered through the introduction of the directional yield criterion in the elasto-plastic behavior (Figure 8.16). Its effect on the deformation characteristics of the rock mass is not considered. If there is any yielding in a given element, the excess stress is computed and the iteration scheme for the elasto-plastic behavior is generally implemented. If there is more than one discontinuity set, the excess stresses are computed for the discontinuity set, which yields the largest value.

8.11.3 Smeared crack element

The smeared crack element method within the FEM was initially proposed by Rashid (1968) and adopted by Pietruszczak and Mroz (1981) in media having weakness planes or developing fracture planes. This method evaluates the equivalent stiffness matrix of the element, and it allows the directional plastic yielding within the element (Figure 8.17).

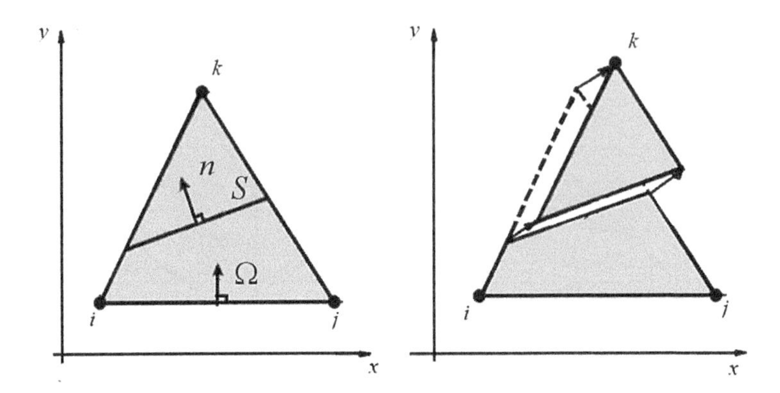

Figure 8.17 An illustration of the smeared crack element.

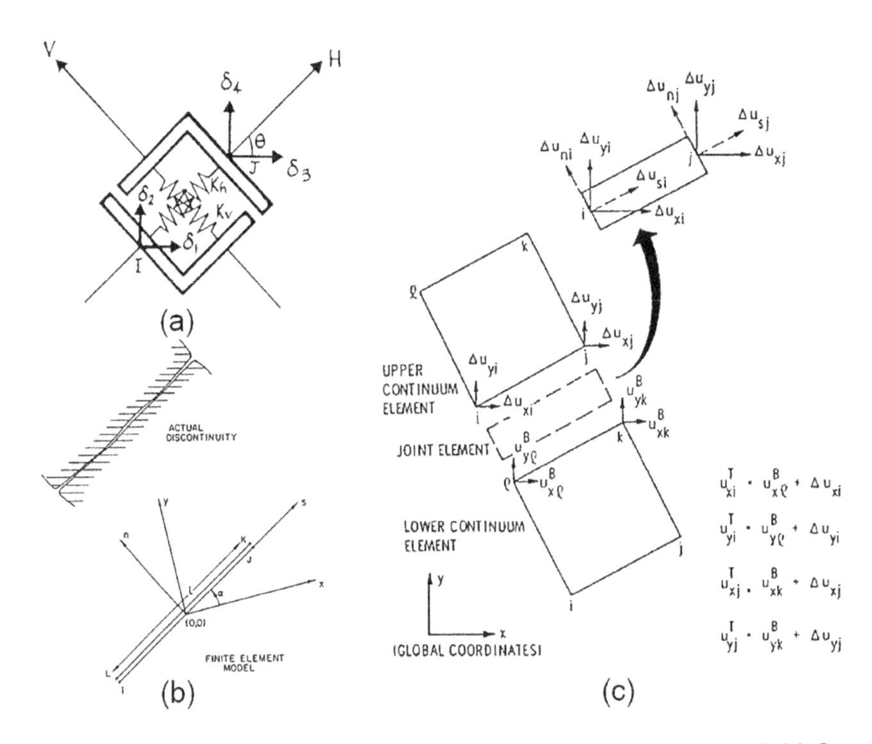

Figure 8.18 Illustrations of various elements in the finite element method. (a) Contact element. (b) Goodman's joint element. (c) Ghaboussi's interface element.

8.11.4 Finite element method with joint or interface element (FEM-J)

Finite element techniques using contact, joint or interface elements have been developed for representing discontinuities among blocks in rock masses (Figure 8.18). The simplest approach for representing joints is the contact element, which was originally developed for bond problems between steel bars and concrete. The contact element is a two-node element having normal and shear stiffnesses. Goodman et al. (1968) proposed a four-node joint element for joints. This model is simply a four-node version of the contact element of Ngo and Scordelis (1967) and it has the following characteristics. In a two-dimensional domain, joints are assumed to be tabular with zero thickness. They have no resistance to the net tensile forces in the normal direction, but they have high resistance against compression. Joint elements may deform under normal pressure, especially if there are crushable asperities. The shear strength is presented by a bilinear Mohr–Coulomb envelope. The joint elements are designed to be compatible with solid elements. Ghaboussi et al. (1973) proposed a four-node interface element for joints. This model is a further improvement of the joint element by assigning a finite thickness to joints.

8.11.5 Discrete finite element method (DFEM)

This model is recently used to model block systems by Aydan and Mamaghani (e.g. Aydan et al. 1996; Tokashiki et al. 1997) by assigning a finite thickness to contact element and employing an updated Lagrangian scheme to deal with large block movements. The contact element can easily deal with sliding and separation movements. The fundamental principles of the DFEM are illustrated in Figure 8.19.

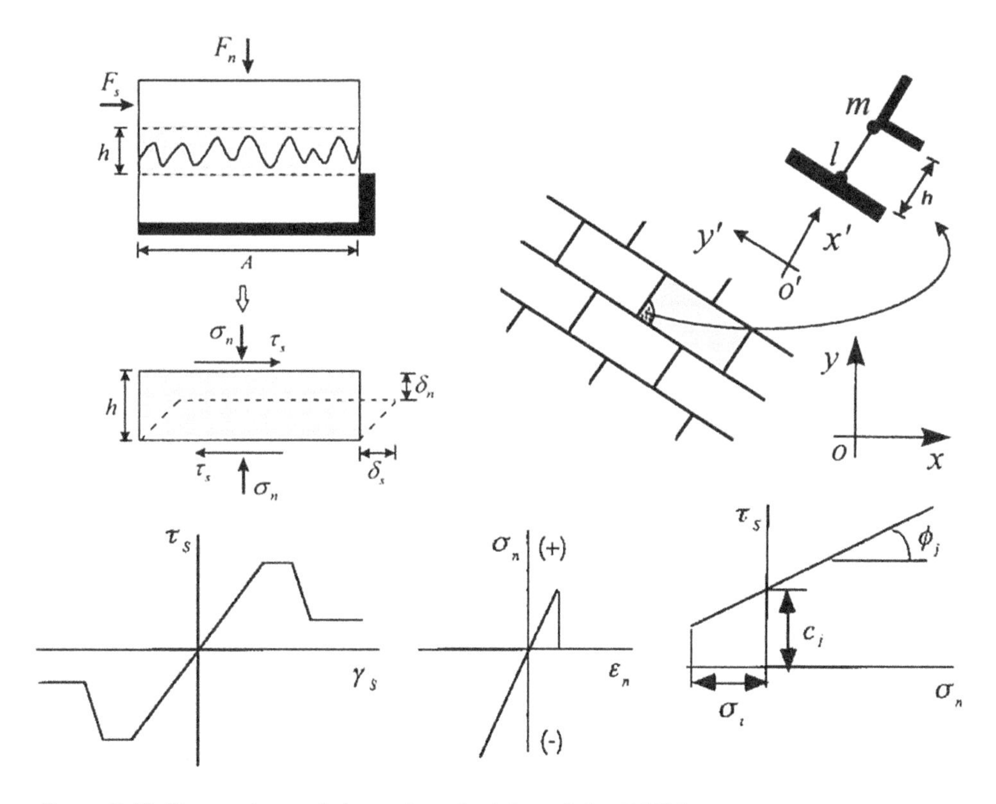

Figure 8.19 Illustrations of the main principles of the DFEM.

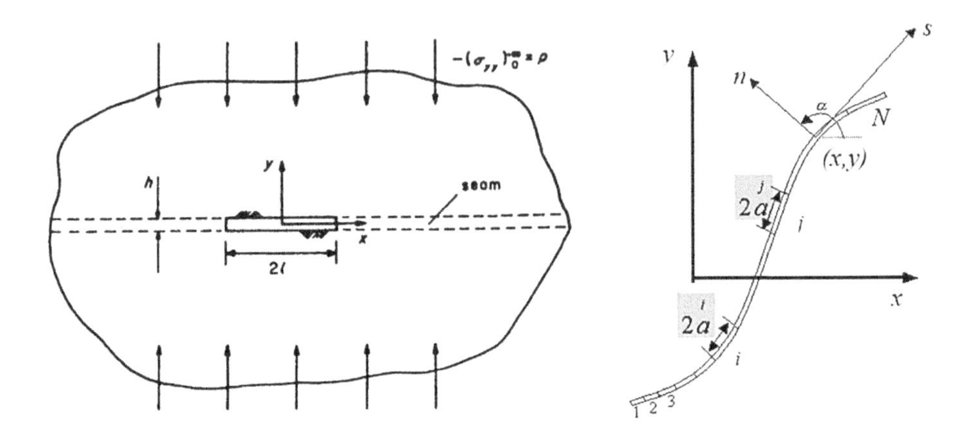

Figure 8.20 An illustration of the principle of the displacement discontinuity method.

8.11.6 Displacement discontinuity method (DDM)

This technique is generally used together with the BEM. The discontinuities are modeled as a finite length segment in an elastic medium with a relative displacement. In other words, the discontinuities are treated as internal boundaries with prescribed displacements. As an alternative approach to the technique of Crouch and Starfield (1983), Crotty and Wardle (1985) used interface elements to model discontinuities and the domain is discretized into several subdomains as shown in Figure 8.20. This technique has been used to simulate large-scale discontinuities such as faults.

8.11.7 Discrete element method (DEM)

The DEM (rigid block models) for jointed rocks was developed by Cundall in 1971. In Cundall's model, problems are treated as dynamic ones from the very beginning of formulation. It is assumed that the contact force is produced by the action of springs, which are applied whenever a corner penetrates an edge. Normal and shear stiffness were introduced between the respective forces and displacements in his original model. Furthermore, to account for slippage and separation of block contacts, plastic yielding is introduced. For the simplicity of calculation of contact forces due to the overlapping of the block, the blocks are assumed not to change their original configurations. To solve the equations of the whole domain, the equilibrium equations of blocks are never assembled into a large equation system. They are solved through a step-by-step procedure, which he called the marching scheme (Figure 8.21). This solution technique has two main merits:

1. Storage memory of computers can be small (note that computer technology was not so advanced during the late 1960s); therefore, it could run on a microcomputer.
2. The separation and slippage of contacts can be easily considered as the global matrix representing block connectivity is never assembled. If a large assembled matrix is used, such matrix will result in zero or very nearly zero diagonals, which subsequently cause a singularity problem or ill-conditioning of the matrix system.

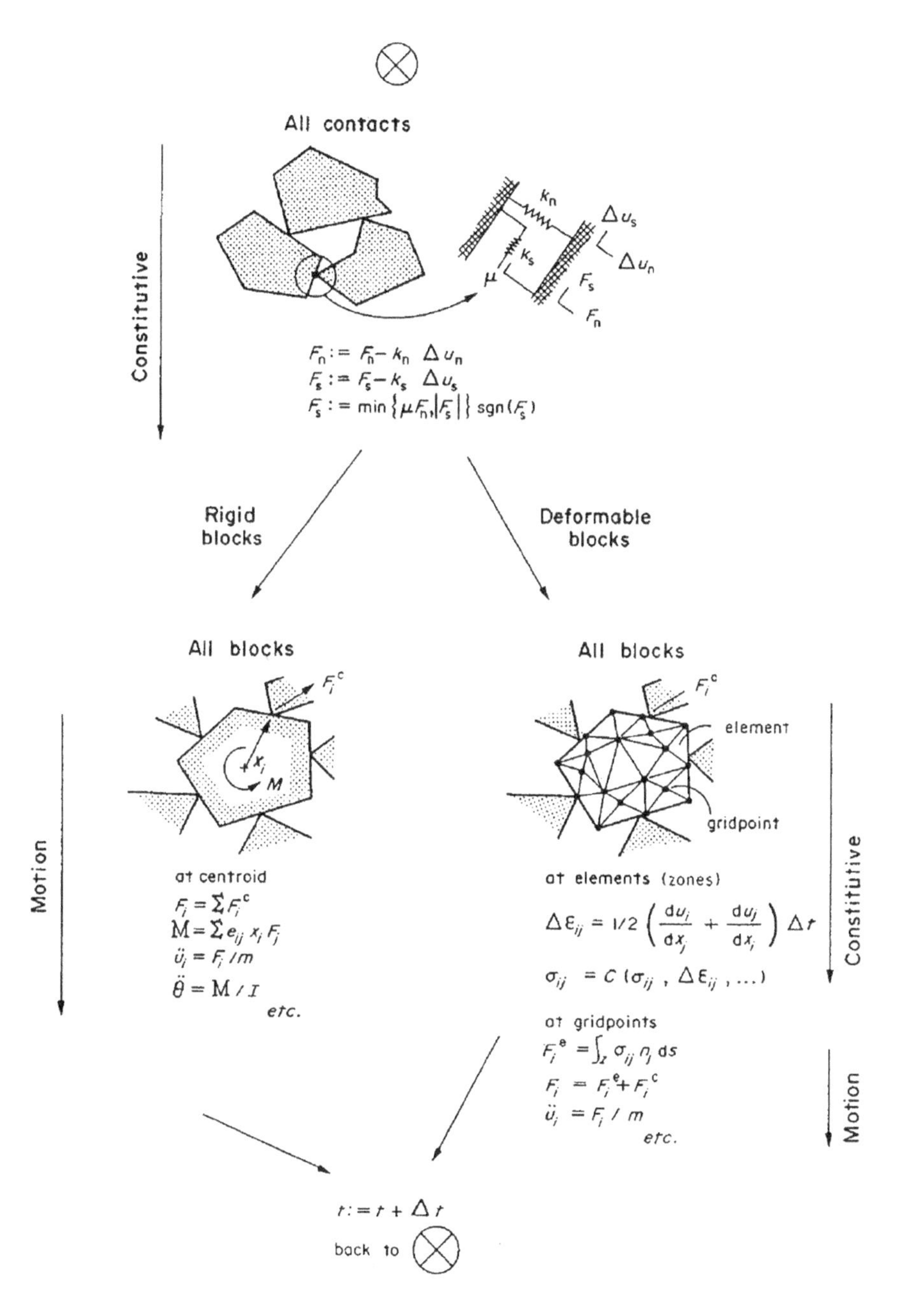

Figure 8.21 An illustration of the solution scheme of the discrete element method.

As the governing equation is of hyperbolic type, the system could not become stabilized even for static cases unless a damping is introduced into the equation system. In recent years, the original model was modified by considering the deformability of intact blocks and their elasto-plastic behavior.

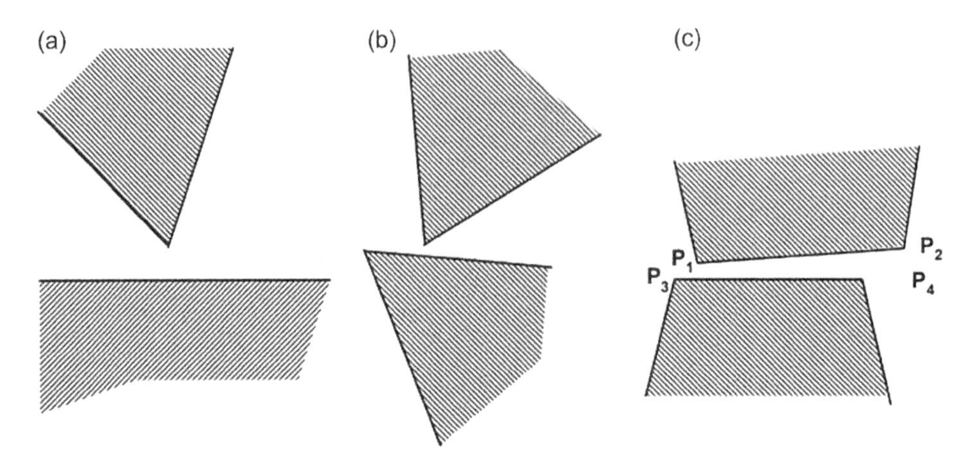

Figure 8.22 Contact conditions considered in the DDA. (From Shi 1993.)

8.11.8 Discontinuous deformation analysis method (DDA)

Shi (1988) proposed the DDA method. Intact blocks were assumed to be deformable and are subjected to constant strain and stress due to the order of the interpolation functions used for the displacement field of the blocks, which is the simplest form of interpolation function used in the FEM (see Section 8.10). In the original model, the inertia term was neglected so that the damping becomes unnecessary. For dynamic problems, although the damping is not originally introduced into the system, the large time steps used in the numerical integration in time-domain results in artificial damping. Although the fundamental concept is not very different from Cundall's model, the main difference results from the solution procedure adopted in both methods. In other words, the equation system of blocks and their contacts is assembled into a global equation system in Shi's approach. Recently, Ohnishi et al. (1995) introduced an elasto-plastic constitutive law for intact blocks and gave an application of this method to rock engineering structures. The original DDA has been recently modified to count actual dynamic loading and the various block contact conditions. It seems that it is adopting the procedures used in the DEM (Figure 8.22).

References

Aydan, Ö. 2017. *Time Dependency in Rock Mechanics and Rock Engineering*, London, CRC Press, Taylor & Francis Group, 241 p.

Aydan, Ö. 2018. *Rock Reinforcement and Rock Support*, London, CRC Press, Taylor & Francis Group, 486 p.

Aydan, Ö. and T. Kawamoto 2001. The stability assessment of a large underground opening at great depth. *17th International Mining Congress and Exhibition of Turkey*, IMCET, Ankara, 1, 277–288.

Aydan, Ö., A. Ersen, Y. Ichikawa and T. Kawamoto 1985. Temperature and thermal stress distributions in mass concrete shaft and tunnel linings during the hydration of concrete (in Turkish). *The 9th Mining Science and Technology Congress of Turkey*, Ankara, 355–368.

Aydan, Ö., R. Güloğlu and T. Kawamoto 1986. Temperature distributions and thermal stresses in tunnel linings due to hydration of cement (in Japanese). *Tunnels and Underground*, 17(2), 29–36.
Aydan, Ö., I.H.P. Mamaghani and T. Kawamoto 1996. Application of discrete finite element method (DFEM) to rock engineering. *North American Rock Mechanics Symposium*, Montreal, 2, 2039–2046.
Aydan, Ö., F. Uehara and T. Kawamoto 2012. Numerical study of the long-term performance of an underground powerhouse subjected to varying initial stress states, cyclic water heads, and temperature variations. *International Journal of Geomechanics*, ASCE, 12(1), 14–26.
Baudendistel, M., H. Malina and L. Müller 1970. Einfluss von Discontinuitaten auf die Spannungen und Deformationen in der Umgebung einer Tunnelröhre. *Rock Mechanics*, 2, 17–40.
Crotty, J.M. and L.J. Wardle 1985. Boundary integral analysis of piece-wise homogenous media with structural discontinuities. *International Journal of Rock Mechanics and Mining Sciences*, 22(6), 419–427.
Crouch, S.L. and A.M. Starfield 1983. *Boundary Element Methods in Solid Mechanics*, London, Allen & Unwin.
Cundall, P.A. 1971. The measurement and analysis of acceleration in rock slopes. Ph.D. Thesis, University of London (Imperial College).
Ghaboussi, J., E.L. Wilson and J. Isenberg 1973. Finite element for rock joints and interfaces. *Journal of the Soil Mechanics and Foundations Division, ASCE*, 99(10), 833–848.
Goodman, R.E., R. Taylor and T.L. Brekke 1968. A model for the mechanics of jointed rock. *Journal of the Soil Mechanics and Foundations Division, ASCE*, 94(3), 637–659.
Kawamoto, T. and Ö. Aydan 1999. A review of numerical analysis of tunnels in discontinuous rock masses. *International Journal of Numerical and Analytical Methods in Geomechanics*, 23, 1377–1391.
Kreyszig, E. 1983. *Advanced Engineering Mathematics*, New York, John Wiley & Sons.
Ngo, D. and A.C. Scordelis 1967. Finite element analysis of reinforced concrete beams. *Journal of American Concrete Institute*, 64(3), 152–163.
Ohnishi, Y., T. Sasaki and M. Tanaka 1995. Modification of the DDA for elasto-plastic analysis with illustrative generic problems. *35th US Rock Mechanics Symposium*, Lake Tahoe, 45–50.
Owen, D.R.J. and E. Hinton 1980. *Finite Element in Plasticity: Theory and Practice*, Swansea, Pineridge Press Ltd.
Pietruszczak, S. and Z. Mroz 1981. Finite element analysis of deformation of strain-softening materials. *International Journal for Numerical Methods in Engineering*, 17, 327–334.
Rashid, Y.R. 1968. Ultimate strength analysis of prestresses concrete pressure vessels. *Nuclear Engineering and Design*, 7, 334–344.
Reddy, J.N. 2006. *An Introduction to the Finite Element Method*. 3rd ed., New York, McGraw-Hill.
Segerlind, L.J. 1976. *Applied Finite Element Analysis*, New York, Wiley.
Shi, G.H. 1988. Discontinuous deformation analysis: a new numerical model for the statics and dynamics of block systems. Ph.D. Thesis, Department of Civil Engineering, University of California, Berkeley, 378 p.
Shi, G.H. 1993. *Block System Modeling by Discontinuous Deformation Analysis*, Southampton, and Boston, MA, Computational Mechanics Publications.
Speigel, M.R. 1994. *Theory and Problems of Calculus of Finite Differences and Difference Equations*, Schaum's outline series, New York, McGraw-Hill, 259 p.
Tokashiki, N., Ö. Aydan, Y. Shimizu and T. Kawamoto 1997. The assessment of the stability of a very old tunnel by discrete finite element method (DFEM). *Numerical Methods in Geomechanics*, NUMOG VI, Montreal, 495–500.
Verruijt, A. 1982. *Groundwater Flow*. 2nd ed., London, Macmillan Press Ltd, 144 p.
Zienkiewicz, O.C. 1977. *The Finite Element Method*. 3rd ed., New York, McGraw-Hill.
Zienkiewicz, O.C., S. Valliappan and I.P. King 1968. Stress analysis of rock as a 'no-tension' materials. *Geotecnique*, 18(1), 56–66.

Chapter 9

Applications of approximate methods in geo-engineering problems

In this chapter, several applications of approximate methods to some typical structures and problems encountered in geo-engineering are described and brief discussions are given. The examples involve the phenomena of groundwater seepage, heat flow, static, time-dependent, and dynamic responses of slopes, foundations and underground structures excavated in rock mass. Furthermore, the examples of applications that are concerned with the effects of degradation and discontinuities on the response and stability of rock engineering structures and the effect of reinforcement for stabilization are also presented and briefly discussed.

9.1 Applications in continuum

9.1.1 The stress state of earth and earth's crust

In-situ stresses are of great importance for the design and stability assessment of rock engineering structures as well as for understanding and predicting earthquakes. The stress state of the earth and its crust has been investigated by several researchers in the past. Jeffreys and Bullen (1940) considered that the stress state of the earth was hydrostatic and they calculated the pressure of the earth by considering the variation of density and gravitational acceleration. Anderson and Hart (1976) modified this approach by considering recent findings. Nadai (1950) also derived some formula for radial and tangential stresses by assuming that the density of the earth is constant and the gravitational acceleration varies linearly with depth.

Aydan (1995) analyzed the stress state of the earth by modeling it as a spherical body consisting of multiple co-centric layers exhibiting a thermo-elasto-plastic behavior under pure gravitational acceleration using the finite element method (Figure 9.1). The computational results of Aydan (1995) are briefly introduced in this section. The co-centric layers constituting the earth is isotropic within each layer while the overall structure is anisotropic. Considering the distribution of material properties (elastic modulus, Poisson's ratio and density) and the variation of gravity through the earth (Fowler 1990), the rock mass is assumed to behave thermo-elasto-plastically obeying the yield criterion of Aydan (see Section 5.3.3).

Figure 9.2 shows the distribution of radial and tangential stresses in the earth. From his study, the following conclusions were drawn for the stress state of the earth:

a. If the spherical structure of the earth is considered, it is possible to explain why the horizontal stress is larger than the vertical stress near the ground surface.

In other words, the large horizontal stress is due to the gravity not due to presumed tectonic forces resulting from unknown sources.

b. For the spherically symmetric earth, the tangential stress (lateral stress) is always the maximum principal stress and the radial stress (vertical stress) is the minimum principal stress irrespective of the mechanical behavior of rocks.

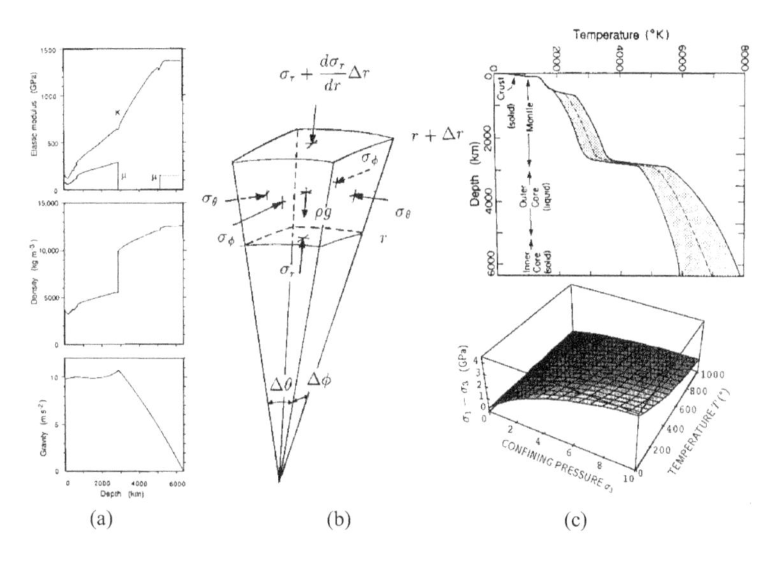

Figure 9.1 Illustration of the spherical model of the earth and material properties (from Aydan 1993, Aydan and Paşamehmetoğlu 1994 and Fowler 1990). (a) Variation of material properties and gravitational acceleration with depth; (b) control-volume for a spherical earth; (c) temperature distribution and thermo-plastic yield surface.

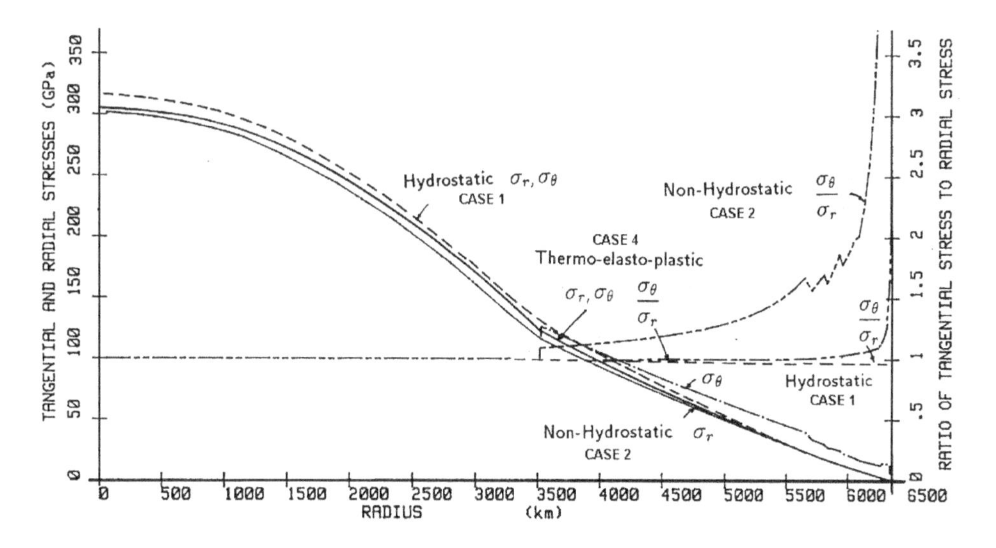

Figure 9.2 The stress state of the earth (from Aydan 1995).

c. The crust and the mantle are in a plastic state, which may have some important implications in rock mechanics and rock engineering as well as in geo-science. In other words, the earth is not an elastic body as presumed in many different studies.

Aydan's approach provided very valuable information and it is a first approximation to the stress state of the earth. Although the gravitational pull is the governing element in shaping the stress state of the earth, slight variations in the computed stress state from the gravitational model are caused by rotation around its axis as well as the sun and mantle convection due to the non-uniformity in the thermal field resulting from the subduction of the cooler plates into the hot mantle. As the spherical symmetry condition does not strictly hold for the earth, better estimations may be done using actual geometry and distributions of materials constituting the earth. Nevertheless, it should be noted that our knowledge of the constitutive parameters of the earth's constituting materials is still insufficient.

Figure 9.3 compares the stress measurements by the acoustic emissions (AE) method together with available in-situ stress measurements worldwide (Aydan 1995) and some empirical relations for horizontal stress ratio over the overburden pressure developed by Hoek and Brown (1980) and Aydan and Paşamehmetoğlu (1994) as well as those from the thermo-elasto-plastic finite element analyses by Aydan (1995). It is interesting to note that the stress measurements by the AE method for Turkey are quite consistent with measurements using other methods and empirical formulas (Aydan and Kawamoto 1997).

9.1.2 Evaluation of the tunnel face effect

Advancing tunnels utilizing support systems consisting of rockbolts, shotcrete, steel ribs and concrete lining are three-dimensional complex structures and it is a dynamic process. However, tunnels are often modeled as a one-dimensional axisymmetric

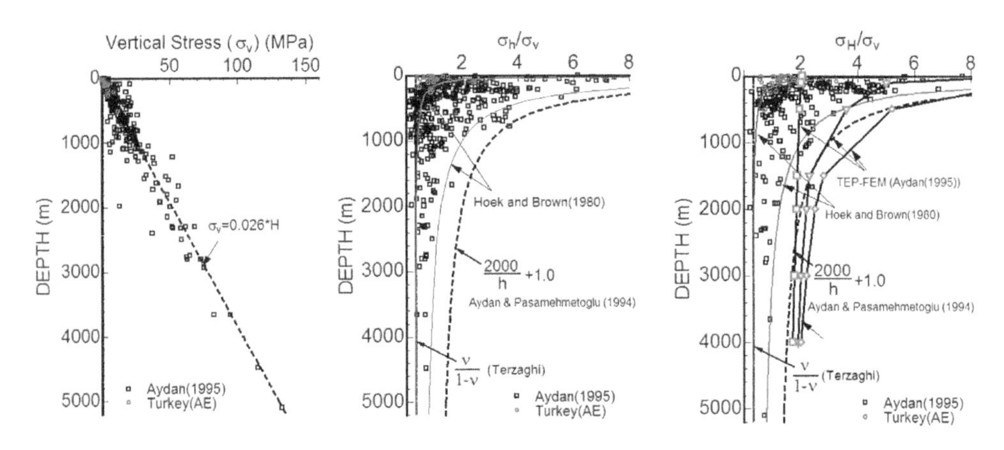

Figure 9.3 Comparison of in-situ stress measurements with various methods (from Aydan et al. 2016).

structure subjected to a hydrostatic initial stress state as a static problem. The effect of tunnel face advance on the response and design of support systems is often replaced through an excavation stress release factor determined from pseudo–three-dimensional (axisymmetric) or pure three-dimensional analyses as given below:

$$f = \frac{e^{-bx/D}}{1/B + e^{-bx/D}} \tag{9.1}$$

where x is the distance from the tunnel face and the values for coefficients B and b suggested by Aydan (2011) are 2.33 and 1.7, respectively.

Figure 9.4a illustrates an unsupported circular tunnel subjected to an axisymmetric initial stress state. The variation of displacement and stresses along the tunnel axis was computed using the elastic axisymmetric finite element method, which is regarded as a pseudo-three-dimensional analysis. The radial displacement at the tunnel wall is normalized by the largest displacement and is shown in Figure 9.4b. As seen from the figure, the radial displacement takes place in front of the tunnel face. The displacement is about 28%–30% of the final displacement. Its variation terminates when the face advance is about +2D. Almost 80% of the total displacement takes place when the

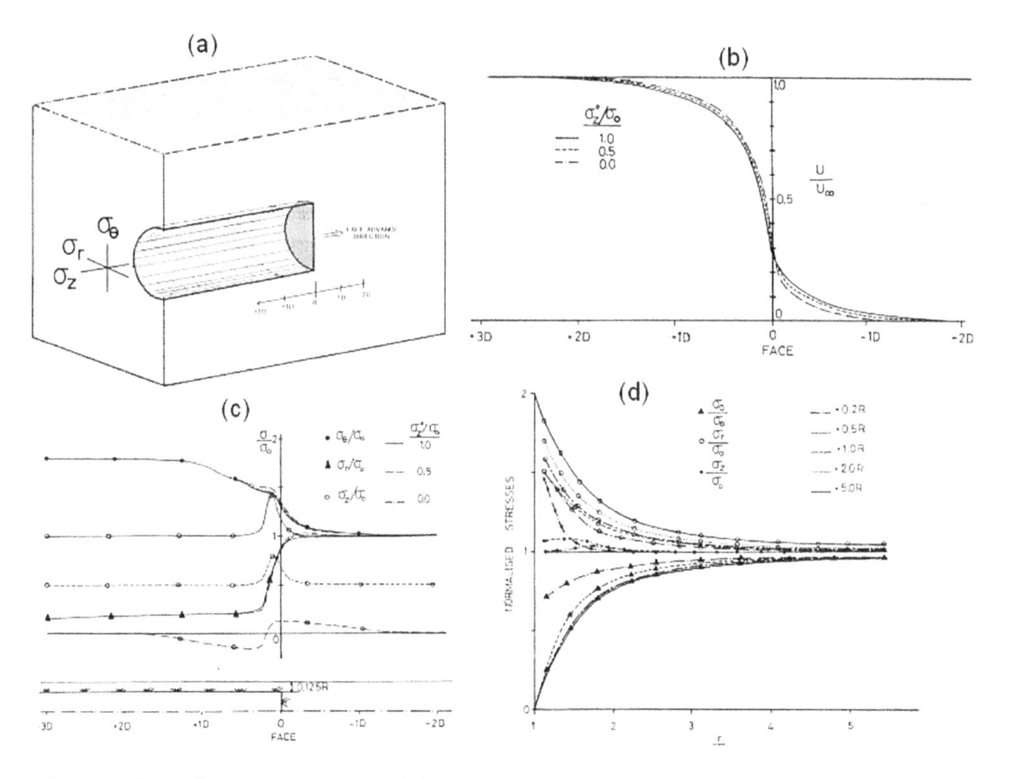

Figure 9.4 (a) Computational model for elastic finite element analysis; (b) normalized radial displacement of the tunnel surface; (c) normalized stress components along the tunnel axis at a distance of 0.125R; (d) the variation of stresses along the *r*-direction at various distances from the tunnel face.

tunnel face is about +1D. The effect of the initial axial stress on the radial displacement is almost negligible.

Figure 9.4c shows the variation of radial, tangential and axial stress around the tunnel at a depth of 0.125R. As noted from the figure, the tangential stress gradually increases as the distance increases from the tunnel face. The effect of the initial axial stress on the tangential stress is almost negligible. The radial stress rapidly decreases in the close vicinity of the tunnel face and the effect of the initial axial stress on the radial stress is also negligible. The most interesting variation is associated with the axial stress distribution. The axial stress increases as the tunnel face approaches, and then it gradually decreases to its initial value as the face effect disappears. This variation is limited to a length of 1R (0.5D) from the tunnel face. It is also interesting to note that if the initial axial stress is nil, even some tensile axial stresses may occur in the close vicinity of the tunnel face.

Figure 9.4d shows the stress distributions along the r-axis of the tunnel at various distances from the face when the initial axial stress is equal to initial radial and tangential stresses. As noted from the figure, the maximum tangential stress is 1.5 times the initial hydrostatic stress and it becomes twice as the distance from the tunnel face is +5R, which is almost equal to theoretical estimations for tunnels subjected to a hydrostatic initial stress state. The stress state near the tunnel face is also close to that of the spherical opening subjected to a hydrostatic stress state. The stress state seems to change from the spherical state to the cylindrical state (Aydan 2011). It should be noted that it would be almost impossible to simulate exactly the same displacement and stress changes of 3D analyses in the vicinity of tunnels by 2D simulations using the stress-release approach irrespective of the constitutive law of surrounding rock as a function of distance from the tunnel (Aydan et al. 1988; Aydan and Geniş 2010).

9.1.3 Three-dimensional simulation of the excavation of a railway tunnel supported with forepoles, rockbolts, shotcrete and steel ribs

Aydan et al. (1988) carried out a three-dimensional analysis to simulate the excavation of a shallow railway tunnel. These analyses were carried out using the super-computer of Nagoya University, which had a storage capacity of 10 GB in 1986. As the rock was highly weathered, forepoles and rockbolts together with shotcrete and steel ribs were used as a support system. Material properties used in the analysis are given in Tables 9.1 and 9.2.

The forepoles in this particular case were also simulated by the three-dimensional (3D) rockbolt element developed by Aydan (1989). Figure 9.5 shows the geology (a) and the block diagram (b) of the analyzed domain. Figure 9.6 shows the comparison of the development of calculated axial stress in forepoles with those measured on-site in

Table 9.1 Material properties of ground

Formations	*E (MPa)*	ν	*c (kPa)*	ϕ (°)
Alluvial deposit	50	0.3	50	35
V. weathered shale	200	0.3	50	30
Weathered shale	800	0.25	950	35

Table 9.2 Dimensions and material properties of support members

Support members	*Dimensions*	*E (GPa)*	ν	σ_c *or* σ_t *(MPa)*
Shotcrete	Thickness $t = 10$ mm	5	2	10
Steel sets	Cross-section $A = 21$ cm^2	210	0.3	300
Rockbolts and forepoles	Diameter of bolt $D_b = 25$ mm	210	0.3	300
Grout	Diameter of hole $D_h = 37$ mm	5	0.25	8

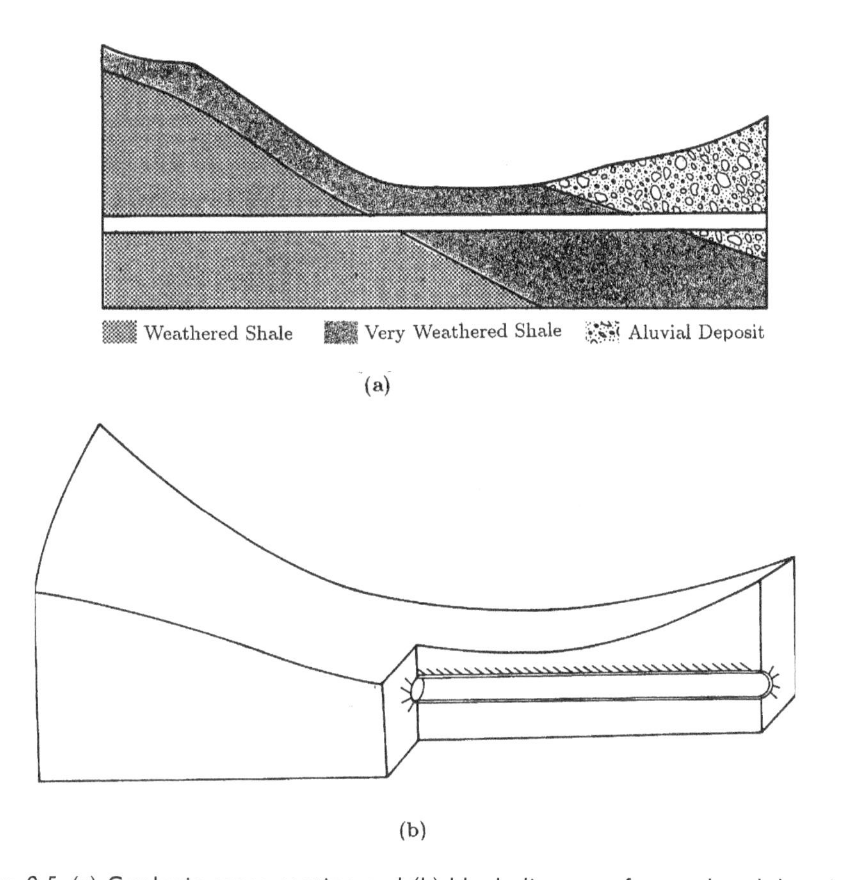

Figure 9.5 (a) Geologic cross-section and (b) block diagram of an analyzed domain.

relation to the distance from the tunnel face (a) and axial stress distributions in bolts (b), respectively. The symbols FP-1, FP-2 and FP-3 in Figure 9.6 denote the forepoles at the crown positioned at 0, 3, and 6 m from the tunnel face. The symbols B1–B8 denote the rockbolts, installed around the tunnel perimetry in positions shown in Figure 9.6b. As noted from the figure, the measured response of forepoles is very well predicted. It is interesting to note that the forepoles are effective when the net axial stress in the forepoles is tensile, and their efficiency decreases as the character of the axial stress tends to become compressive.

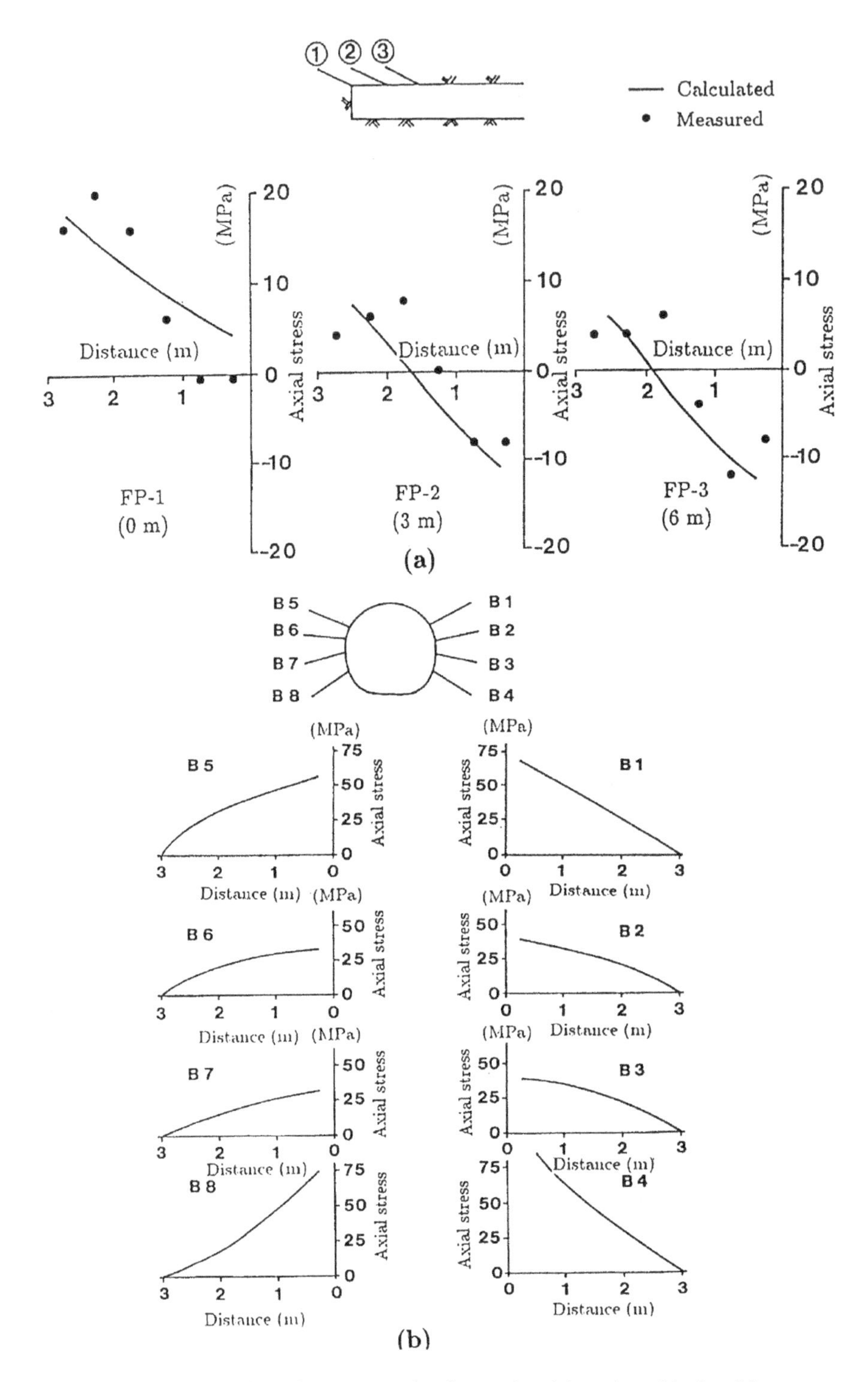

Figure 9.6 Axial stress distributions in the forepoles (a) and rockbolts (b).

9.1.4 Effect of bolting pattern in underground excavations

Aydan and Kawamoto (1991) reported a detailed study on the support systems consisting of rockbolts, shotcrete and steel ribs. The effect of floor bolting on the states of stress and deformation of the surrounding rock was analyzed by Aydan and Kawamoto (1991) by considering two specific cases:

- With floor bolting and
- Without floor bolting.

The shape of the opening was horse-shoe type and the opening was situated 44 m below the ground surface and the lateral stress-coefficient was assumed to be 1.0. Material properties used in the analyses are given in Table 9.3 and the behavior of the surrounding medium was assumed to be elastic-perfectly plastic. Figure 9.7 shows the bolting pattern (a), the displacement of opening perimetry (b), the distribution of contours of safety factor about the opening (c) and axial stress distributions in selected rockbolts (d). As expected, the displacement of the lower half of the opening perimetry in the case of floor bolting is drastically decreased compared with that in the case of no-floor bolting. It is interesting to note that the plastic region about the opening with floor bolting is much less than that without floor bolting. The distribution of safety factor contours about the opening has a wavy form near the opening perimetry. The safety factor contours tend to go away from the opening perimetry at the mid-distance between two bolts while they approach the opening perimetry near the rockbolts. This is actually the arching action between rockbolts due to bolting as noted in the model tests by Rabcewicz (1957). It is also interesting to note that the all-around bolting results in more uniform axial stress distributions as compared with those in the case of no-floor bolting.

9.1.5 Numerical studies on the indentation (impression) experiment

Indenters are used to infer the mechanical properties of intact rock. The stress state is also close to that beneath foundations. The finite element method is used for studying stress–strain responses of various objects under different loading regimes (Aydan et al. 2016). An axisymmetric finite element analysis of cylindrical indenter with a diameter of 3 mm under elastic behavior was carried out (Figure 9.8). Axisymmetric simulations can be regarded as pseudo-three-dimensional modeling. The material properties on the indenter and rock are given in Table 9.4. The applied pressure on the indenter was 10 kgf/cm^2 (0.981 MPa) and the top and side of the model were assumed to be free to move while the central vertical line and bottom of the model can move vertically and radially, respectively.

Table 9.3 Material properties of steel bar and grout annulus

E_b (GPa)	μ_b	σ_t^b (MPa)	σ_s^b (MPa)	K_t	K_s	E_{ga} (GPa)	μ_{ga}	c^p (MPa)	c^r (MPa)	ϕ^p (°)	ϕ^r (°)
210	0.3	500	288	0.1	0.1	5	0.2	15	1.5	35	35

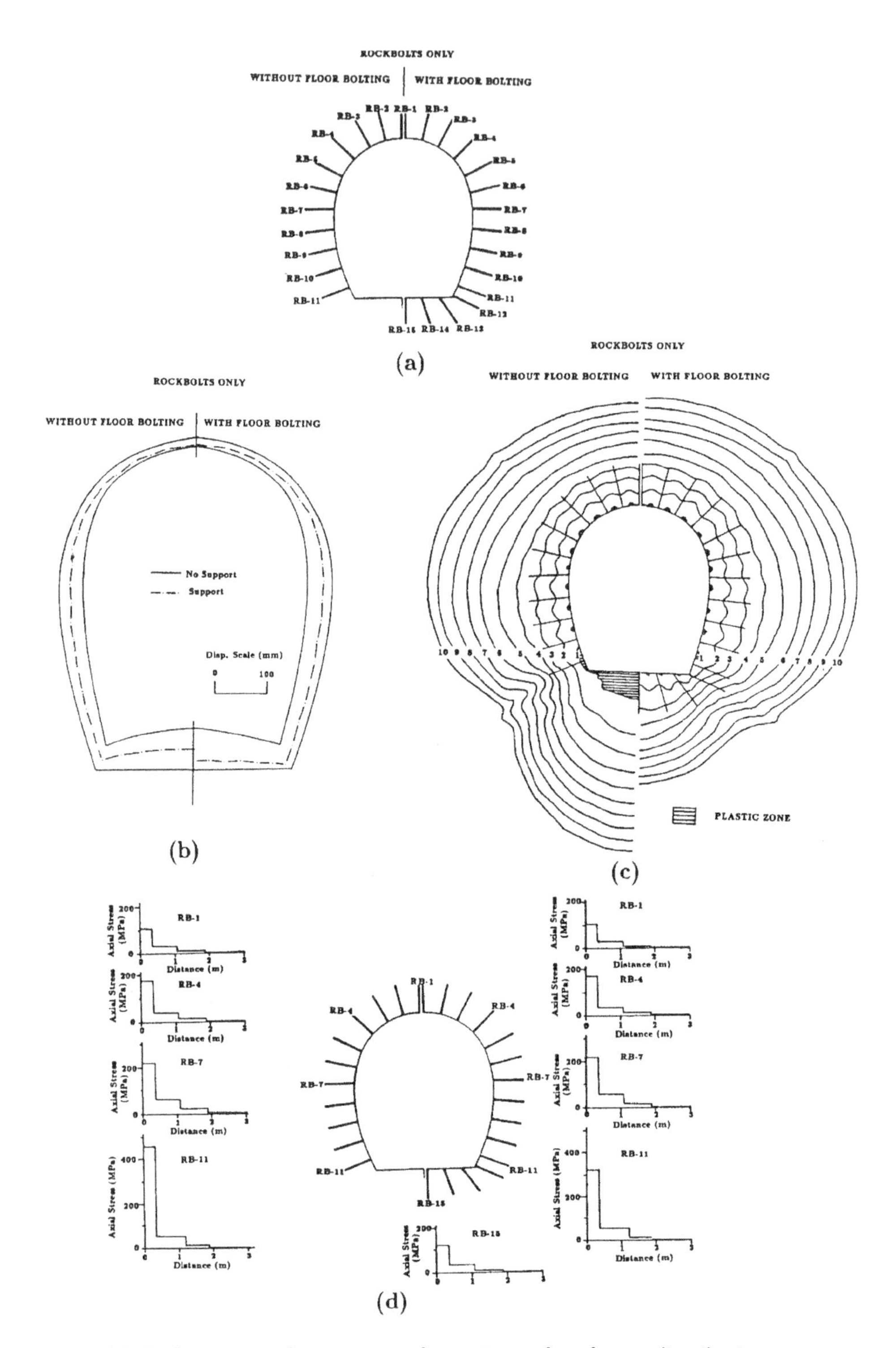

Figure 9.7 Deformation of perimetry of opening, safety factor distribution contours and axial stress distributions in rockbolts. (a) Support pattern; (b) deformation of tunnel perimeter; and (c) Safety contours and plastic zone, (d) axial stress in rockbolts.

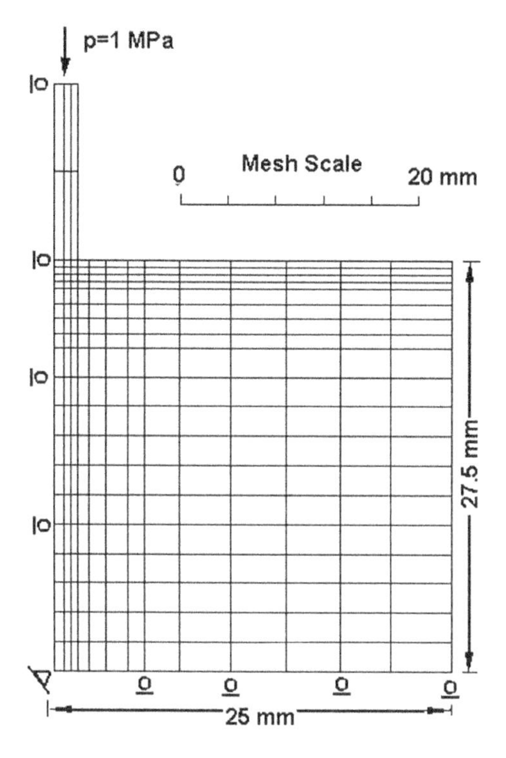

Figure 9.8 Finite element models for impression tests.

Table 9.4 Material properties used in finite element analyses

Material	*Elastic modulus (GPa)*	*Poisson's ratio*
Indenter	200	0.30
Rock	10	0.25

The stress and strain field induced in the impression experiments is close to the compression of the rock under a rigid indenter. Timoshenko and Goodier (1970) developed the following relation for a circular rigid indentation of elastic half-space problem:

$$\frac{\delta}{D} = \frac{\pi}{4}\frac{1-v^2}{E}p \text{ with } p = \frac{4F}{\pi D^2} \tag{9.2}$$

where F, v and E are applied load, Poisson's ratio and elastic modulus of rock.

Aydan et al. (2008a) showed that the following relation with the use of a spherical cavity approach:

$$\frac{\delta}{D} = \frac{1+v}{2E}p \tag{9.3}$$

where v is Poisson's ratio of rock and D is the diameter of the indenter.

Figure 9.9 shows the computed deformed configuration, minimum principal stress contours (tension is assumed to be positive) and maximum shear stress distribution. The minimum principal stress and maximum shear stress contours resemble pressure bulbs as expected. However, the distributions are not uniform just below the indenter. Nevertheless, the pressure bulbs become spherical beyond a distance equivalent to the radius of the indenter. Table 9.5 compares the computed average displacement responses of the indenter from the finite element analysis and Eqs. (9.2) and (9.3). As noted in Table 9.5, the finite element analysis yields smaller displacement compared to the theoretical derivations. The reason for the discrepancy is due to the differences in

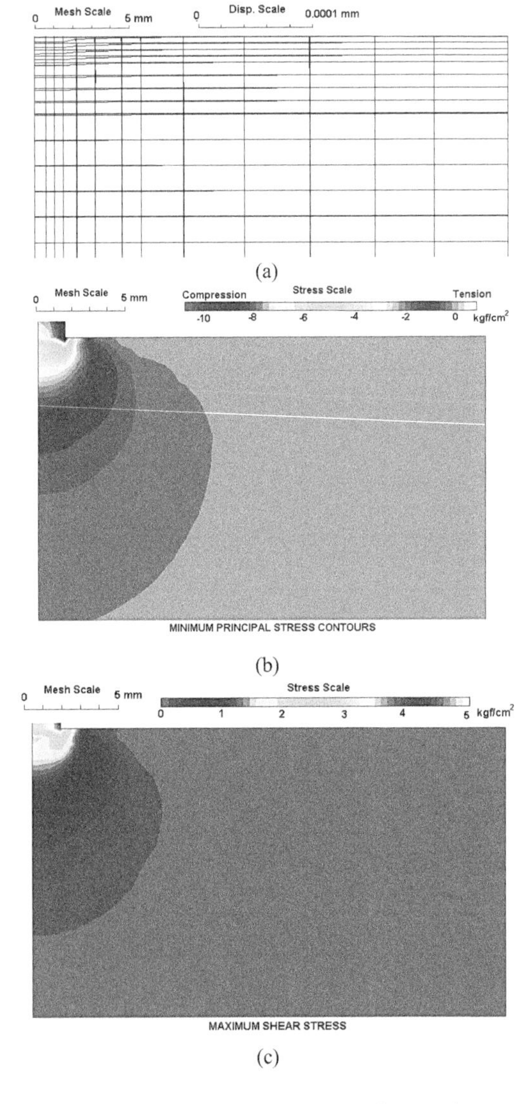

Figure 9.9 Computed results from the axisymmetric finite element analysis for indentation (impression) test. (a) Deformed configuration, (b) minimum principal stress contours and (c) maximum shear stress.

Table 9.5 Computed average impression displacement

Model	*Computed displacement (mm)*
FEM	1.815×10^{-4}
Equation (9.2)	2.209×10^{-4}
Equation (9.3)	1.875×10^{-4}

boundary conditions. While the domain is finite in the FEM analysis, it is a half-space in the derivations of Eqs. (9.2) and (9.3). Nevertheless, it is interesting to note that Eq. (9.3) yields reasonably close results to those from the FEM, as the ratio of diameter and length of the domain is greater than 16 times the indenter radius.

9.1.6 The evaluation of the long-term response of an underground cavern

A pumped-storage scheme consists of two reservoirs and an underground powerhouse and it was constructed about 30 years ago. The upper reservoir consists of a 125 m high rockfill-dam with a storage capacity of 5,780,000 m^3. The lower reservoir has a gravity dam with a height of 44 m. The underground powerhouse is 55 m long, 22 m wide and 39 m high and it has two turbines (Aydan et al. 2012). The maximum water level variation may reach 45 m in 12 hours at the full capacity. As it is expected that the deformability and strength of every geo-material have some time-dependent characteristics, the time-dependent behavior of the underground powerhouse would naturally occur following its construction. The time-dependency of cohesion (c) of rock mass was assumed to obey the functional form given by Eq. (9.4) by considering experimental results on the igneous rocks (Aydan and Nawrocki 1998).

$$\frac{c(t)}{c_o} = 1.0 - 0.0282\log(t^*) \tag{9.4}$$

where c_0 is the cohesion obtained from a short-term experiment with a duration t_0, $t^* = \frac{t}{t_o}$, and t is the time.

The elasto-visco-plastic finite element method described in Chapter 8 was used to analyze the long-term response of rock mass around the underground cavern (Aydan et al. 2012). The visco-elastic response during the excavation stage was modeled according to the Zener type rheological model. Figures 9.10 and 9.11 show the deformed configurations and plastic zone formation around the cavern at the time of construction and after 30 years for the lateral stress coefficient of 1.0, which was used in the initial design. Compared with the deformation of the cavern shown in Figure 9.10a, the deformation becomes larger after 30 years. Furthermore, a plastic zone develops in the vicinity of the sidewall after 30 years, which was not observed during the excavation stage (Figure 9.10). The increase of inward deformation and the formation of the plastic zone would impose additional loads on the support system, which may also undergo some degree of degradation, eventually resulting in the rupture of rock-anchors and the fracturing of shotcrete.

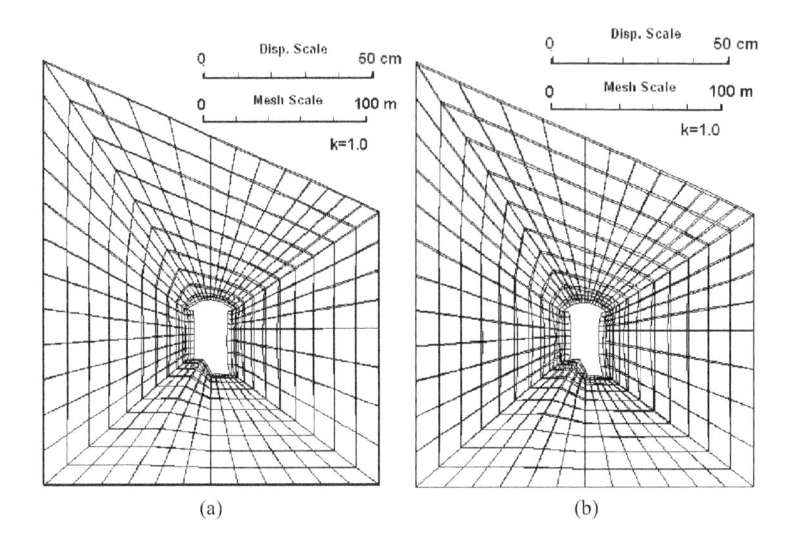

Figure 9.10 Comparison of deformed configuration after excavation and 30 years. (a) Just after construction and (b) after 30 years.

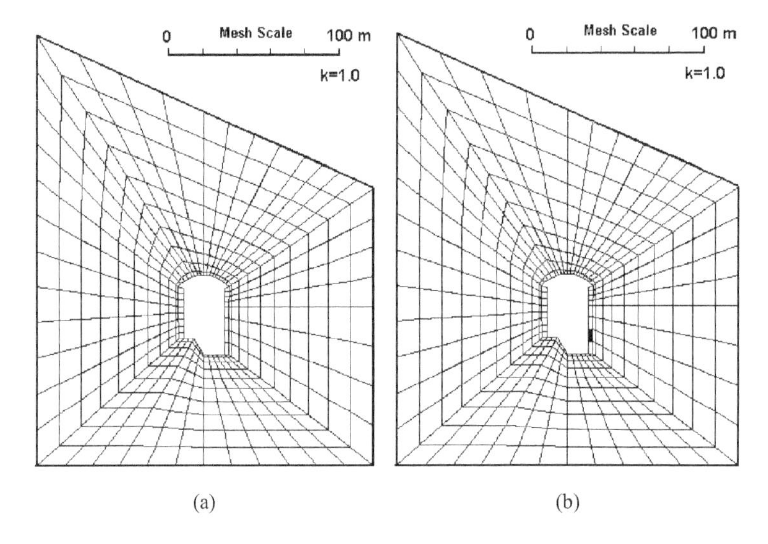

Figure 9.11 Comparison of plastic zone formation around the cavern after and 30 years after the construction. (a) Just after construction and (b) 30 years after construction.

9.1.7 Long-term stability of the Derinkuyu underground city, Cappadocia, Turkey

There are many antique semi-underground and underground settlements in the Cappadocia Region of Turkey. Aydan and Ulusay (2013) reported an integrated study on the long-term response and stability of the Derinkuyu underground city. For the analyses of long-term response and stability of this underground city, a visco-elasto-plastic

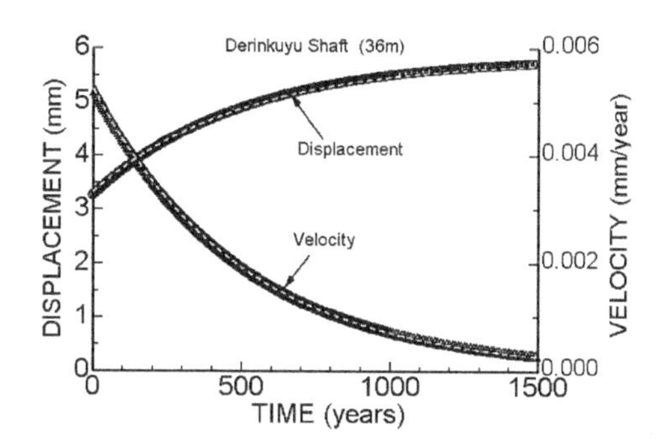

Figure 9.12 Comparison of exact (closed-form) solutions with the axisymmetric finite element solution.

behavior of surrounding rock was considered. The constitutive law is based on the model described in Chapter 5. However, the behavior of rock mass at the excavation stage was modeled using a Zener type rheological model. Figure 9.12 compares the finite-element analysis based on the elasto-visco-plastic model with results of closed-form solutions for a one-dimensional axisymmetric situation for the shaft of the Derinkuyu underground city at the level of 36 m below the ground surface.

A series of visco-elasto-plastic finite element analyses were carried out to assess the short- and long-term stability of a vertical shaft and a hall on the seventh floor of the Derinkuyu underground city (Aydan and Ulusay 2013). The vertical shaft is still in operation and it is used as one of the ventilation shafts connecting all floors of the Derinkuyu underground city to the ground surface. The cross-section of the shaft is circular. Therefore, the problem was treated as an axisymmetric problem. The shaft was 37.5 m deep and the hall was at the depth of 40 m. The vertical stress was assumed to be equivalent to the overburden and the lateral stress coefficient was taken 0.5 based on initial in-situ stress predictions (Aydan et al. 1999; Watanabe et al. 1999; Aydan and Ulusay 2013). The Mohr–Coulomb yield function was replaced with the Drucker–Prager yield criterion by using the theoretical relations between two yield criteria while considering the strain rate dependency. Relevant mechanical properties used in the analyses are given in Table 9.6, which were determined from short- and long-term laboratory tests on rock samples and in-situ characterization of the rock mass (Ito et al. 2016).

The analysis on the shaft showed that there should be no yielding occurring around the shaft soon after the excavation. The computations were carried out up to 1500 years, which corresponds to the present time if this underground city was assumed to be excavated 1500 years ago. Figure 9.13 shows the maximum shear stress distribution

Table 9.6 Physical and mechanical properties used in finite element analyses

ρ (kN/m^3)	λ_h (MPa)	μ_h (MPa)	λ_m (MPa)	μ_m (GPa)	$\lambda^* = \mu^*$ (GPa year)	ϕ (°)
18	96	96	76	76	25.189	22

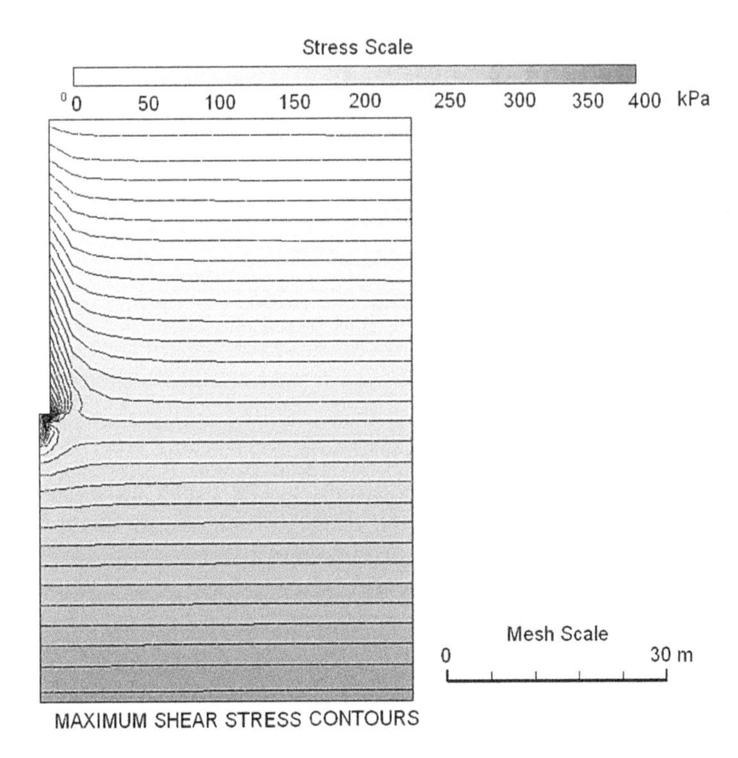

Figure 9.13 Computed maximum shear stress distributions around shaft using elasto-visco-plastic FEM.

contours at the time of 1500 years following the excavation and the displacement and velocity responses of the shaft at the level of 36 m from the ground surface. Computations also indicated that yielding of rock mass should not also occur at present. As the behavior of surrounding rock mass is visco-elastic, the deformation of the shaft should have been nearly converged to its final value and the stress state in the surrounding rock mass should also be the same as that at the time of excavation.

The hall on the seventh floor of the Derinkuyu underground city was analyzed next. The hall has an arch-shaped roof and its shape is close to a horseshoe. The width is about 4.5 m and it is 20 m long. Therefore, the problem was treated as a two-dimensional plain-strain problem. Although the hall has three pillars along the center line, they were neglected in computations as they had some thoroughgoing discontinuities. The analysis showed that no yielding occurs soon after the excavation. The computations were carried out up to 1500 years, which corresponds to the present time. Figure 9.14 shows the deformed configurations of the domain analyzed 1500 years after the excavation and the displacement and velocity response of the hall. Computations also indicated that yielding of rock mass should not occur until the present time. As the behavior of surrounding rock mass is visco-elastic, the deformation of the hall should have been nearly converged to its final value and the stress state in the surrounding rock mass should also be the same as that at the time of excavation.

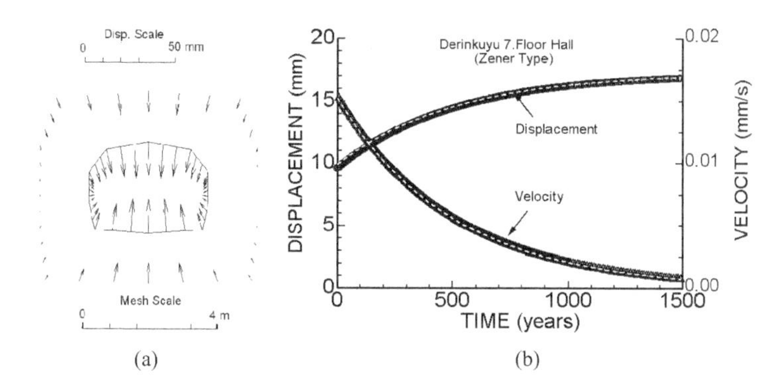

Figure 9.14 Deformation and velocity responses of the hall at the seventh floor of the Derinkuyu underground city. (a) Deformation vectors at 1500 years and (b) displacement and velocity response.

9.1.8 Stability analyses of Tomb of Pharaoh Amenophis III, Luxor, Egypt

Three-dimensional elasto-plastic numerical analyses of the tomb were carried out using the FLAC code developed by ITASCA (2007) and it is based on the finite difference method (Figure 9.15). Two cases, in which the properties of rock mass are assumed to be equivalent to those of intact rock under dry and wet conditions as given in Table 9.7, are only presented (see Aydan et al. 2008b for details). The stress state of the tomb was obtained from the fault striation method of Aydan (2000). The computational results indicated that if rock mass is assumed to be dry, there should be no plastic zone development in rock mass around the tomb (Figure 9.16). However, if rock becomes saturated, the computational results indicated that the damage in the walls between J-room and Jd-room has a great influence on the overall stability of the J room and adjacent rooms. Saturation of rock mass from time to time due to floods in the region for 3400 years may also have a negative effect on the stability of the tomb.

9.1.9 Dynamic response of a large underground cavern

Geniş and Aydan (2007) carried out a series of numerical studies for the static and dynamic stability assessments of a large underground opening for a hydroelectric powerhouse. The cavern is in granite under high initial stress conditions and approximately 550 m below the ground surface (Aydan and Kawamoto 2001). The area experienced in 1891 the largest inland earthquake in Japan. In the numerical analyses, the amplitude, frequency content and propagation direction of waves were varied (Figure 9.17). The numerical analyses indicated that the yield zone formation is frequency and amplitude dependent (Aydan et al. 2011). Furthermore, the direction of wave propagation has also a large influence on the yield zone formation around the cavern. When maximum ground acceleration exceeds 0.6–0.7 g, it increases plastic zones around the opening. Thus, there will be no additional yield zone around the cavern if the maximum ground acceleration is less than these threshold values.

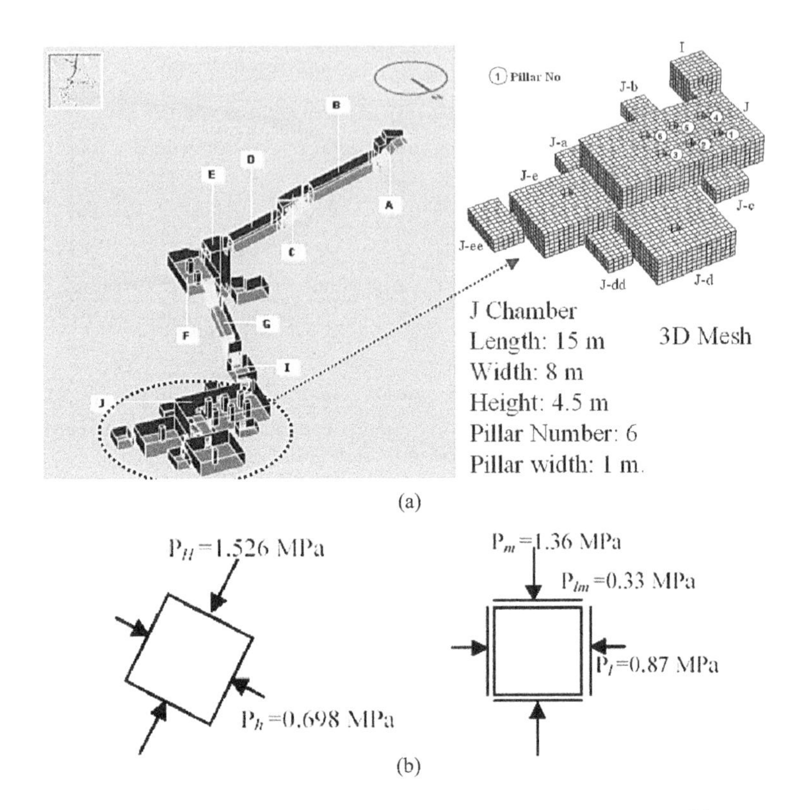

Figure 9.15 (a) Geometry and numerical model of the tomb and (b) in-situ stress state.

Table 9.7 Rock mass properties used in numerical analyses

σ_c *(MPa)*	ϕ	*E (GPa)*	*C (MPa)*	σ_t *(MPa)*	v	γ *(kN/m^3)*	*Remarks*
20	35	2	5.2	2	0.25	20	Case 1: dry
5	30	0.8	1.44	0.5	0.25	20	Case 2: wet

9.1.10 Response and stability of abandoned room and pillar mine under static and earthquake loading

There are many abandoned lignite mines in Mitake town in Gifu Prefecture of Japan. Following the 2011 Great East Japan earthquake, there is a great concern on the response and stability of these abandoned lignite mines as they are beneath the urbanized areas (Aydan and Tano 2012). Using material properties given in Table 9.8, Figure 9.18 shows the distribution of minimum principal stress (tension is positive) for an abandoned room and pillar mine beneath Kyowa Secondary School in Mitake Town of Gifu Prefecture, Japan under static loading condition. The static analyses showed that the pillars should behave elastically under static loading conditions as the induced

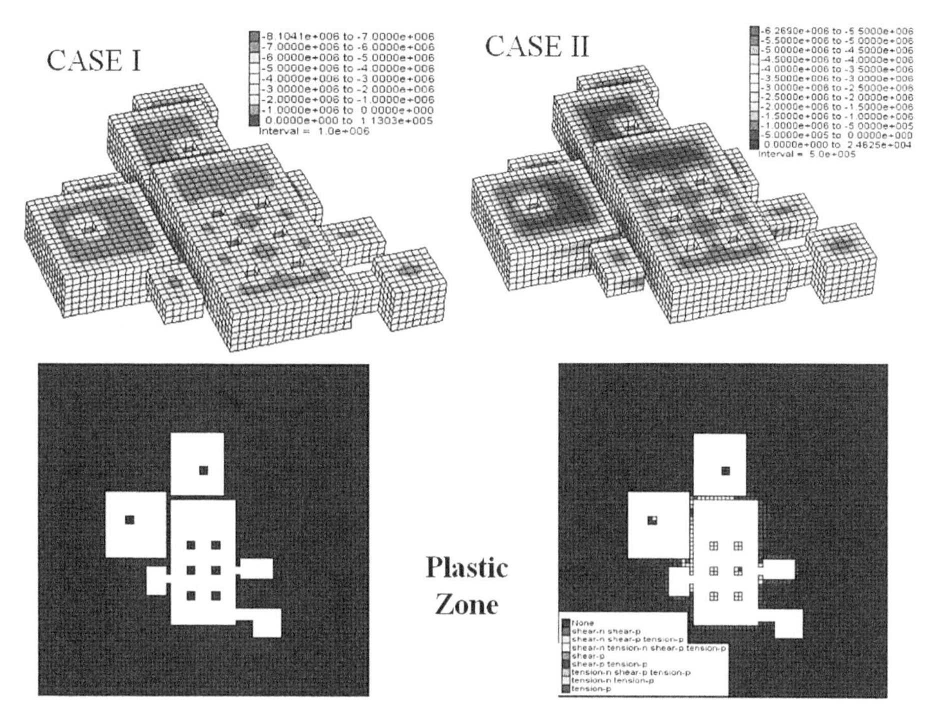

Figure 9.16 Stress distribution and plastic zone development around the tomb.

stresses are less than the surrounding rock masses. However, the pillar stress is highest at the deepest location as expected.

This area would be subjected to the anticipated Nankai–Tonankai–Tokai earthquake in the future and there is a great concern about it. The authors have been involved with the stability assessment of the abandoned lignite mine beneath Kyowa Secondary School during the anticipated Nankai–Tonankai–Tokai earthquake (Aydan and Tano 2012; Geniş and Aydan 2013, 2020). Material properties of investigated ground are given in Table 9.8. The authors carried out 1D, 2D and 3D dynamic simulations for an estimated base ground motion data at Mitake Town obtained from the method of Sugito et al. (2000). Figure 9.19 illustrates the numerical model of the ground and abandoned lignite mine beneath the Kyowa Secondary School. Figure 9.20 shows the computed responses from 1D and 3D numerical analyses. It is interesting to note that responses from 1D and 3D analyses are quite similar to each other.

A three-dimensional elasto-plastic numerical analysis of abandoned lignite mine beneath the Kyowa Secondary school (Figure 9.19) was analyzed using the estimated ground motion record (Figure 9.21), based on the methods developed by Sugito et al. (2000) and Aydan (2012) for the anticipated Nankai–Tonankai–Tokai mega earthquake. The Nankai earthquake terminated at 43 seconds and the Tonankai earthquake started and terminated at 75 seconds. The last earthquake is the Tokai earthquake and it terminated at about 125 seconds. The failure of the pillars started at the deepest site

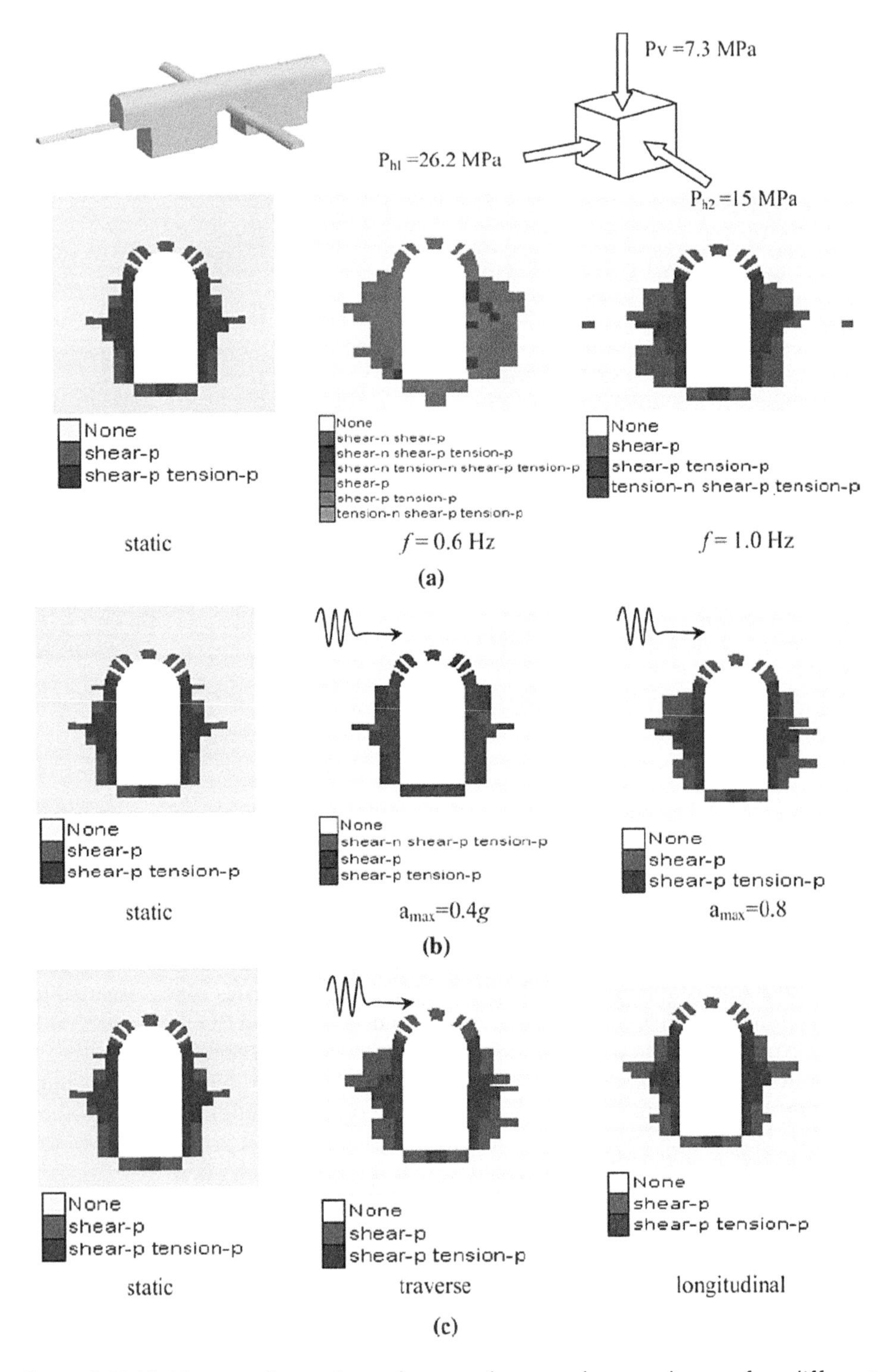

Figure 9.17 Yield zone formation of an underground powerhouse for different cases of input ground motions (from Aydan et al. 2011). (a) $a_{max} = 0.8$ g; (b) $f = 1$ Hz; and (c) $a_{max} = 0.8$ g and $f = 1$ Hz.

Table 9.8 Material properties of layers

Layer	γ *(kN/m³)*	*E (MPa)*	ν	*c (MPa)*	ϕ *(°)*
Topsoil	19	270	0.35	0.0	38
Upper Mst-Sst	19	750	0.3	0.7	25
Lignite	14	400	0.3	0.66	45
Lower Mst-Sst	19	1073	0.3	1.00	45
Chert	19	3647	0.3	3.00	45

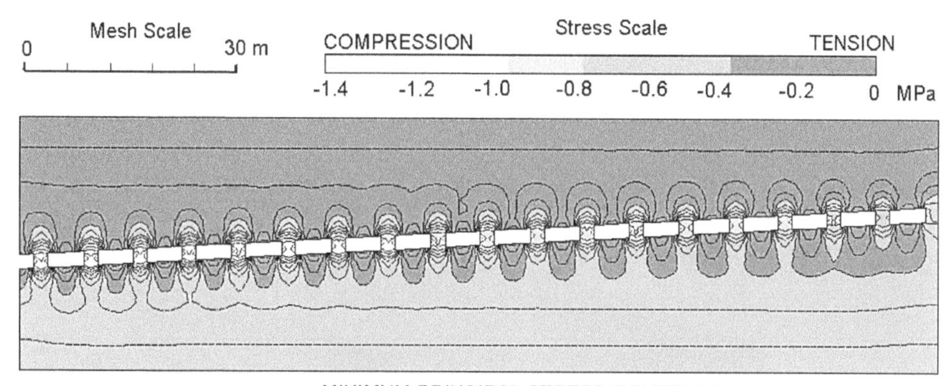

Figure 9.18 Contours of minimum principal stress beneath the Kyowa Secondary School.

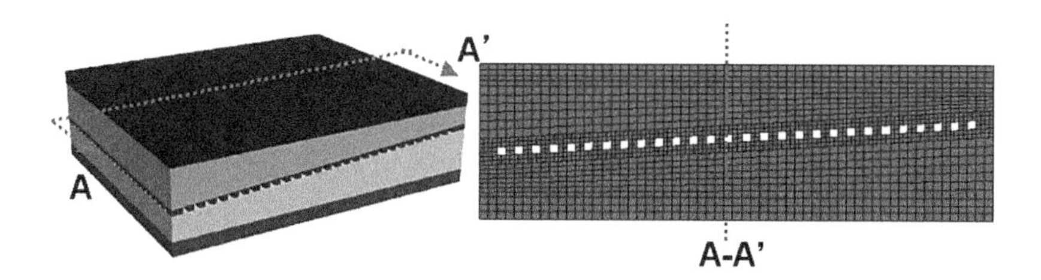

Figure 9.19 Illustration of models used in numerical analyses and selected section.

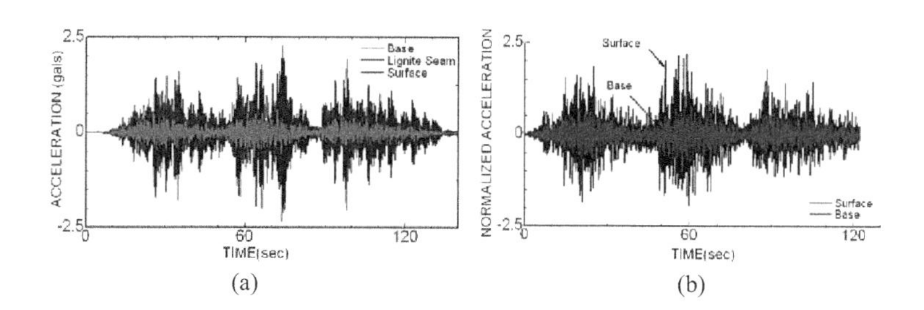

Figure 9.20 Acceleration responses at selected section from 1D to 3D numerical analyses. (a) 1D analysis and (b) 3D analysis.

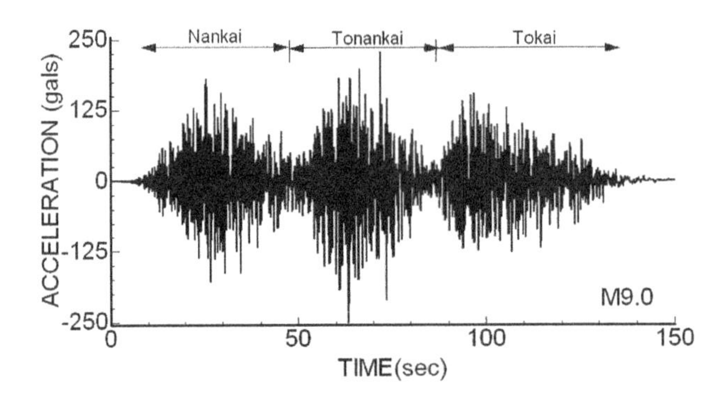

Figure 9.21 Input base acceleration record.

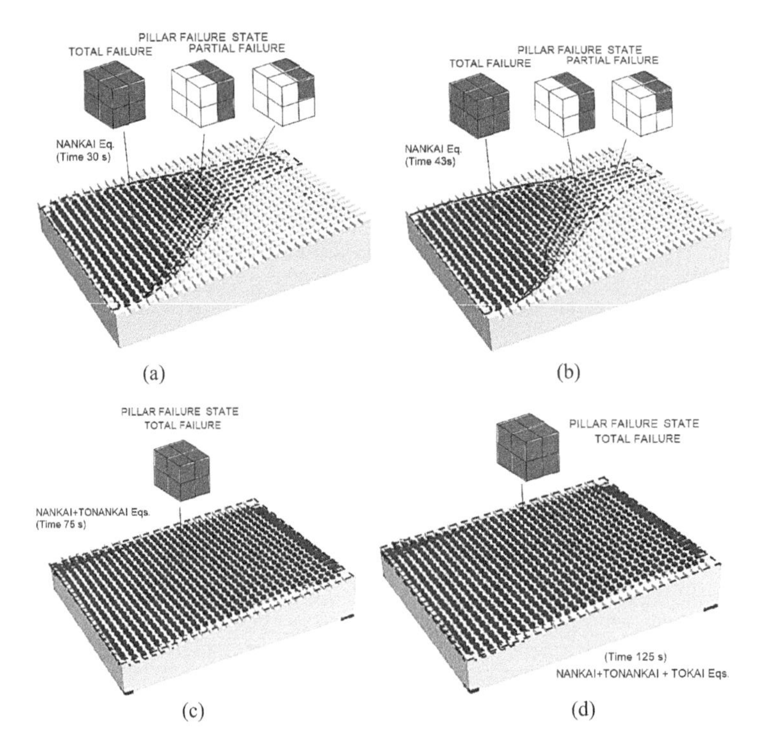

Figure 9.22 Failure state of pillar of abandoned mine at different time steps (a) 30s; (b) 43s; (c) 75s; (d) 125s.

and propagated toward shallower parts as estimated (Figure 9.22). The failure state of pillars can be broadly classified as a total failure and partial failure as illustrated. When the Nankai earthquake terminated, about 60% of the pillars were already totally or partially yielded. The Tonankai earthquake was the nearest one to the Mitake town and all pillars were in a total failure state. The Tokai earthquake had no further effect on the failure state.

9.1.11 *Modal analyses of shafts at the Horonobe Underground Laboratory*

Sato et al. (2019) reported dynamic responses of the Horonobe Underground Laboratory in Hokkaido, Japan during the 2018 Soya and 2018 Iburi earthquakes. A series of 3D finite element modal analyses were carried out for four conditions, which are namely, no shafts, single shaft, double shafts, and triple shafts. Table 9.9 gives the material properties used in numerical analyses while Table 9.10 compares the Eigenvalues for four different conditions and Figure 9.23 shows the displacement response for Mode 1.

Table 9.9 Material properties

Material	*Unit weight (kN/m^3)*	*E (GPa)*	*Poisson's ratio*
Rock mass	26.5	0.600	0.37
Concrete	23.5	11.042	0.20

Table 9.10 Eigenvalues for modes 1, 2 and 3

	No shaft	*Single*	*Double*	*Triple*
Mode 1(s)	1.763	1.752	1.203	1.199
Mode 2(s)	1.645	1.635	1.889	1.172
Mode 3(s)	1.564	1.554	1.117	1.111

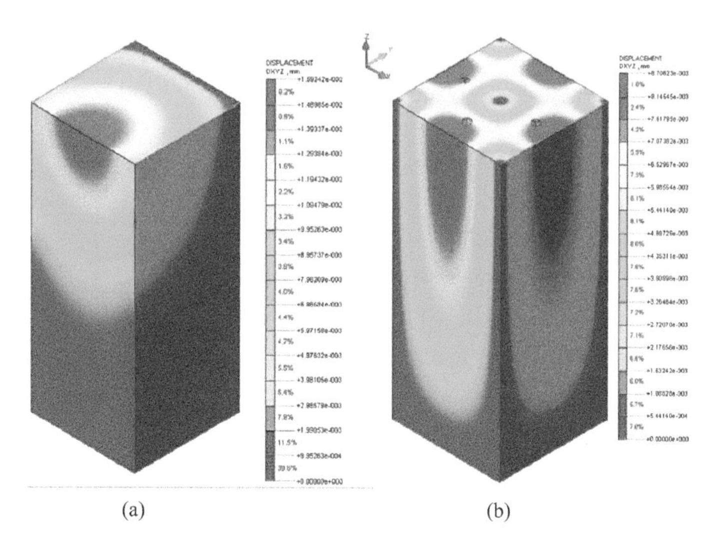

Figure 9.23 Displacement response for Mode 1. (a) No shaft and (b) triple shafts.

9.1.12 Temperature and stress distributions around an underground opening

The example given here is concerned with temperature fluctuation in rock mass in an underground opening subjected to temperature variation applied on the surface of the opening (see details in Aydan et al. 2012). Temperature fluctuation in the opening was based on actual measurements and it was subjected to ±10° yearlong sinusoidal temperature variation. The temperature distribution is computed from a FEM program based on the theory presented in Chapters 4 and 8 of this book. Thermo-mechanical properties used in computations are given in Table 9.11. Figure 9.24 shows the temperature distribution and associated maximum tangential and radial stress distribution in the close vicinity of the cavern when it is modeled as an axisymmetric cylindrical cavern with a diameter of 32 m. The tangential stress variation in the close vicinity of the cavern is the highest and its value is about 0.4 MPa. This implies that the rock mass adjacent to the cavern wall will be subjected to ±0.4 MPa thermal stress cycles.

Next two-dimensional pseudo coupled finite element analyses were carried out with the application of the same temperature variation on the cavern surface shown in Figure 9.24. The finite element mesh used is shown in Figure 9.25.

Figure 9.26 shows responses of stresses and displacements for RHS and LHS walls and crown in rock mass around the underground cavern. In Figure 9.26b, the measured displacement responses are also plotted. Figure 9.27 shows the temperature and

Table 9.11 Thermo-mechanical properties of surrounding rock mass

Unit weight (kN/m^3)	*Elastic modulus (GPa)*	*Poisson's ratio*	*Cohesion (MPa)*	*Friction angle (°)*	*Thermal diffusivity (m^2/day)*	*Thermal expansion coefficient (1/°C)*
26	5–10	0.25	3	40	0.1	1.0×10^{-5}

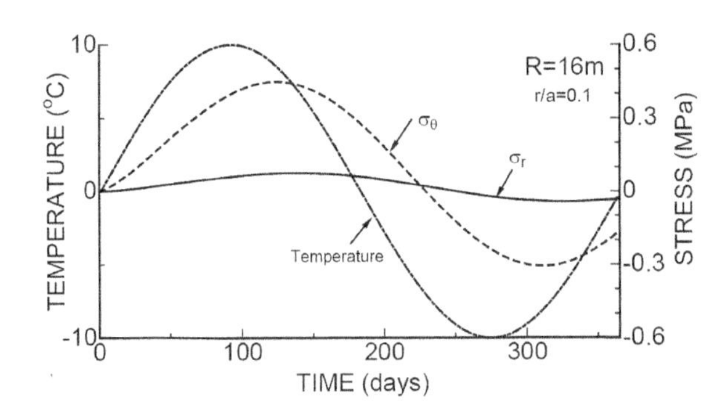

Figure 9.24 Applied temperature at the surface of the cavern and associated tangential and radial stress changes if the cavern is modeled as a cylindrical opening with a diameter of 32 m (from Aydan et al. 2012).

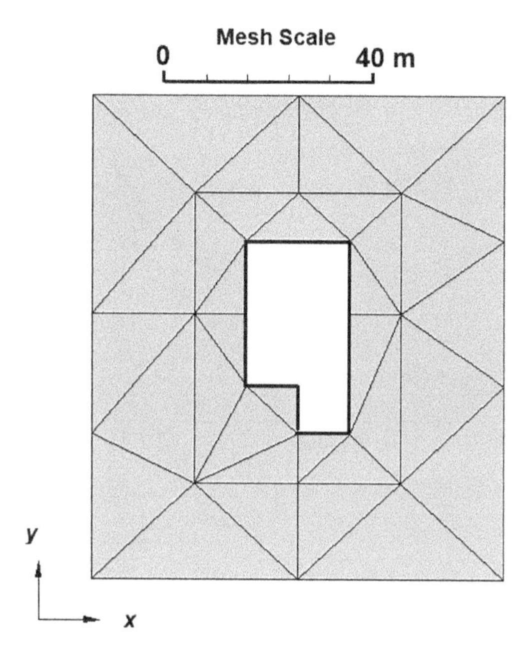

Figure 9.25 Finite element mesh used in pseudo-coupled thermo-mechanical analyses.

maximum shear stress distributions at Days 99 and 272. Days 99 correspond to the highest temperature while Days 272 correspond to the lowest temperature in the cavern. As noted from the figure, temperature distributions are reversed while the maximum shear stress remains almost the same at both extreme values of temperature fluctuations.

9.1.13 Water-head variations in rock mass around an underground cavern

Computations on water-head variations in rock mass around an underground cavern in the association of daily water level variations in a pumped-storage scheme, which is briefly mentioned in Section 9.1.6, were carried out using a two-dimensional coupled seepage analyses, and the mesh used is shown in Figure 9.28 (Aydan et al. 2012). The initial water-level was assigned using Dupuit's equation. The cavern was assumed to be unlined. Figure 9.29 shows the water-head distribution when the water is drawn from the upper reservoir while Figure 9.30 shows the water-head distribution when the water is pumped back to the upper reservoir. This process repeats itself as the water levels of reservoirs are cyclically varied. The large variations of water heads in rock mass occur in the close vicinity of the reservoirs and their effect is quite negligible around the cavern. These results further imply that there is no need to consider the effective stress changes around the caverns of the pumped storage schemes in long term. Nevertheless, they may have important implications on the slopes of the reservoirs due to large variations of effective stresses.

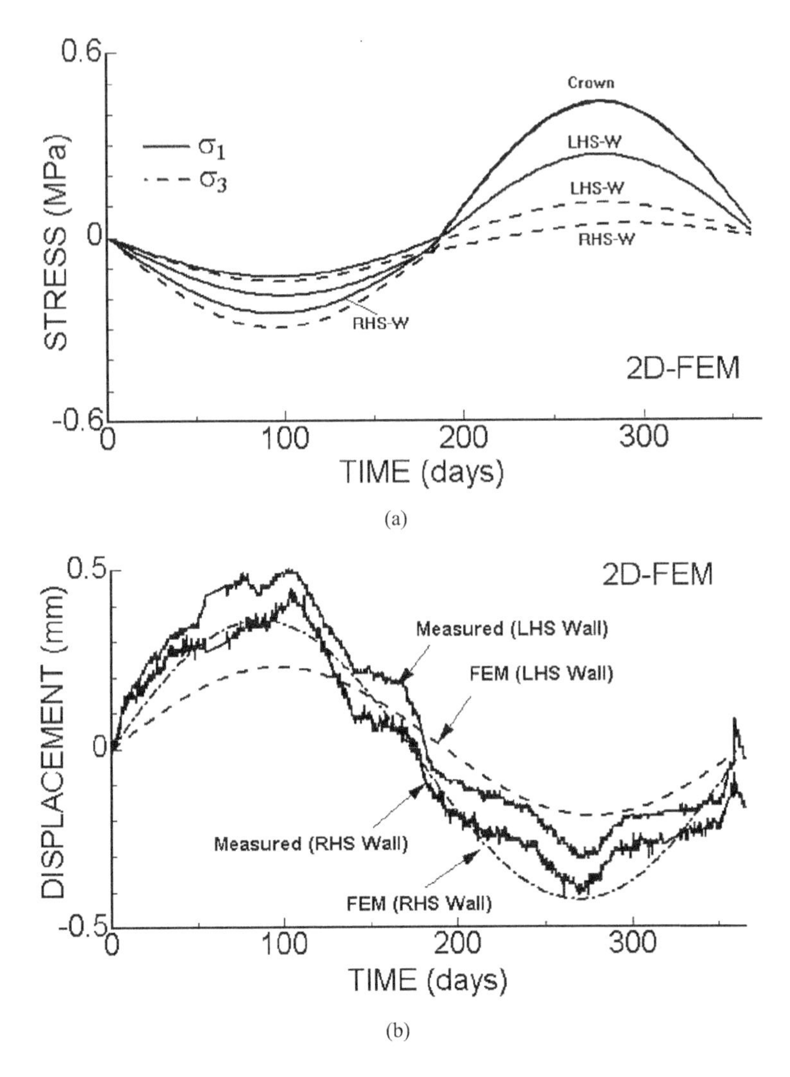

Figure 9.26 Computed stress and displacement responses at selected locations (from Aydan et al. 2012). (a) Principal stress variations at selected locations and (b) computed and measured displacement responses at selected locations.

9.1.14 Breakout formation in boreholes in sedimentary rocks due to moisture loss

It is known that some saturated sedimentary rocks fracture when they lose their water content during the drying period. This phenomenon is sometimes called slaking. This process involves volumetric variations during the drying and wetting periods. When sedimentary rocks lose their water content, they shrink volumetrically. On the other hand, they swell when the water content increases. In other words, it is a cyclic

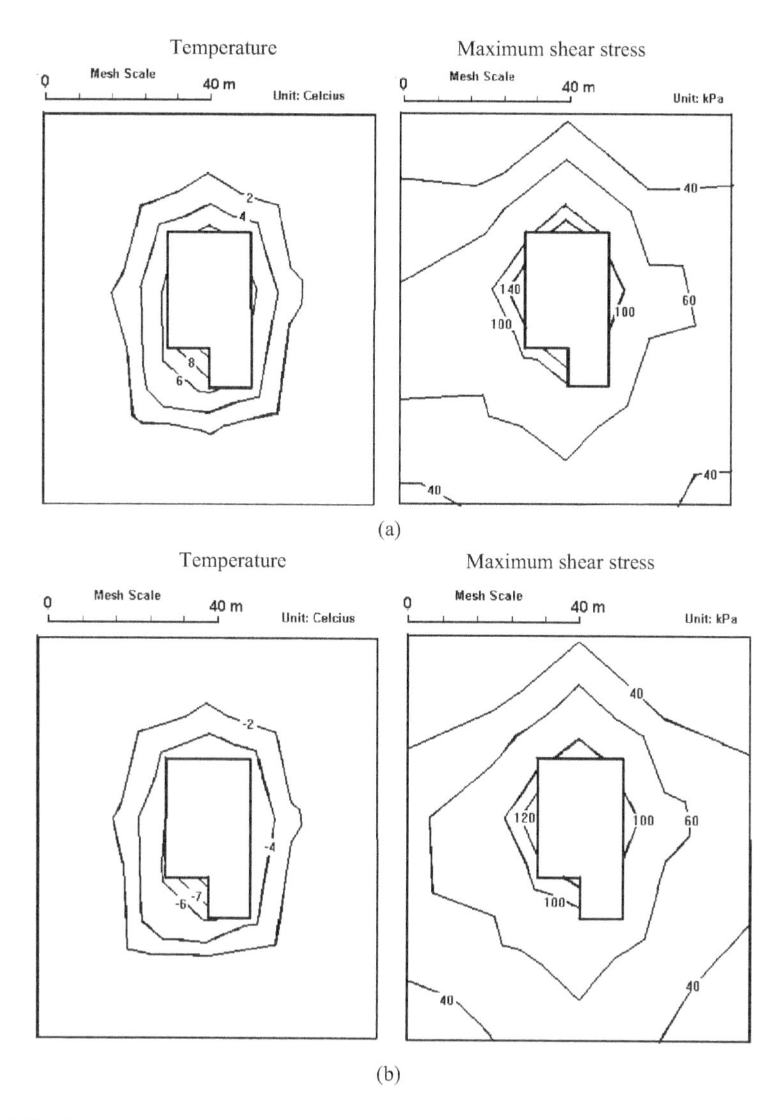

Figure 9.27 Temperatures and maximum shear stress distributions (from Aydan et al. 2012). (a) Days 99 (highest temperature) and (b) days 272 (lowest temperature).

shrinkage and swelling process. This process has been theoretically explained by Aydan in his recent book (Aydan 2017). However, the fundamental studies were done by Aydan and his coworkers during 1994 and 2006 (Aydan 2003; Aydan et al. 1993, 1994, 2006).

A very interesting phenomenon was observed by Aydan et al. (2006) at the Tono mine where circular boreholes were drilled for measuring in-situ stress by the overcoring method. Breakouts at the top of the boreholes were observed after about 6 months. The initial interpretation was that it may be related to creep failure. However, if it

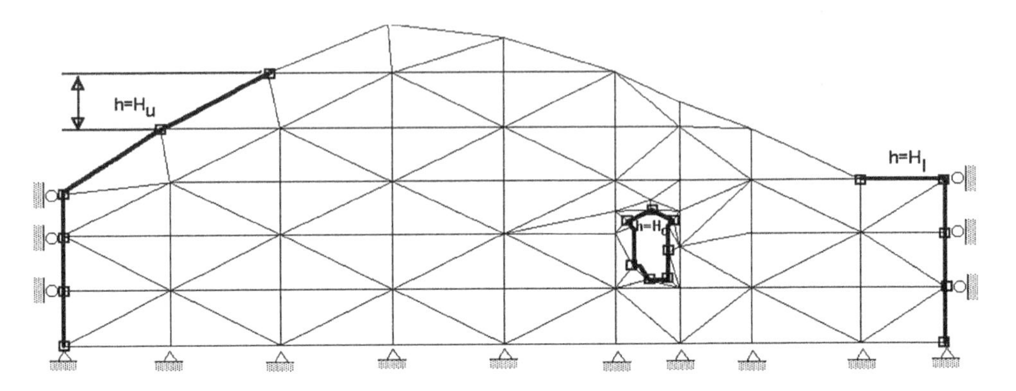

Figure 9.28 Finite element mesh for coupled hydro-mechanical analyses (from Aydan et al. 2012).

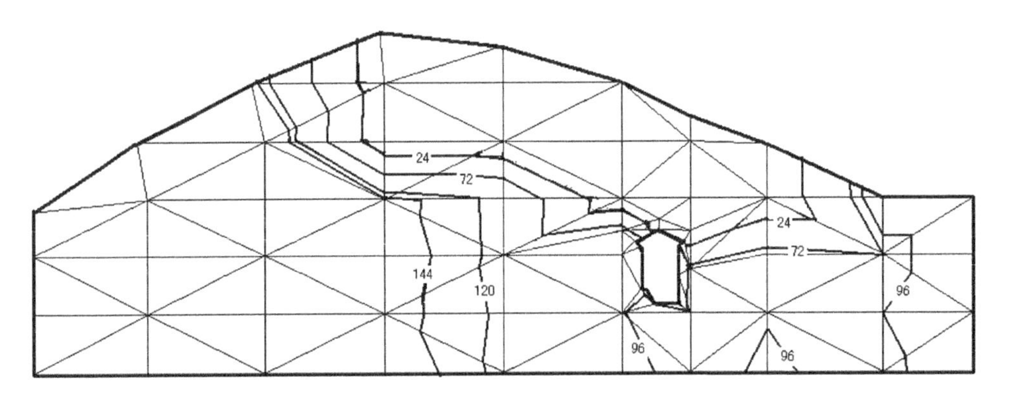

Figure 9.29 Computed water-head distributions for 45 m water level decrease of the upper reservoir (from Aydan et al. 2012).

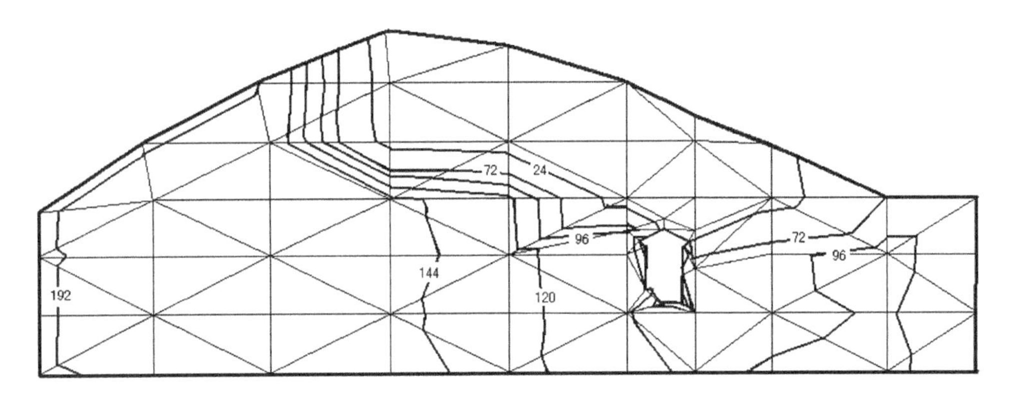

Figure 9.30 Computed water-head distributions for 45 m refill of the upper reservoir (from Aydan et al. 2012).

was due to creep failure, it should have been symmetrically taking place. This led Aydan and his coworkers (Aydan et al. 2006) to model the phenomenon as a coupled process of volumetric change as a result of water content variation and associated stress changes. The first analyses were concerned with the simulations of displacement, strain, and stress field around a circular borehole in a hydrostatic stress field to check the leading idea was correct or not. Experiments indicated that there are fine-grain and coarse-grain sandstones with different swelling characteristics. Specifically, the effects of sandstone type and diameter of borehole were analyzed. Figure 9.31 shows the computed results for displacement, water content and stress fields for fine

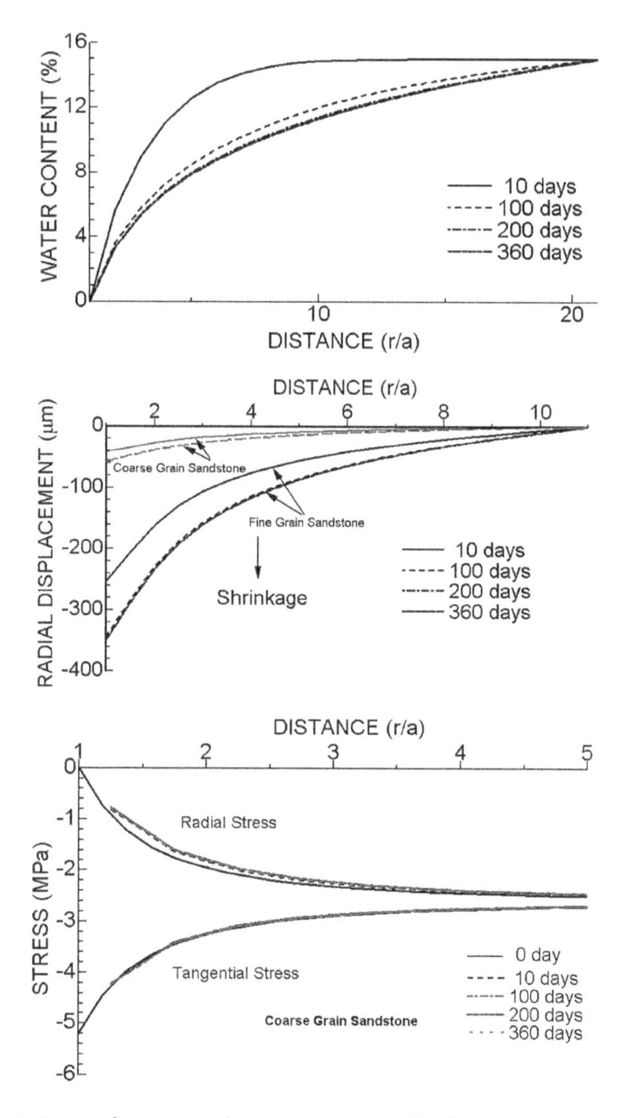

Figure 9.31 Variations of computed water content, displacement, stress fields for fine and coarse grain sandstones (from Aydan et al. 2006).

and coarse grain sandstones for a borehole with a diameter of 200 mm under the overburden of the adit at selected time steps.

As the water-migration characteristics of both fine and coarse sandstones were the same, the resulting water content migration distributions with time were the same. However, displacement, strain and stress fields were entirely different for each sandstone type. Since the volumetric variation of fine-grain sandstone as a function of water content is much larger than that of coarse grain sandstone, the shrinkage of the borehole in fine-grain sandstone is larger than that in coarse-grain sandstone. Consequently, radial stress in the close vicinity of the borehole wall becomes tensile in fine-grain sandstone. This, in turn, implies that there would be fractures parallel to the borehole wall if the tensile strength of the rock was exceeded. Furthermore, such fractures would only occur in the vicinity of boreholes in fine-grain sandstone, as it is observationally confirmed in-situ.

The next computation was concerned with a circular borehole under two-dimensional in-situ stress fields as shown in Figure 9.32. It is observed that the bottom of the borehole was wet or covered with water in-situ. To consider this observation in computations, the boundary conditions for water content migration field and displacement field were assumed as illustrated in Figure 9.32. The other properties were the same as those used in axisymmetric simulations.

The computed displacement field and associated yielding zone are shown in Figure 9.33. As noted from the figure, the bottom of the borehole heaves and the crown of the borehole shrinks upward. In other words, the upper part of the borehole expands outward due to water content loss. The displacement and stress fields of the surrounding rock are entirely different at the lower and upper parts of the borehole.

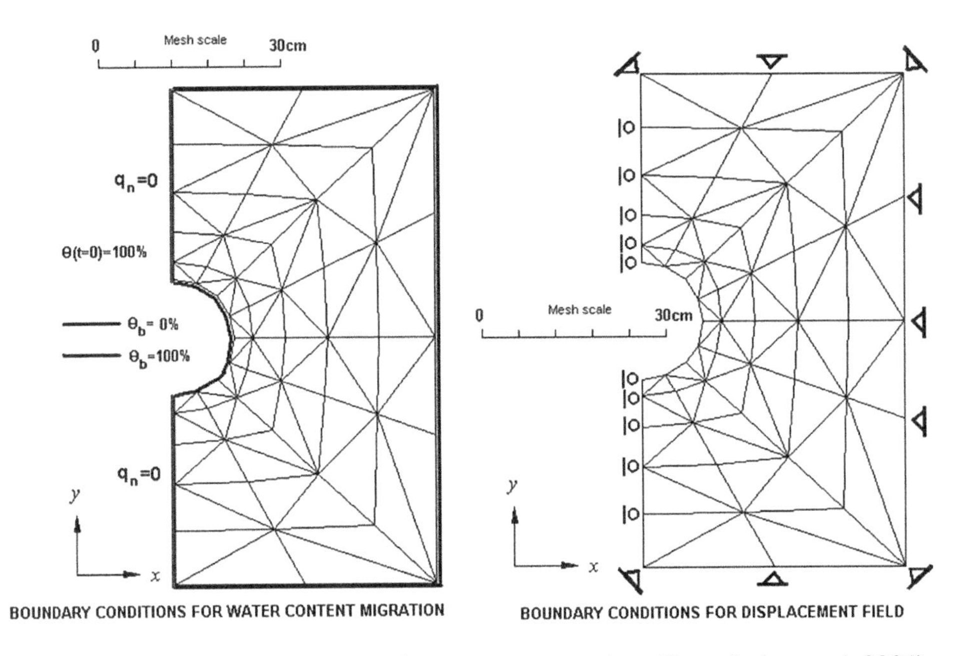

Figure 9.32 Assumed boundary conditions in computations (from Aydan et al. 2006).

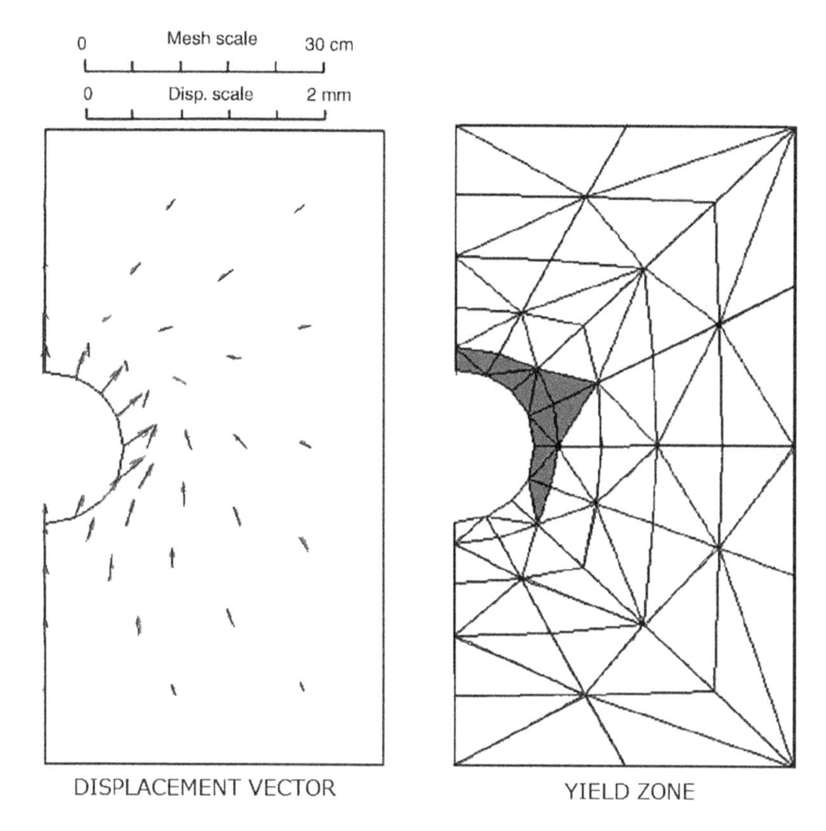

Figure 9.33 Computed displacement field and yield zone (from Aydan et al. 2006).

As a result of this fact, yielding occurs only in the upper part of the borehole. This computational result is in accordance with actual observations. The yielding zone is not depleted in this computation. However, if the yielding zones were depleted in the computation region, the process would repeat itself after depletion of the yielded zone at each time.

9.2 Applications in discontinuum

9.2.1 *Earthquake fault rupture simulation*

Earthquakes are of great importance in the fields of science and engineering of earthquakes. Although there are different approaches for this purpose, there is no explicit consideration of fault rupture, specifically. For the first time, Toki and Muira (1985), Toki and Sawada (1988) and Tsuboi and Muira (1996) utilized Goodman-type joint elements in 2D-FEM to simulate both rupture process and ground motions (Figure 9.34). Fukushima et al. (2010) utilized this method to simulate ground motions caused by the 2000 Tottori earthquake (Figure 9.35). Later Mizumoto et al. (2005) extended the same method to three-dimensions.

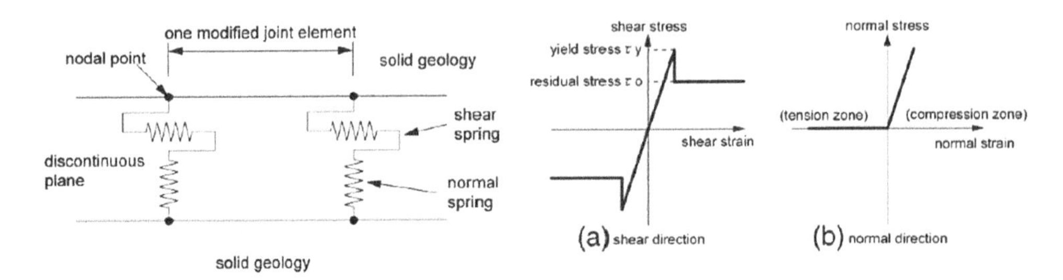

Figure 9.34 Representation of joint elements for faults and their constitutive law (from Fukushima et al. 2010). (a) Shear direction and (b) normal direction.

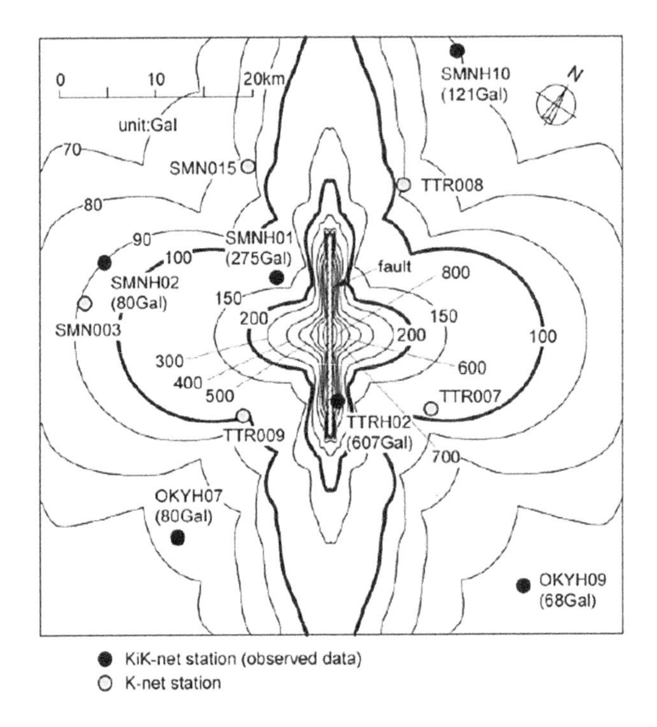

Figure 9.35 Comparison of computed and observed maximum ground acceleration for the 2000 Tottori earthquake (from Fukushima et al. 2010).

Iwata et al. (2016) recently investigated the strong motions induced by the 2014 Nagano–Hokubu earthquake. The model is based on a 3D-FEM version with the consideration of fault plane using joint elements. Figure 9.36a shows the fault parameters and Figure 9.36b shows the 3D mesh of the earthquake fault and its vicinity. Figure 9.37a shows the time histories of surface acceleration at distances of 1 and 2 km from the surface rupture in the 3D-FEM model. The rupture time is about 7–8 seconds. The maximum acceleration is higher on the east-side (hanging-wall) than that on the west-side (footwall), which is close to the general trend observed in strong motion records.

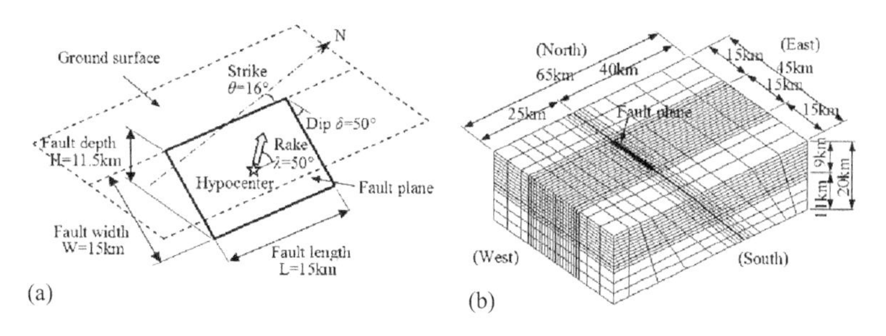

Figure 9.36 Fault model (a) and 3D FEM mesh (b) (from Iwata et al. 2016).

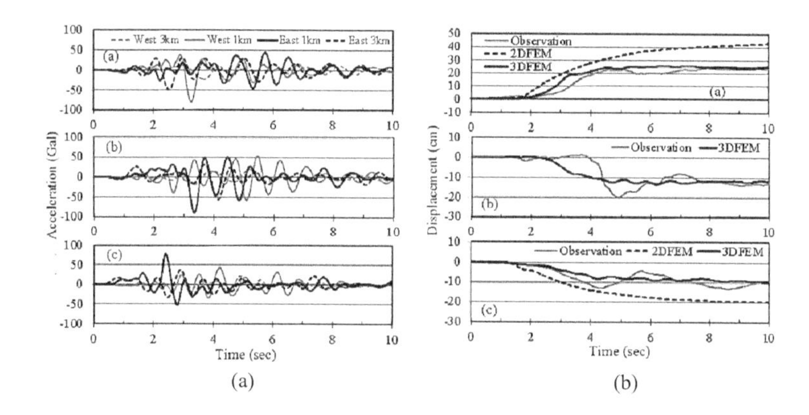

Figure 9.37 Acceleration (a) and displacement (b) responses (from Iwata et al. 2016).

Nevertheless, the computed acceleration was less than the measured accelerations. Figure 9.37b shows the time histories of surface displacement at distances of 1 and 2km from the surface rupture. The east side of the fault moves upward with respect to the footwall together with the movement to the north direction and the vertical displacement of the east-side is larger than that of the footwall and the computed results are close to the observations. However, Iwata et al. (2016) concluded that it was necessary to utilize finer meshes for better simulations of ground accelerations, which requires the use of super-computers.

Iwata et al. (2019) recently reanalyzed this earthquake. Figure 9.38 shows the slip distribution on the fault plane. The slip spreads from the hypocenter and reaches the ground surface. The maximum slip amount is 1.69 m; by considering the bending of the fault plane and the elastic velocity structure at a shallow depth, it was further possible to simulate the seismic behavior by the 3D-FEM approach, which are comparable with the observations. In addition, the refinement of the finite element mesh and the utilization of representative properties of the earth's crust resulted in better simulations of the ground accelerations and permanent and transient displacements.

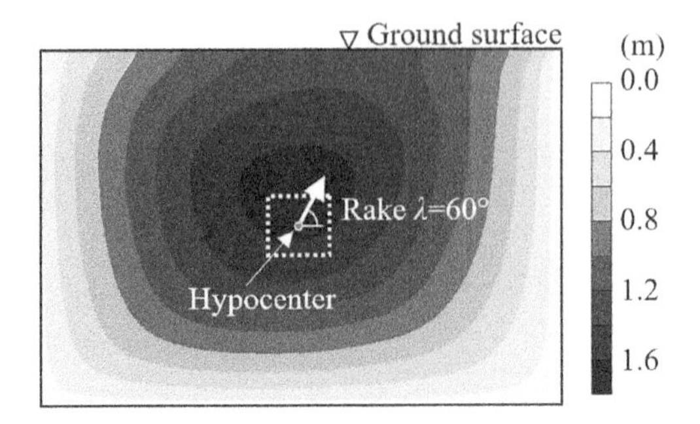

Figure 9.38 Distribution of slip on the fault plane for the 2014 Nagano earthquake (from Iwata et al. 2019).

9.2.2 Pseudo-dynamic analyses on the interaction of structures and earthquake faults

The most important aspect of earthquake engineering is the interaction between structures and fault breaks. For this purpose, a truss structure straddling over the projected fault trace on the ground surface was considered, and normal faulting and thrust faulting conditions are imposed through prescribed displacement at selected points as in the computations using a pseudo-dynamic version of the discrete finite element method (Aydan et al. 1996; Aydan 2002). The material properties for solid and fault are given in Table 9.12. The elastic modulus and the cross-section area of a typical truss were chosen as 90 GPa and 0.1 m^2 and their behavior were assumed to be elastic. Figure 9.39 shows the finite element meshes and boundary conditions used in simulations. Figure 9.40 shows the deformed configurations at computations steps 1 and 10 for normal and thrust type faulting modes. In both cases, the truss structure tilts. While the thrust type faulting causes the contraction of trusses, the normal faulting condition results in the extension of trusses and separation of the supporting members fixed to the ground. Although trusses were assumed to be behaving elastically in these simulations, it is quite easy to implement their elasto-plastic behavior within the discrete finite element method (DFEM).

9.2.3 Dynamic stability conditions of a single rock block

Tokashiki et al. (1997a) reported a very interesting study on the stability of a single block on an incline. The correct kinematic conditions for sliding and toppling of a single block on an incline have been given by Sagaseta (1986) and Aydan et al. (1989). The chart is referred to as the Sagaseta–Aydan chart or briefly S–A chart hereafter.

Table 9.12 Material properties used in discrete finite element method simulations

Material	λ *(MPa)*	μ *(MPa)*	γ *(kN/m^3)*	*c (MPa)*	ϕ *(°)*	σ_t *(MPa)*
Solid	2000	2000	26	–	–	–
Fault	50	50	–	0.0	40	0.0

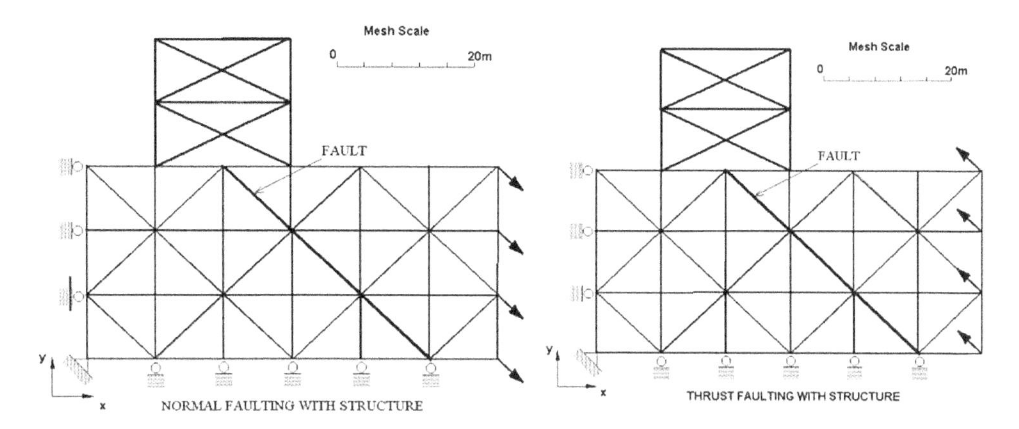

Figure 9.39 Finite element meshes and boundary conditions for fault-structure interaction simulations.

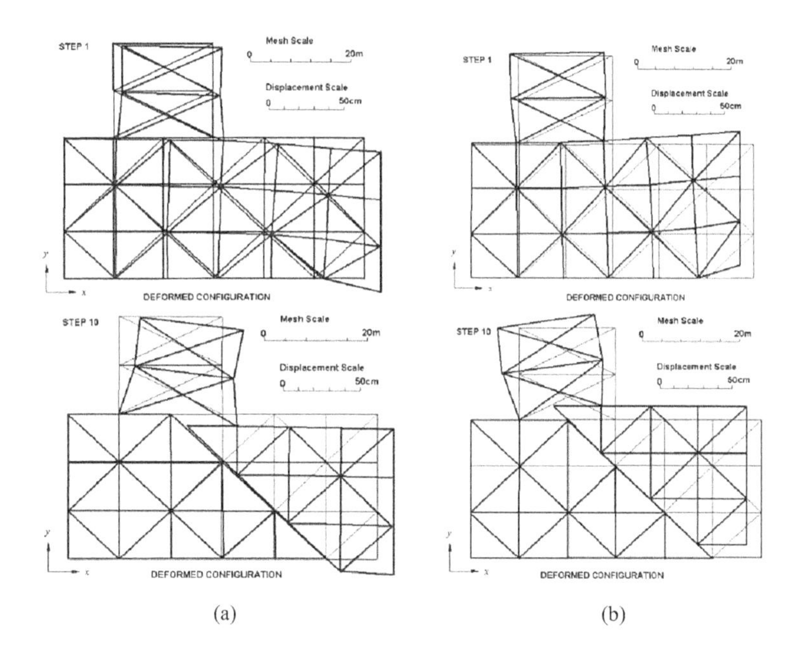

Figure 9.40 Simulations of the fault-structure interaction for normal faulting and thrust conditions. (a) Normal faulting and (b) thrust faulting.

First computed configurations of a square block (4×4m) and a rectangular block (12×4m) on a plane inclined at an angle of $\alpha = 26.57°$ are shown in Figure 9.41. The friction angle between the block and the incline was 25°. The square block slides on the incline (time step $\Delta t = 0.04$ s) while the rectangular block topples (time step $\Delta t = 0.01$ s). These predictions are consistent with the kinematic conditions for the stability of a single block given. In the S–A chart, four modes of behavior, namely, (I) stable, (II)

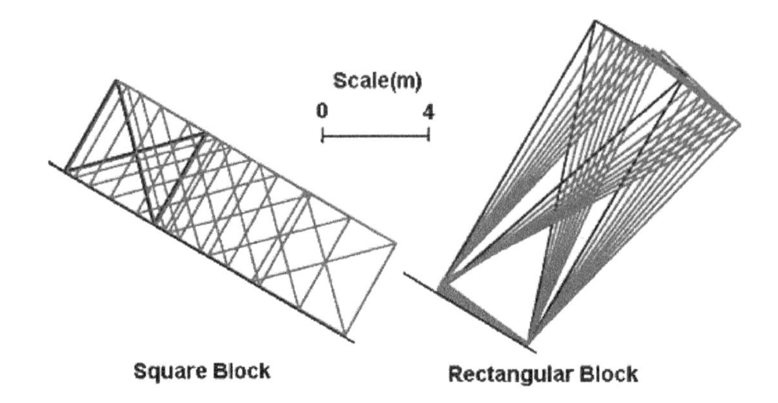

Figure 9.41 Deformed configuration of a square and rectangular block.

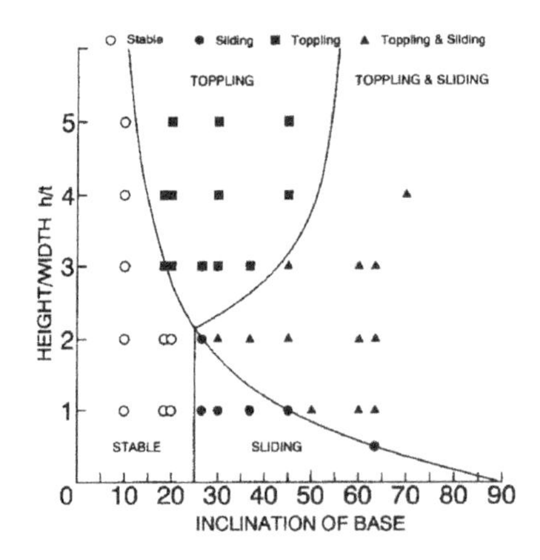

Figure 9.42 Comparison of block failure modes with theoretical stability S–A chart.

sliding only, (III) toppling (forward) only and (IV) sliding and toppling (forward) were observed. The DFEM was applied to study the one-block-on-an-incline case, and the results were compared with those predicted by the S–A chart. For a methodical comparison, the slope angle α and the ratio of block height to width h/t were varied systematically, while the friction angle was fixed at 25°. Different symbols representing different modes of behavior are plotted on the S–A chart for $\phi = 25°$ as shown in Figure 9.42. As seen from these plots, the computational results are in complete agreement with the S–A chart despite the theoretical solutions that do not take into account the deformability of the block.

9.2.4 Stability of a slope against planar sliding

Aydan et al. (1997) analyzed a planar sliding failure of slopes with a tension crack. It was assumed that a block bounded by the slope surface and a vertical discontinuity and a discontinuity inclined at an angle of 33.69°. The cohesion of the failure plane

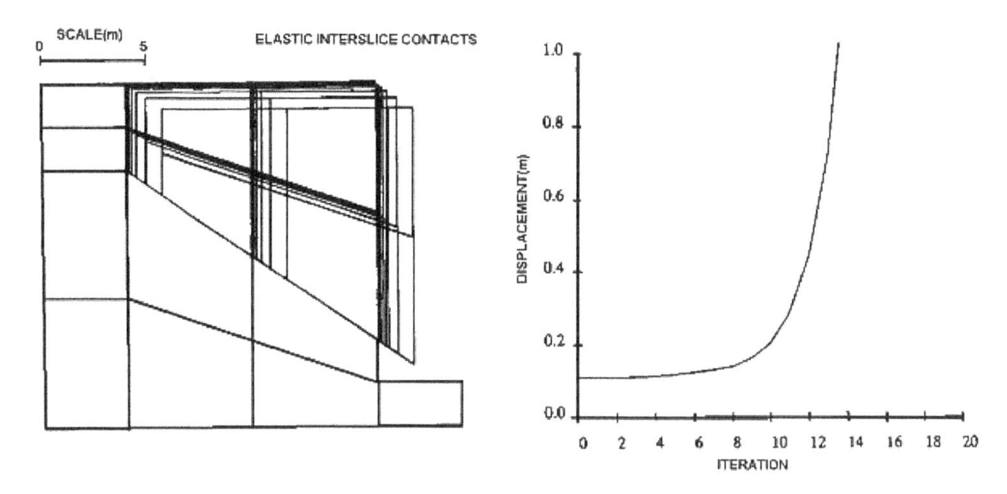

Figure 9.43 Simulation of planar sliding with a tension crack (from Aydan et al. 1997).

Table 9.13 Properties of intact rock and contacts

Rock		*Contacts*				
λ *(MPa)*	μ *(MPa)*	E_n *(MPa)*	G_s *(MPa)*	*c (MPa)*	σ_t *(MPa)*	ϕ *(°)*
10	10	0.3	0.3	0	0	20

and the unit weight of the region susceptible to sliding were assumed to be 0 and 23 kN/m^3, respectively. Its friction angle was selected as 20°. The safety factors computed from limiting equilibrium methods were the same, that is, 0.5459. The reason for such a coincidence is that the failure plane has the same inclination for each slice and the resistance and disturbance forces along this plane are not affected by the interslice forces as theoretically shown by Aydan et al. (1989). The stability of the slope was analyzed by the DFEM and the results are shown in Figure 9.43. In this figure, deformed configurations and the absolute displacement of the crest of the slope are plotted. The material properties used in numerical analyses are given in Table 9.13. As expected, separation at the vertical discontinuity and sliding along the inclined discontinuity occurred. The displacement of the sliding block increased as the iteration number increased.

9.2.5 Stability of rock slope against columnar toppling

The example described herein is concerned with the columnar toppling failure of rock slopes. The material properties used in numerical analyses are given in Table 9.14. It was assumed that a basal plane inclined at an angle of 12° dipping into slope and discontinuities dipping into the mountainside with an inclination of 70° exist within the slope. Figure 9.44 shows the configuration of the slope at each iteration step and the vertical displacement of the slope crest. As expected, each rock column tends to rotate clockwise and the slope fails in toppling mode.

Table 9.14 Properties of intact rock and contacts

Properties of blocks			*Properties of contact*							
λ *(MPa)*	μ *(MPa)*	γ *(kN/m³)*	E_n^b *(MPa)*	G_s^b *(MPa)*	E_n^j *(MPa)*	G_s^j *(MPa)*	c *(MPa)*	σ_t *(MPa)*	ϕ^b *(°)*	ϕ^j *(°)*
	10	25	0.6	0.6	0.1	0.1	0	0	20	10

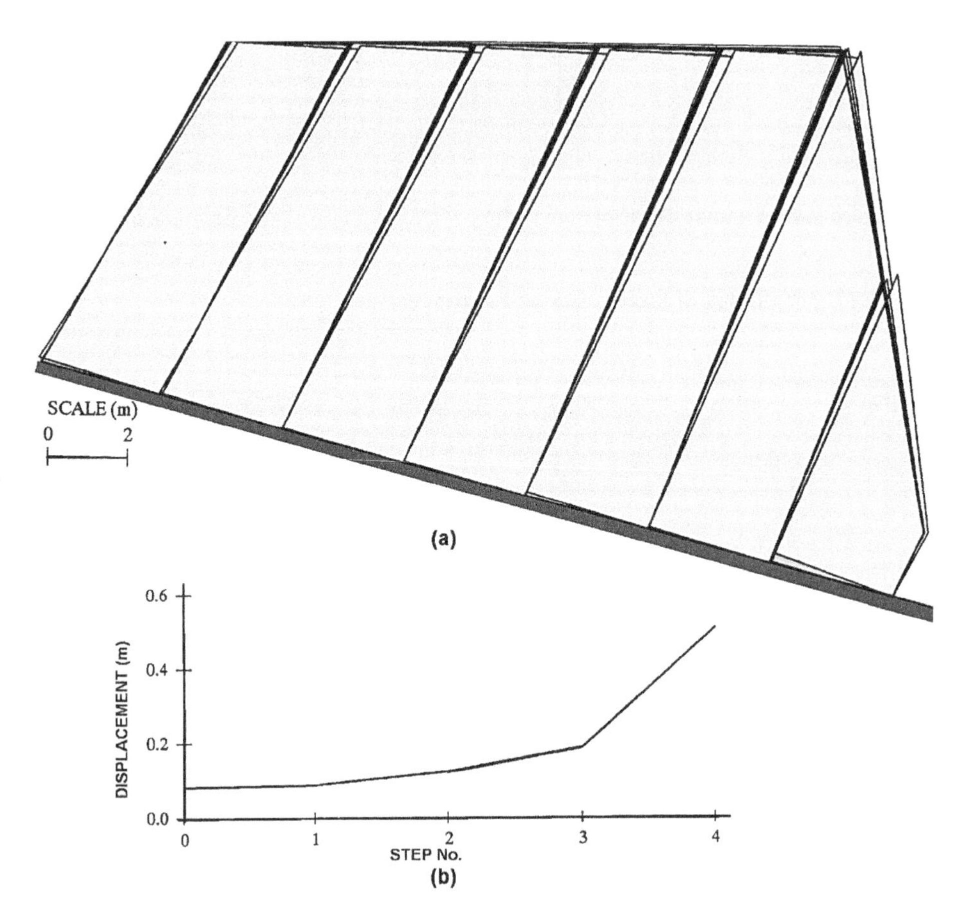

Figure 9.44 The simulation of columnar toppling (from Aydan et al. 1996). (a) Deformed configuration at different time steps; (b) deformation of a selected point.

9.2.6 Stability of rock slope against flexural toppling and its stabilization

The stability of rock slopes against flexural toppling was formulated by Aydan and Kawamoto (1987, 1992). This was the first time for mathematical treatment of this instability mode under static and dynamic conditions with/without reinforcement.

To investigate the effect of rockbolting, it is assumed that two layers inclined at various angles with respect to horizontal are subjected to a gravitational force field (Figure 9.45a) and material properties given in Table 9.15 were utilized in computations and the thin-band element (Aydan 1989; Aydan et al. 1990; Appendix 6) was used to simulate the interface behavior. Four different case studies were made to see the differences between the various bolting patterns (Figure 9.45b). Figure 9.46a-1 and a-2 show the fiber stress distributions at the base of the layers for four different cases for two inclinations 45° and 75°. It is also quite interesting to note that there is no interaction between two layers of the same height implying that no load is transferred to the lower layer from the upper layer. As noted from the figures, the behavior of a multi-layered column approaches that of a monolithic column of equivalent thickness through rockbolting.

Figure 9.46b-1, b-2, c-1 and c-2 show the shear stresses in bolts at the discontinuity plane and the axial stress distributions in the bolts, respectively. The distribution of stresses in bolts differs depending upon the location. It is expected that the

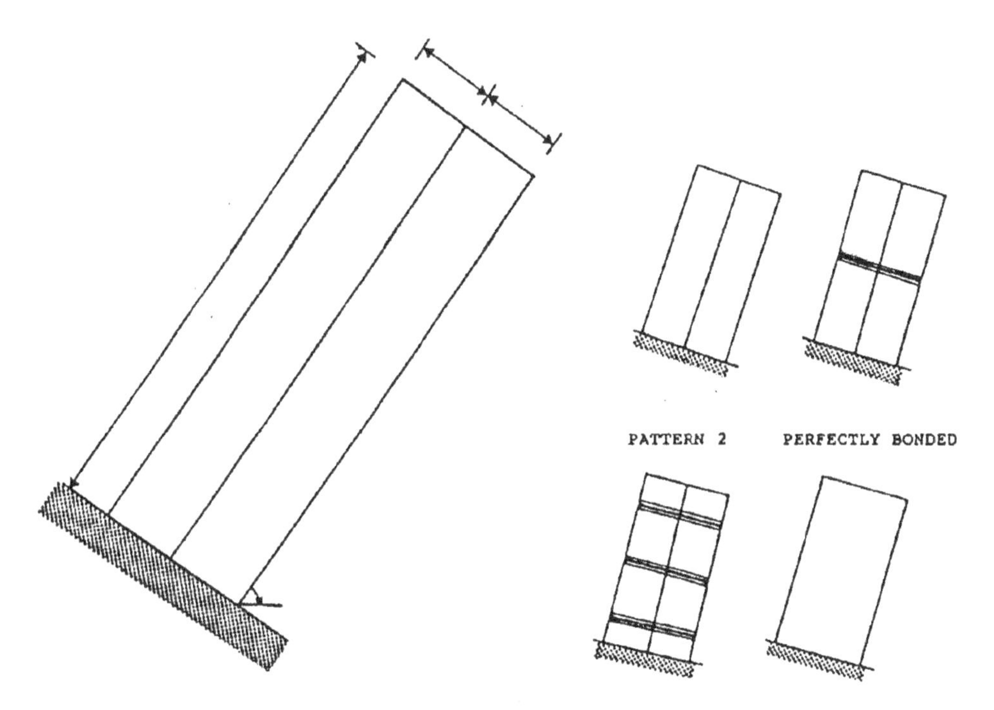

Figure 9.45 Physical model and reinforcement pattern for two layers subjected to flexural toppling failure.

Table 9.15 Mechanical properties and dimensions

Materials	σ_t *(MPa)*	γ *(kN/m³)*	ϕ *(°)*	*H (m)*	*t (m)*	*d (mm)*	Ω *(°)*	θ *(°)*	*n*
Rock	0.5	25	–	20	0.5	–	–	–	–
Discontinuity	–	–	30	–	0.01	–	–	–	–
Rockbolts	300	–	–	–	–	25	0.56	30	2

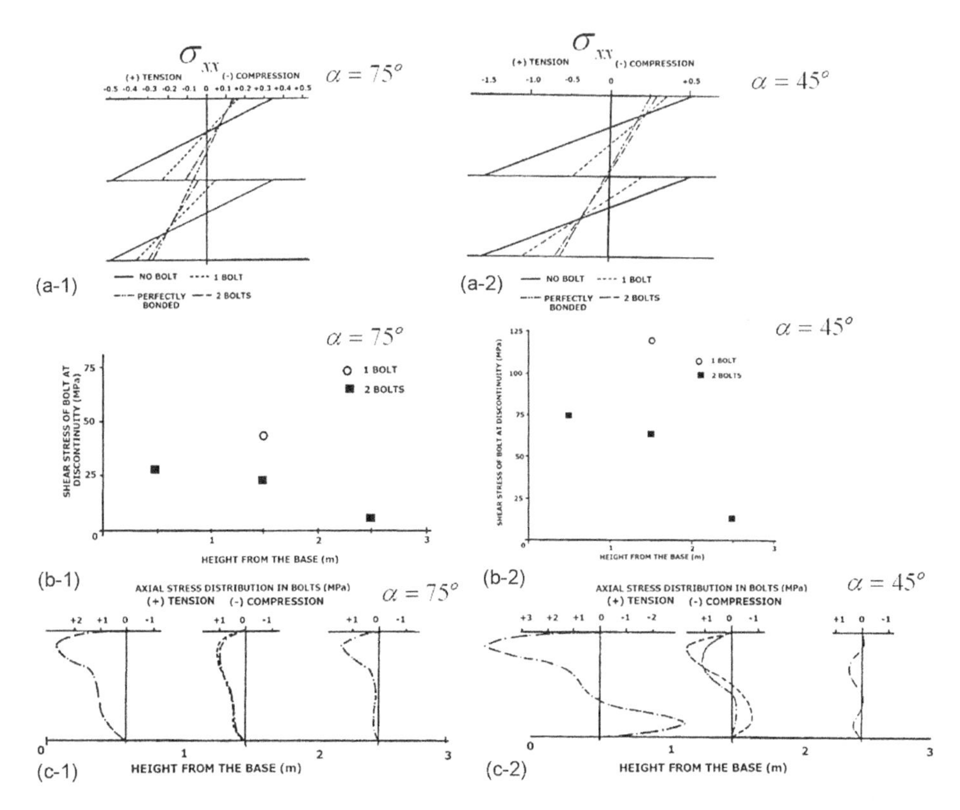

Figure 9.46 Shear reinforcement against a flexural toppling failure. (a1) and (a2) Bending stress at the fixed side f the layers; (b1) and (b2) axial stress of bolts at discontinuities; (c1) and (c2) axial stress distributions in bolts.

reinforcement offered by rockbolts by thickening the layers is quite effective. Furthermore, the layers act independently while interslice forces are nil.

In the next example, a 4-m high rock-slope with a potential failure mode of flexural toppling failure was considered. The layers were dipping mountainside with an inclination of 80°. A horizontally installed rockbolt stitching of the first three layers was assumed to be installed. Material properties given in Table 9.15 were used. Figure 9.47 shows the DFEM mesh with/without bolt and deformed configurations of the slope at the computation step of 10. As noted from the figure, the rockbolt greatly restrain the deformation of layers as expected. Furthermore, the third and second layers exert forces to the front layer and the contact condition is very close to the top of the layer, which confirms the assumption of Aydan and Kawamoto (1987, 1992) used for analyzing slopes and underground openings against flexural toppling failure.

9.2.7 Retrofitting of unlined tunnels

The Unten tunnel in the Nakijin region of Okinawa Island (Japan) was an unlined single lane roadway tunnel. Following the collapse of the Toyohama tunnel in Hokkaido Island in 1996, the authorities were ordered to check the safety of all roadway

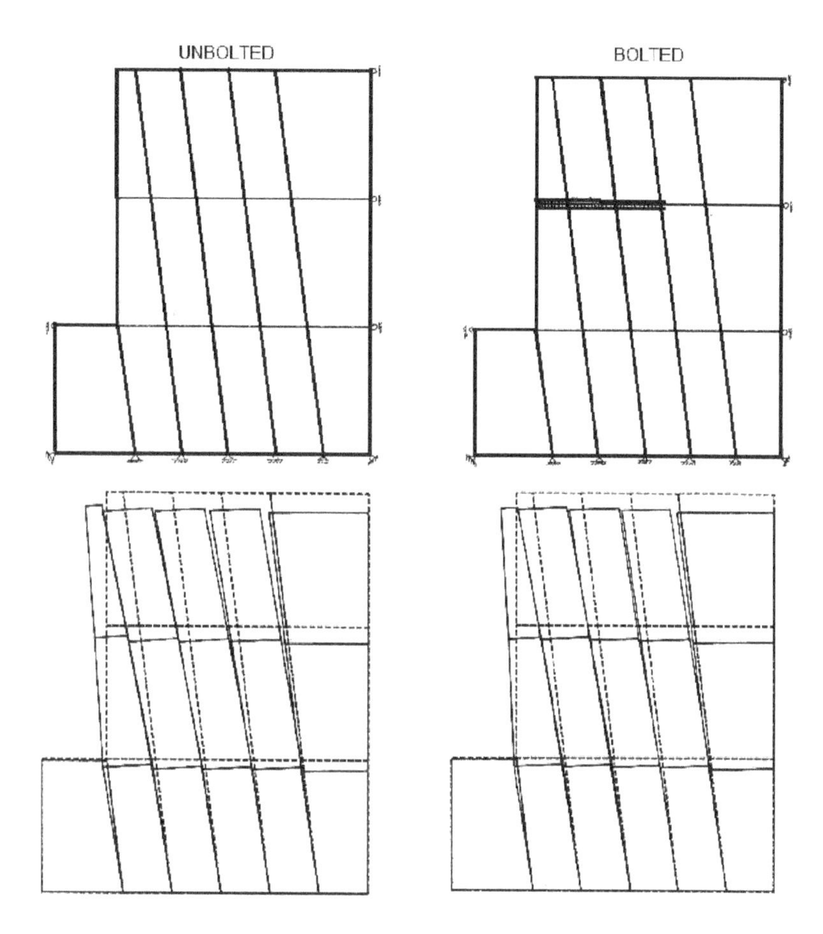

Figure 9.47 DFEM mesh and deformed configurations of unbolted and bolted rock slope models against flexural toppling failure.

tunnels in Japan. The Unten tunnel was designated as unsafe after the checking procedure and it was decided to close to traffic. However, the strong demand to keep the tunnel open to traffic by residents resulted in the reassessment of the stability of the tunnel and its retrofitting.

The site investigations revealed several thoroughgoing discontinuities as shown in Figure 9.48. Model experiments using the base friction apparatus indicated that the tunnel might be unstable if the frictional properties of discontinuities decrease with time. The reduction of friction angle of discontinuities was achieved by introducing the double layer Teflon sheets along discontinuities in the model tests.

The DFEM was used to assess the stability of the tunnel (see Tokashiki et al. 1997b, for the details of numerical analyses). The DFEM analyses also indicated that the tunnel might become unstable if the frictional properties of discontinuities were drastically reduced. For retrofitting of the tunnel, glass-fiber rockbolts were installed and reinforced concrete lining was constructed. Figure 9.49 shows

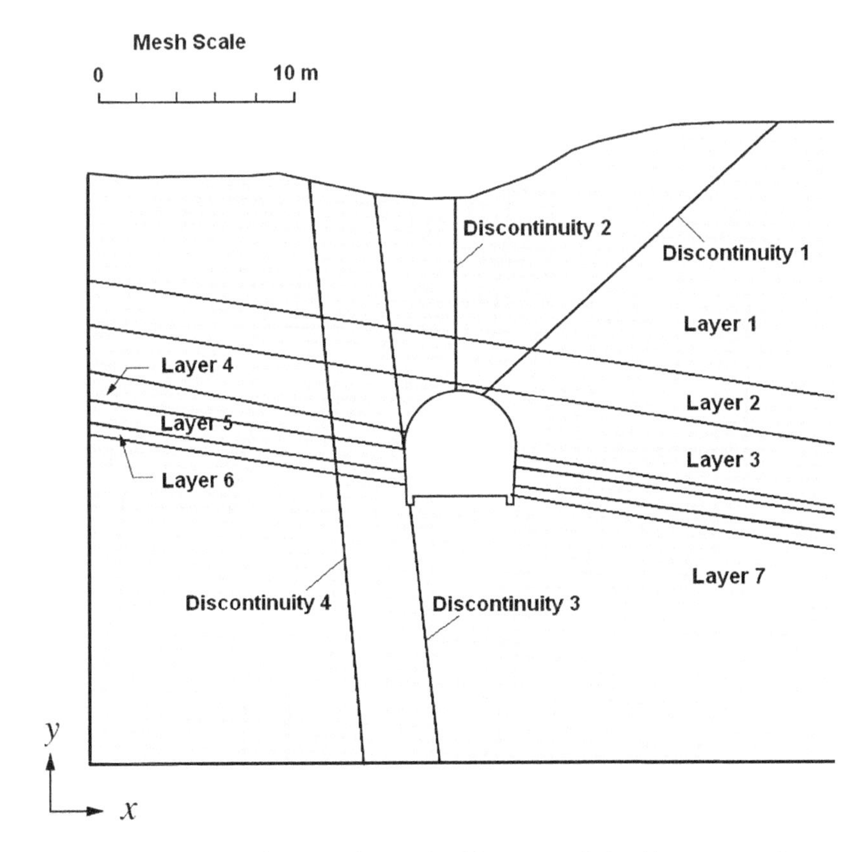

Figure 9.48 Distribution of major discontinuities around the Unten tunnel.

the deformed configurations of the tunnel without and with retrofitting. As noted from the figures, the tunnel should be stable if the selected measures of retrofitting are employed.

9.2.8 Analysis of backfilling of abandoned mines

9.2.8.1 Stability of pillars

A series of analyses using DFEM was carried out on the supporting effect of backfilling on a fractured pillar of an abandoned lignite mine. The overburden was 6m, the pillar height was 4m and the width was 2m. Table 9.16 gives material properties used in the analyses. Figure 9.50 shows the finite element meshes used in analyses. The pseudo-dynamic version of DFEM analyses was used in the analyses and computations were carried out up to 12 steps.

Figure 9.51 shows the deformed configuration of non-backfilled and backfilled pillars at the computation step of 12. Figure 9.52 shows the settlement of the pillar. As noted from the figures, when the pillar is not supported by backfilling, sliding occurs

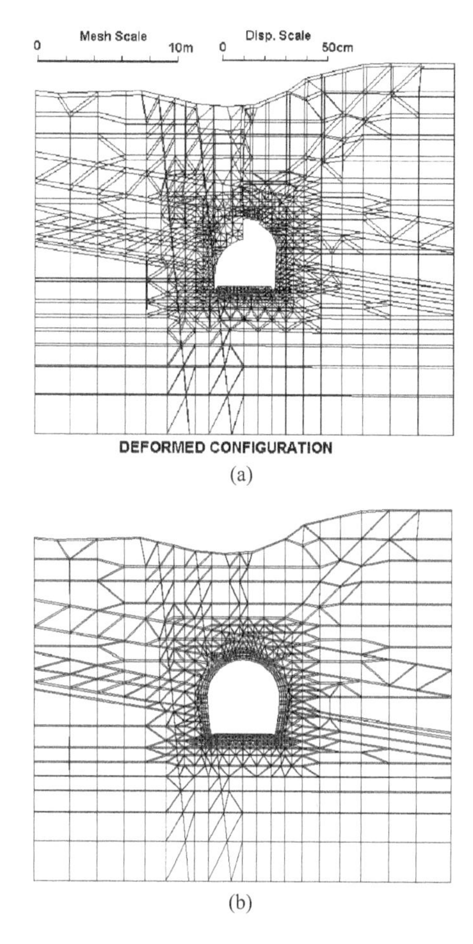

Figure 9.49 Deformation of surrounding rock mass with/without the measures of retrofitting. (a) No retrofitting and (b) with retrofitting.

Table 9.16 Material properties used in DFEM analyses

Material	$\lambda(E_n)$ *(MPa)*	$\mu(E_s)$ *(MPa)*	γ *(kN/m³)*	*c (kPa)*	ϕ *(°)*	σ_t *(kPa)*
Top layer	200	200	22			
Lignite	200	200	14			
Lower layer	300	300	22			
Backfilling	100	100	14			
Fracture plane	200	10	–	10	10	5

along the fracture plane and a vertical crack appears in the middle of the pillar. However, if the pillar is supported by backfilling, the sliding and separation of fracture planes are restrained and the pillar is stabilized by the backfilling material. Although the backfilling material is quite soft, its supporting effect is remarkable.

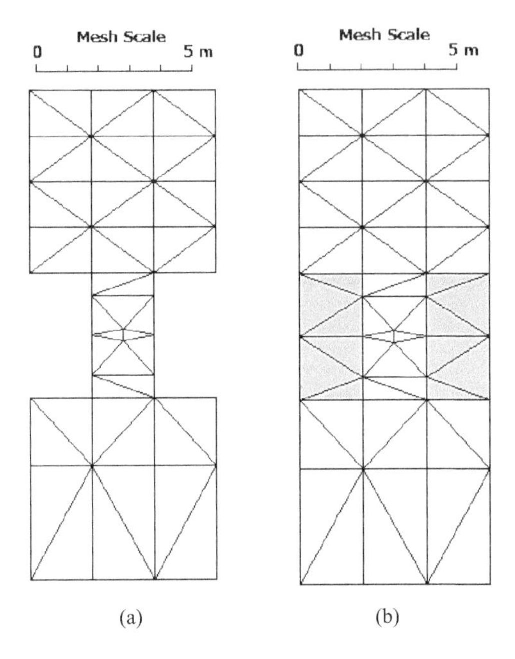

Figure 9.50 Finite element meshes used in DFEM analyses. (a) No backfilling and (b) backfilled.

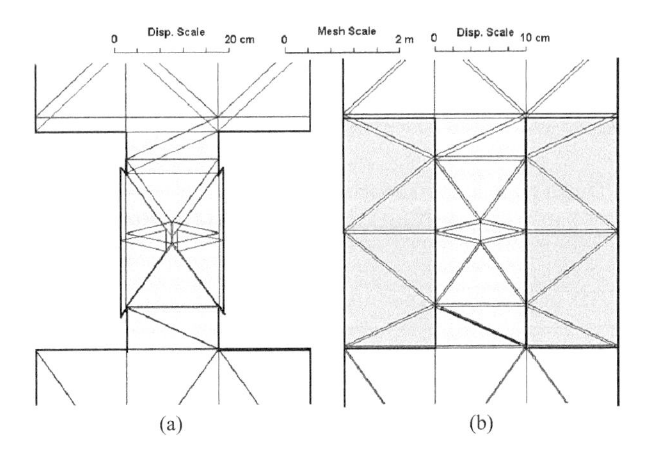

Figure 9.51 Deformed configurations of the pillar for non-backfilling and backfilling cases. (a) No backfilling and (b) backfilled.

9.2.8.2 The stability of an abandoned room and pillar mine next to a steep cliff

An abandoned lignite mine with an overburden of 4 m next to a steep cliff was considered as shown in Figure 9.53. The pillar height was assumed to be 2 m with a width of 1 m. The computations were carried out up to 12 steps and the deformed configurations

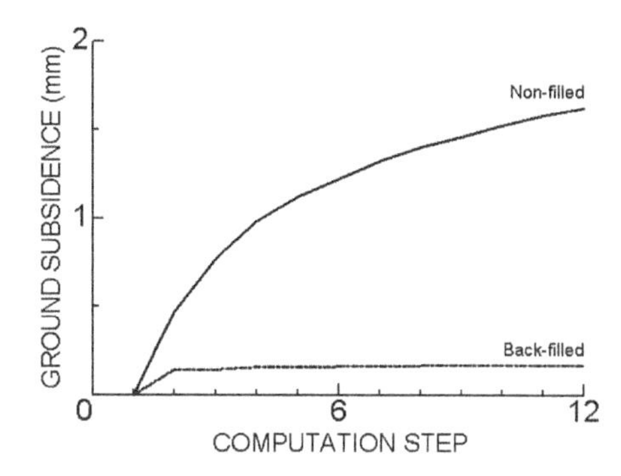

Figure 9.52 Settlement of the pillar for non-backfilling and backfilling cases.

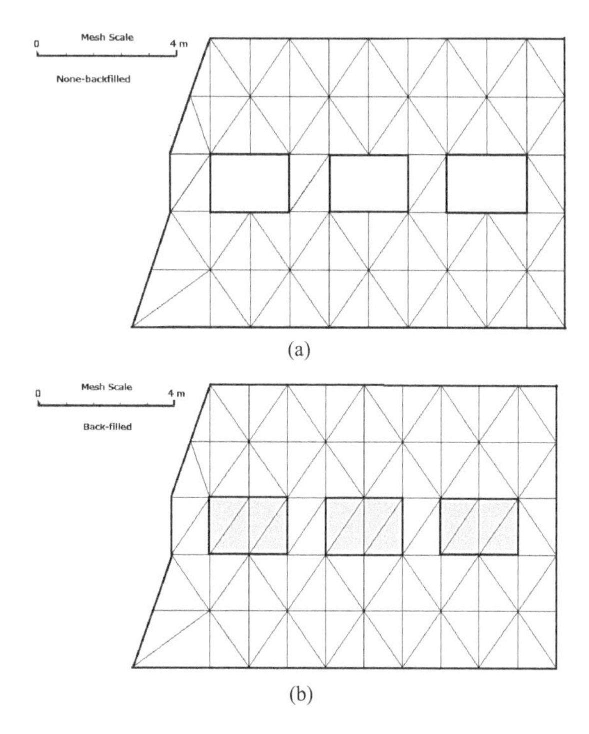

Figure 9.53 Abandoned lignite mine with an overburden of 4 m next to a steep cliff.

of the abandoned mine for two cases are shown in Figure 9.54. The material properties used were the same as those used in the previous analyses. The pillars near the cliff were assumed to have fracture planes, which extended to the ground surface from the second pillar. When the results of analyses were compared, the settlement of the abandoned mine was restrained and the abandoned mine was stable while the abandoned mine is unstable and the settlement continues. These analyses again clearly show the

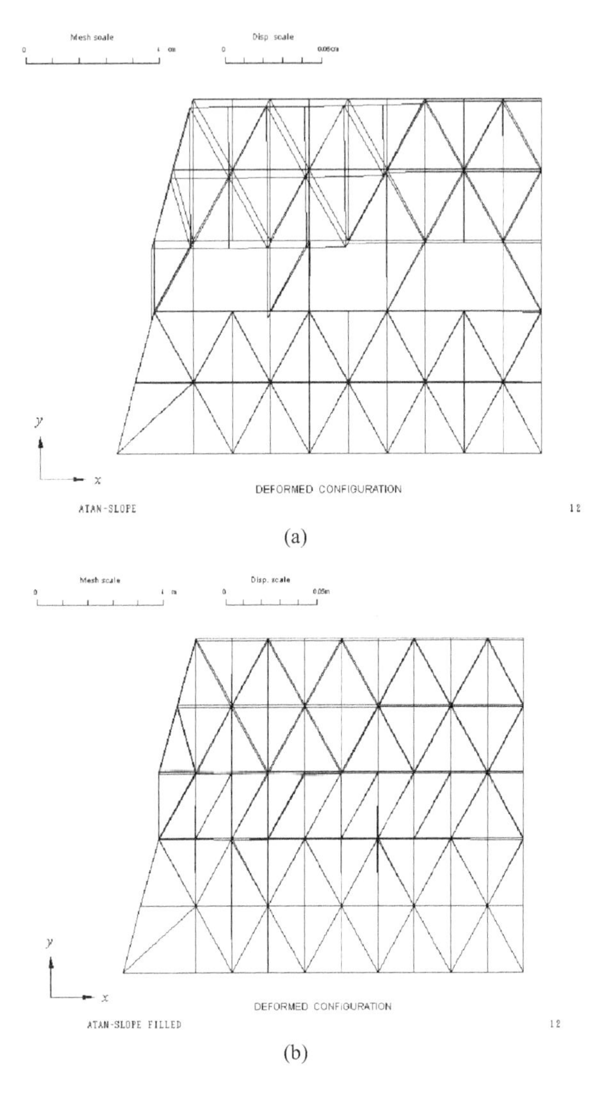

Figure 9.54 Deformed configuration of the pillar for non-backfilling (a) and backfilling (b) cases at computation step 12.

effect of backfilling of abandoned mines even though its material properties are relatively soft and weak.

9.2.9 Simulation of creep-like deformation of the Babadağ landslide by DFEM

Kumsar et al. (2015) described a numerical model for creep-like ground deformations at the Gündoğdu district of Babadağ, which could not be evaluated by classical sliding-type models as the failure process would take place in few seconds. On the other hand, ground deformations have been taking place for decades in the district since the 1940s. The fundamental concept has been implemented in the DFEM, in which cyclic softening and hardening of the contact zone is modeled. The flow-chart of the implementation of the DFEM code for this situation is shown in Figure 9.55.

Figures 9.56 and 9.57 show the maximum shear stress contours and deformed configurations for selected time steps for rainfall data starting from May 2011 continuing

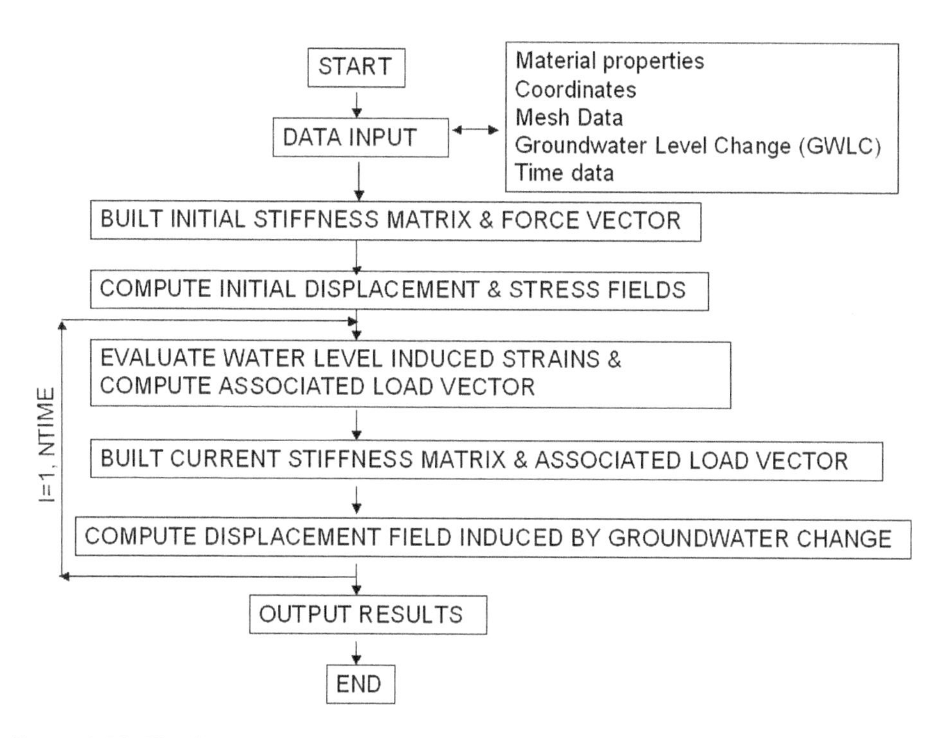

Figure 9.55 The flow-chart of DFEM-CSH.

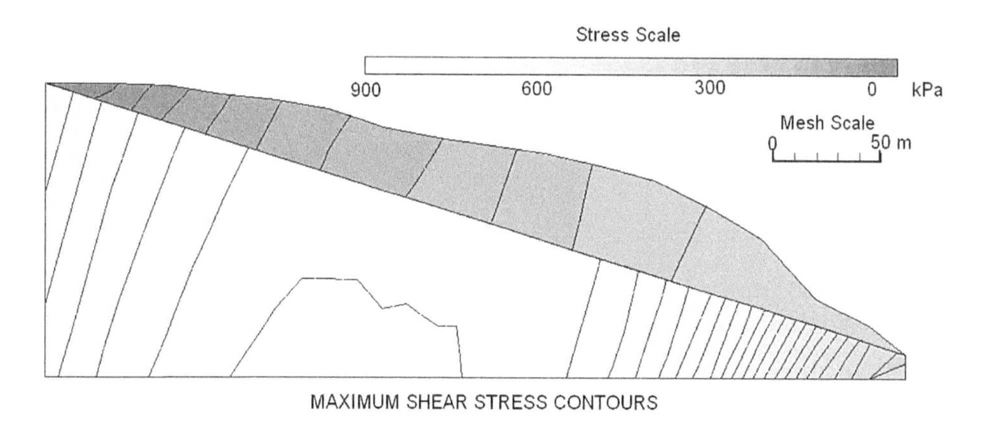

Figure 9.56 Computed maximum shear stress contours.

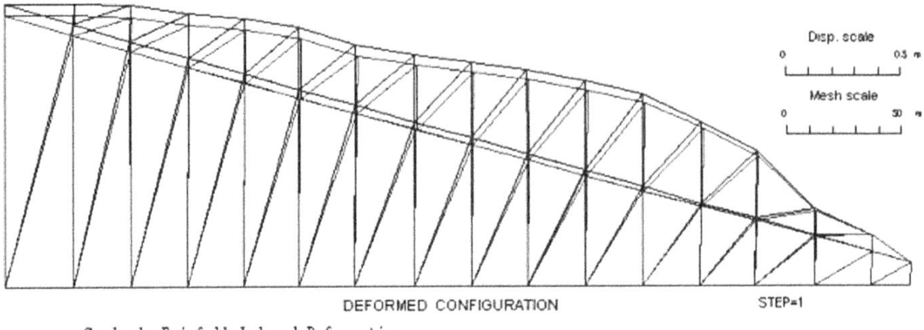

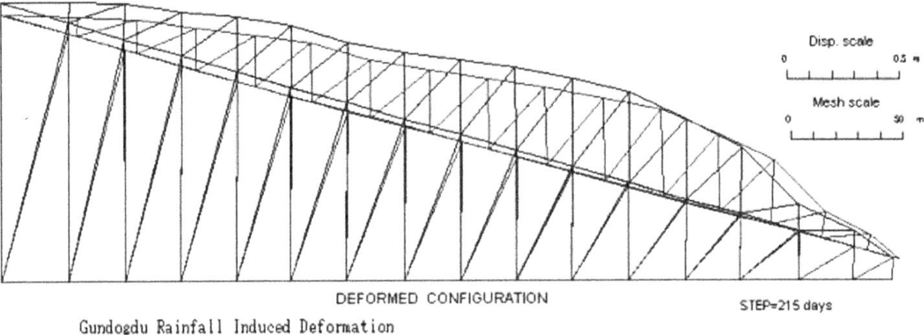

Figure 9.57 Deformed configurations of the analyzed domain at time steps 1 and 215 days.

into 2012. The yielding of the contact zones did not occur. However, the deformation of the body took place upon each cycle of saturation and drying. As noted in Figure 9.57, a very large displacement of the unstable zone does occur.

Figure 9.58 shows the response of three points at the rear, toe and middle-top of the potentially unstable body. As noted from the figure, the displacements differ and it is not purely a rigid-body-like ground deformation. The maximum ground deformation occurs at the middle top of about 150 mm on day 215.

Figure 9.59 shows horizontal ground deformation at a given section for different time-steps. It is very interesting to notice that the overall ground deformation resembles those measured from the pipe-strain gauge in the field as reported by Kumsar et al. (2015).

9.2.10 Simulation of creep-like deformation of a rock block at the Nakagusuku Castle

A multi-parameter monitoring system was also initiated at the Nakagusuku Castle in 2013. The system at the castle was the first attempt regarding the masonry structures in the world. The monitoring initiated in December 2013 has been continuing. During the period of measurements, some earthquakes occurred and long-term creep-like

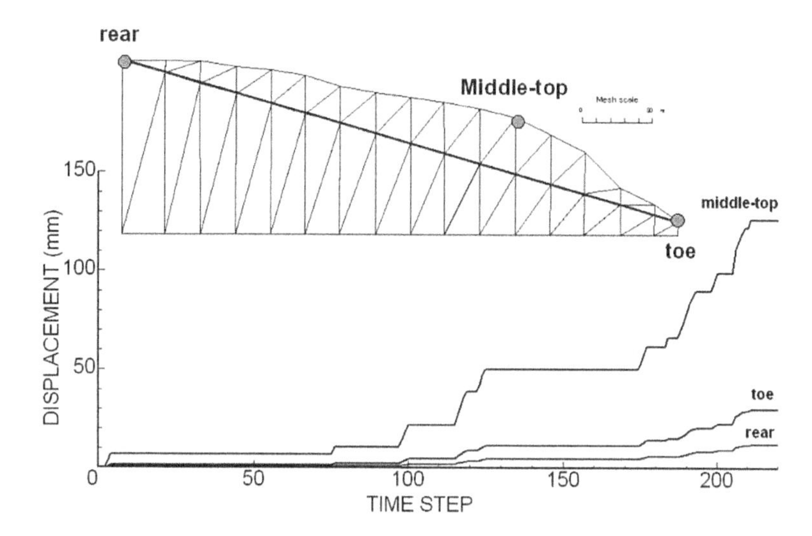

Figure 9.58 Displacement responses of three selected points in the analyzed domain.

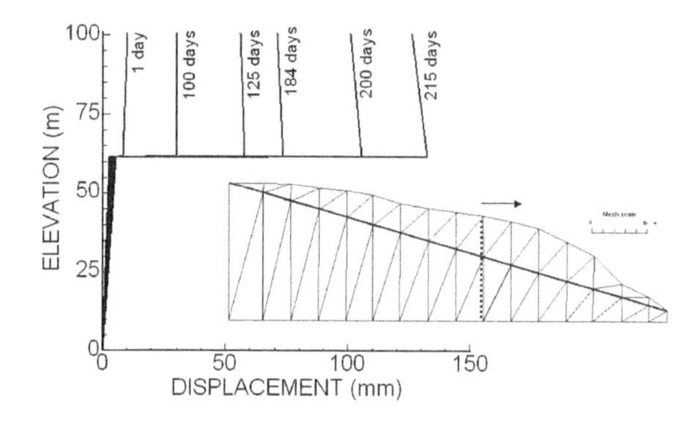

Figure 9.59 Ground deformation at a given section for different time-steps.

separation of a huge crack in the Ryukyu Limestone layer extending to the Shimajiri formation layer beneath has been taking place. Instruments were installed in the upper part of the crack as illustrated in Figure 9.60. An earthquake with a moment magnitude of 6.5 occurred at 5:10 AM on 2014 March 13 (JST) in the East China Sea at a depth of 120 km on the western side of the Okinawa Island. Another earthquake occurred at 11:27 AM on the same day near the Kumejima Island. Although the magnitude of the earthquake was intermediate and far from the location, some permanent displacement occurred as seen in Figure 9.61.

A series of analyses using DFEM (Aydan et al. 1996; Kumsar et al. 2015) together with the implementation of softening and hardening process of weathered rock mass in relation to rainfall (Aydan 2017) were carried out. Table 9.17 gives material properties

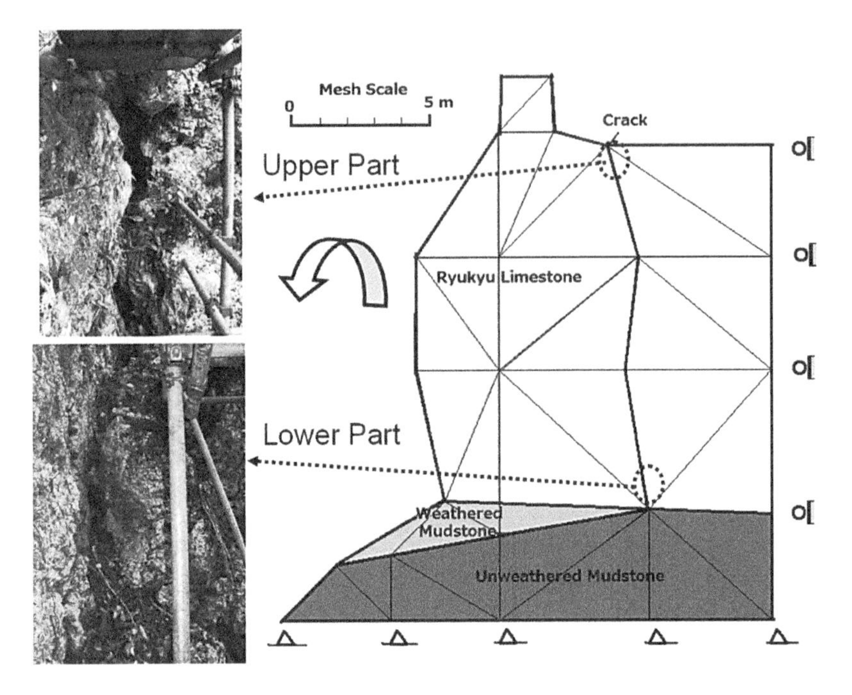

Figure 9.60 Views of monitoring location.

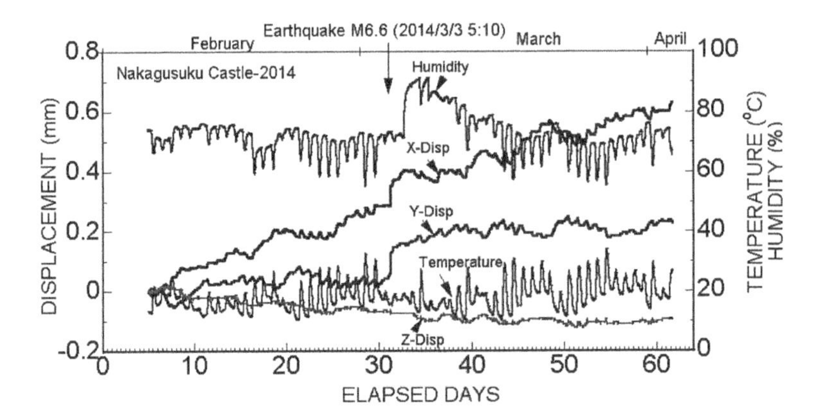

Figure 9.61 Monitoring results during February to March 2014.

and Figure 9.60 shows the mesh used in analyses. Computed results are compared with measurements in Figure 9.62. As noted in Figure 9.62, the computational model can simulate both permanent displacements induced by the earthquake as well as the rainfall-induced softening process of the weathered rock mass.

Table 9.17 Material properties used in DFEM analyses

Material	*λ (MPa)*	*μ (MPa)*	*γ (kN/m³)*	*c (kPa)*	*φ (°)*	*σ_t (MPa)*
Ryukyu limestone	900	900	25	–	–	–
Mudstone	400	400	22	–	–	–
Weathered mudstone	300	100	20	–	–	–
Discontinuity-L	100	20	–	10	10	1
Discontinuity-V	0.02	0.02	–	10	10	1

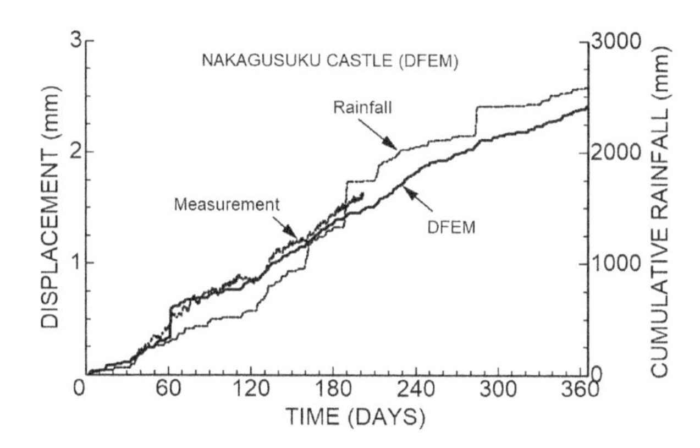

Figure 9.62 Mesh used in DFEM analysis and comparison with measured responses.

References

Anderson, L.D. and R.S. Hart 1976. An earth model based on free oscillations and body waves. *Journal of Geophysical Research* 81, 1461–1475.

Aydan, Ö. 1989. The stabilisation of rock engineering structures by rockbolts. Doctorate Thesis, Nagoya University, Nagoya, Japan.

Aydan, Ö. 1993. A consideration on the stress state of the earth due to the gravitational pull. *The 7th Annual Symposium on Computational Mechanics*, Tokyo, 243–252.

Aydan, Ö. 1995. The stress state of the earth and the earth's crust due to the gravitational pull. *35th US Rock Mechanics Symposium*, Lake Tahoe, 237–243.

Aydan, Ö. 2000. A stress inference method based on structural geological features for the full-stress components in the earth' crust. *Yerbilimleri*, 22, 223–236.

Aydan, Ö. 2002. Actual observations and numerical simulations of surface fault ruptures and their effects on engineering structures. *The Eight Workshop on Earthquake Resistant Design of Lifeline Facilities and Countermeasures against Liquefaction*, December 15–18, Tokyo, Japan, 228–237.

Aydan, Ö. 2003. The moisture migration characteristics of clay-bearing geo-materials and the variations of their physical and mechanical properties with water content. *2nd Asian Conference on Saturated Soils* (UNSAT-ASIA 2003).

Aydan, Ö. 2011. Some issues in tunnelling through rock mass and their possible solutions. *First Asian Tunnelling Conference*, Tehran, ATS11–15, 33–44.

Aydan, Ö. 2012. Ground motions and deformations associated with earthquake faulting and their effects on the safety of engineering structures, in R. Meyers, ed., *Encyclopedia of Sustainability Science and Technology*, New York, Springer, 3233–3253.

Aydan, Ö. 2017. *Time Dependency in Rock Mechanics and Rock Engineering*, London, CRC Press, Taylor & Francis Group, 241 p.

Aydan, Ö. and T. Kawamoto 1987. Toppling failure of discontinuous rock slopes and their stabilisation (in Japanese). *Journal of Japan Mining Society*, 103(1197), 763–770.

Aydan, Ö. and T. Kawamoto 1991. A comparative numerical study on the reinforcement effects of rockbolts and shotcrete systems. The 5th *International Conference* on Computer Methods and Advances in Geomechanics, Cairns, 2, 1443–1448.

Aydan, Ö. and T. Kawamoto 1992. The stability of slopes and underground openings against flexural toppling and their stabilisation. *Rock Mechanics and Rock Engineering*, 25(3), 143–165.

Aydan, Ö. and G. Paşamehmetoğlu 1994. In-situ measurements and lateral stress coefficients in various parts of the earth. *Kaya Mekaniği Bülteni*, 10, 1–17.

Aydan, Ö. and T. Kawamoto 1997. The general characteristics of the stress state in the various parts of the earth's crust. *International Symposium on Rock Stress*, Kumamoto, 369–373.

Aydan, Ö. and P. Nawrocki 1998. Rate-dependent deformability and strength characteristics of rocks. *International Symposium on* the Geotechnics of Hard Soils-Soft Rocks, Napoli, 1, 403–411.

Aydan, Ö. and T. Kawamoto 2001. The stability assessment of a large underground opening at great depth. *17th International Mining Congress and Exhibition of Turkey*, IMCET, Ankara, 1, 277–288.

Aydan, Ö. and M. Geniş 2010. Rockburst phenomena in underground openings and evaluation of its counter measures. *Journal of Rock Mechanics*, TNGRM (Special Issue, No. 17), 1–62.

Aydan, Ö. and H. Tano 2012. The observations on abandoned mines and quarries by the Great East Japan Earthquake on March 11, 2011 and their implications. *Journal of Japan Association on Earthquake Engineering*, 12(4), 229–248.

Aydan, Ö. and R. Ulusay 2013. Geomechanical evaluation of Derinkuyu Antique Underground City and its implications in geoengineering. *Rock Mechanics and Rock Engineering*, 46, 731–754, Springer.

Aydan, Ö., T. Kyoya, Y. Ichikawa, T. Kawamoto, T. Ito and Y. Shimizu 1988. Three-dimensional simulation of an advancing tunnel supported with forepoles, shotcrete, steel ribs and rockbolts. *The 6th International Conference on Numerical Methods in* Geomechanics, Innsbruck, 2, 1481–1486.

Aydan, Ö., Y. Shimizu and Y. Ichikawa 1989. The effective failure modes and stability of slopes in rock mass with two discontinuity sets. *Rock Mechanics and Rock Engineering*, 22(3), 163–188.

Aydan, Ö., Y. Ichikawa and T. Kawamoto 1990. Numerical modelling of discontinuities and interfaces in rock mass. *The 4th Symposium on Computational Mechanics of Japan*, 254–261.

Aydan, Ö., T. Akagi and T. Kawamoto 1993. Theoretical and numerical modeling of swelling phenomenon of rocks in tunnel excavations. *The 2nd Asian-Pacific Conference on Computational Mechanics*, Sydney, 331–336.

Aydan, Ö., T. Ito, T. Akagi and T. Kawamoto 1994. Theoretical and numerical modeling of swelling phenomenon of rocks in rock excavations. *International Conference* on Computer Methods and Advances in Geomechanics, IACMAG, Morgantown, 3, 2215–2220.

Aydan, Ö., I.H.P. Mamaghani and T. Kawamoto 1996. Application of discrete finite element method (DFEM) to rock engineering structures. *NARMS'96*, 2039–2046.

Aydan, Ö., H. Kumsar, R. Ulusay and Y. Shimizu 1997. Assessing limiting equilibrium methods (LEM) for slope stability by discrete finite element method (DFEM). *IACMAG*, Wuhan, 1681–1686.

Aydan, Ö., R. Ulusay and E. Yüzer 1999. Man-made structures in Cappadocia, Turkey and their implications in rock mechanics and rock engineering. *ISRM News Journal*, 6(1), 63–73.

Aydan, Ö., M. Daido, H. Tano, S. Nakama and H. Matsui 2006. The failure mechanism of around horizontal boreholes excavated in sedimentary rock. *50th US Rock mechanics Symposium*, Paper No. 06–130 (on CD).
Aydan, Ö., S. Watanabe and N. Tokashiki 2008a. The inference of mechanical properties of rocks from penetration tests. *5th Asian Rock Mechanics Symposium (ARMS5)*, Tehran, 213–220.
Aydan, Ö., H. Tano, M. Geniş, I. Sakamoto and M. Hamada 2008b. Environmental and rock mechanics investigations for the restoration of the tomb of Amenophis III. *Japan – Egypt Joint Symposium New Horizons in Geotechnical and Geoenvironmental Engineering*, Tanta, Egypt, 151–162.
Aydan, Ö., Y. Ohta, M. Daido, H. Kumsar, M. Genis, N. Tokashiki, T. Ito and M. Amini 2011. Chapter 15: Earthquakes as a rock dynamic problem and their effects on rock engineering structures, in Y. Zhou and J. Zhao, eds., *Advances in Rock Dynamics and Applications*, CRC Press, Taylor & Francis Group, 341–422.
Aydan, Ö., F. Uehara and T. Kawamoto 2012. Numerical study of the long-term performance of an underground powerhouse subjected to varying initial stress states, cyclic water heads, and temperature variations. *International Journal of Geomechanics, ASCE*, 12(1), 14–26.
Aydan, Ö., T. Ito and F. Rassouli 2016. *Chapter 11: Tests on Creep Characteristics of Rocks*, London, CRC Press, 333–364.
Fowler, C.M.R. 1990. *The Solid Earth: An Introduction to Global Geophysics*, Cambridge, Cambridge University Press.
Fukushima, K., Y. Kanaori & F. Miura 2010. Influence of fault process zone on ground shaking of inland earthquakes: verification of Mj = 7.3 Western Tottori Prefecture and Mj = 7.0 West Off Fukuoka Prefecture earthquakes, southwest Japan. *Engineering Geology*, 116, 157–165.
Geniş, M. and Ö. Aydan 2007. Static and dynamic stability of a large underground opening. *Proceedings of 2nd Symposium on Underground Excavations for Transportation*, 317–326.
Geniş, M. and Ö.Aydan (2013). A numerical study on the ground amplifications in areas above abandoned room and pillar mines and longwall old mines. The 2013 ISRM EUROCK International Symposium, Gliwice, 733–737.
Geniş, M. and Ö. Aydan 2020. Dynamic analyses of abandoned mines in Japan during earthquakes. *Journal of Environmental Geotechnics*. doi: 10.1680/jenge.18.00120.
Hoek, E. and E.T. Brown 1980. *Underground Excavations in Rock*, London, Institution of Mining & Metallurgy.
ITASCA 2007. *FLAC3D-Fast Lagrangian Analysis of Continua-User Manual (Dynamic Option) (Version 2.1)*, Minneapolis, MN, Itasca Consulting Group Inc.
Ito, T., T. Akagi, Ö. Aydan, R. Ulusay and T. Seiki 2016. Time-dependent properties of tuffs of Cappadocia, Turkey. EUROCK2016, Ürgüp, 229–234.
Iwata, N., K. Adachi, Y. Takahashi, Ö. Aydan, N. Tokashiki and F. Miura 2016. Fault rupture simulation of the 2014 Kamishiro Fault Nagano Prefecture Earthquake using 2D and 3D-FEM. EUROCK2016, Ürgüp, 803–808.
Iwata, N., R. Kiyota, Ö. Aydan, T. Ito and F. Miura 2019. Effects of fault geometry and subsurface structure model on the strong motion and surface rupture induced by the 2014 Kamishiro Fault Nagano Earthquake. *Proceedings of 2019 Rock Dynamics Summit in Okinawa*, May 7–11, Okinawa, Japan, ISRM (Aydan, Ö., T. Ito, T. Seiki, K. Kamemura and N. Iwata, eds.), 222–228.
Jeffreys, H. and K.E. Bullen 1940. *Seismological Tables*, London, British Association for the Advancement of Science.
Kumsar, H., Ö. Aydan, H. Tano, S.B. Çelik and R. Ulusay 2015. An integrated geomechanical investigation, multi-parameter monitoring and analyses of Babadağ-Gündoğdu Creep-like

landslide. *Rock Mechanics and Rock Engineering*, Special Issue on Deep-seated Landslides, 49(6), 2277–2299.

Mizumoto, T., T. Tsuboi and F. Miura 2005. Fundamental study on failure rupture process and earthquake motions and near a fault by 3D-FEM. *Journal of Japan Society of Civil Engineers*, 780(70), 27–40 (in Japanese with English abstract).

Nadai, A.L. 1950. Theory of flow and fracture of solids, Vol. 2, 623–624.

Rabcewicz, L. 1957. Model tests with anchors in cohesionless material (in German). *Die Bautechnik*, 34(5), 171–173.

Sagaseta, C. 1986. On the modes of instability of a rigid block. *Rock Mechanics and Rock Engineering*, Technical Note, 19, 261–266.

Sato, T., K. Aoyagi, N. Miyara, Ö. Aydan, J. Tomiyama, T. Morita 2019. The dynamic response of Horonobe Underground Research Center during the 2018 June 20 earthquake. *Proceedings of 2019 Rock Dynamics Summit in Okinawa*, May 7–11, Okinawa, Japan, ISRM (Aydan, Ö., T. Ito, T. Seiki, K. Kamemura and N. Iwata, eds.), 640–645.

Sugito, M., Y. Furumoto and T. Sugiyama 2000. Strong motion prediction on rock surface by superposed evolutionary spectra. *12th World Conference on Earthquake Engineering*, 2111/4/A, CD-ROM.

Timoshenko, S.P. and J.N. Goodier 1970. *Theory of Elasticity*. 3rd ed., New York, McGraw-Hill Int. Book Company.

Tokashiki, N., Ö. Aydan, I.H.P. Mamaghani and T. Kawamoto 1997a. The stability of a rock block on an incline by discrete finite element method (DFEM). *IACMAG*, Wuhan, 523–528.

Tokashiki, N., Ö. Aydan, Y. Shimizu and T. Kawamoto 1997b. The assessment of the stability of a very old tunnel by discrete finite element method (DFEM). *NUMOG*, Montreal, 495–500.

Toki, K. and F. Miura 1985. Simulation of a fault rupture mechanism by a two-dimensional finite element method. *Journal of Physics of the Earth*, 33, 485–511.

Toki, K. and S. Sawada 1988. Simulation of the fault rupture process and near filed ground motion by the three-dimensional finite element method. *Proceedings of 9th World Conference on Earthquake Engineering, Japan*, II, 751–756.

Tsuboi, T. and F. Miura 1996. Simulation of stick-slip shear failure of rock masses by a non-linear finite element method. *Journal of Japan Society of Civil Engineers*, 537(35), 61–76 (in Japanese with English abstract).

Watanabe, H., H. Tano, R. Ulusay, E. Yüzer, M. Erdoğan and Ö. Aydan 1999. The initial stress state in Cappadocia. *Proceedings of the 1999 Japan-Korea Joint Symposium on Rock Engineering*, Fukuoka, Japan, 249–260.

Appendix 1

Gauss divergence theorem

A1.1 One-dimensional (1D) Gauss theorem

The Gauss theorem for the 1D case is written as:

$$\int_{\Omega} \frac{\partial f}{\partial x} d\Omega = \int_{\Gamma} f_n d\Gamma \tag{A1.1}$$

The above expression is explicitly written as

$$\int_{\Omega} \frac{\partial f(x)}{\partial x} \Delta x \Delta y \Delta z = \int_{\Gamma} (f \cdot n) |_{x=a}^{x=b} \Delta y \Delta z \tag{A1.2}$$

As $n_a = \cos 180^\circ = -1$, $n_b = \cos 0^\circ = 1$, $f(x = a) = f_a$ and $f(x = b) = f_b$, the above expression can be rewritten as

$$\int_{\Omega} \frac{\partial f(x)}{\partial x} \Delta x \Delta y \Delta z = \int_{\Gamma} f_b \Delta y \Delta z - \int_{\Gamma} f_a \Delta y \Delta z \tag{A1.3}$$

Figure A1.1 illustrates the geometrical interpretation of the Gauss divergence theorem.

A1.2 Three-dimensional Gauss theorem

To get the three-dimensional (3D) version of Eq. (A1.1), let us introduce the replacements

$$\frac{\partial}{\partial x} \rightarrow \nabla = \frac{\partial}{\partial x}\mathbf{e}_x + \frac{\partial}{\partial y}\mathbf{e}_y + \frac{\partial}{\partial z}\mathbf{e}_z \tag{A1.4}$$

$$n \rightarrow \mathbf{n} = n_x \mathbf{e}_x + n_y \mathbf{e}_y + n_z \mathbf{e}_z \tag{A1.5}$$

where $n_x = \cos\alpha$, $n_y = \cos\beta$ and $n_z = \cos\gamma$.

If the integrand is a scalar function (f), Eq. (A1.1) takes the following form in the 3D case:

$$\int_{\Omega} \nabla f d\Omega = \int_{\Gamma} f\mathbf{n} d\Gamma \tag{A1.6}$$

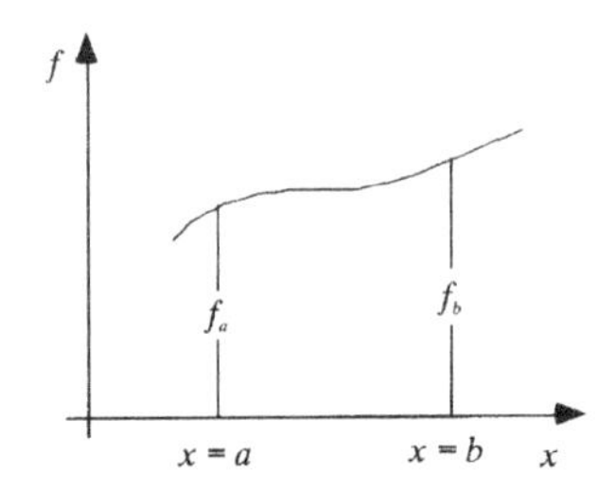

Figure A1.1 Geometrical illustration of the Gauss divergence theorem.

If the integrand is a vector (**v**), Eq. (A1.1) takes the following form in the 3D case:

$$\int_{\Omega} \nabla \cdot \mathbf{v} \, d\Omega = \int_{\Gamma} \mathbf{v} \cdot \mathbf{n} \, d\Gamma \tag{A1.7}$$

If the integrand is a tensor (σ), Eq. (A1.1) takes the following form in the 3D case:

$$\int_{\Omega} \nabla \cdot \sigma d\Omega = \int_{\Gamma} \sigma \cdot \mathbf{n} d\Gamma \tag{A1.8}$$

Appendix 2

Geometrical interpretation of the Taylor expansion

A scalar function ϕ at a given coordinate $x + \Delta x$ can be expressed using the Taylor expansion as (Figure A2.1):

$$\phi_{x+\Delta x} = \phi_x + \frac{\partial \phi}{\partial x}\Delta x + \frac{\partial^2 \phi}{\partial^2 x}\Delta x^2 + \cdots \frac{\partial^{(n)} \phi}{\partial^{(n)} x}\Delta x^n + \cdots \tag{A2.1}$$

The first term on the right-hand side is the value of the function ϕ at position (x). The second term involves the gradient of the function ϕ at position (x) multiplied by the position increment Δx, which corresponds to $\Delta\phi$, which is the increment of the function ϕ. As noted from the figure, there is a deviation in the exact value at $x + \Delta x$. If the higher terms of function ϕ are possible, the use of higher-order derivatives is expected to yield better estimations. However, the linear term is often utilized in the derivation of governing equations in many applications of mechanics. Therefore, the mechanics itself is called linear mechanics.

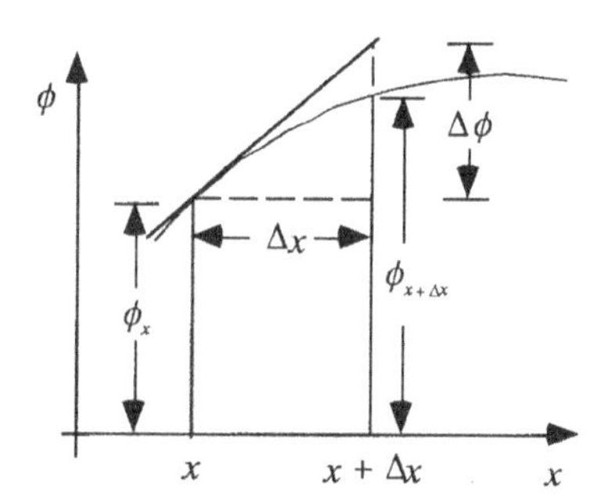

Figure A2.1 Approximation of the function ϕ at $x + \Delta x$.

Appendix 3

Reynolds transport theorem

By inserting the time derivation operator into the integral operator, we get

$$\frac{d}{dt}\int_{\Omega}(\ )d\Omega = \int_{\Omega}\frac{d(\ )}{dt}d\Omega + \int_{\Omega}(\ )\frac{d(d\Omega)}{dt} = \int_{\Omega}\left[\frac{d(\ )}{dt} + (\ )\nabla\cdot\mathbf{v}\right]d\Omega \tag{A3.1}$$

Equation (A3.1) is also known as the Reynolds transport theorem. The time derivative of the infinitely small volume $d\Omega$ takes the following form in 3D and 1D:

In 3D,

$$d\Omega = Jd\Omega_o = \left(\nabla_{x_o}\cdot\mathbf{x}\right)d\Omega_o \tag{A3.2}$$

where J is the Jacobian.

In 1D,

$$dx = \frac{\partial x}{\partial x_o}dx_o = Jdx_o \tag{A3.3}$$

A detailed supplementary explanation for Eq. (A3.3) is given by

$$\frac{d(dx)}{dt} = \frac{d}{dt}\left(\frac{\partial x}{\partial x_o}dx_o\right) = \frac{d}{dt}\left(\frac{\partial x}{\partial x_o}\right)dx_o + \frac{\partial x}{\partial x_o}\frac{d(dx_o)}{dt} \tag{A3.4}$$

The time derivative of the second term is nil as the initial control element length (dx_o) is constant. Thus, we have

$$\frac{d(dx)}{dt} = \frac{d}{dt}\left(\frac{\partial x}{\partial x_o}\right)dx_o \Rightarrow \frac{\partial}{\partial x_o}\left(\frac{dx}{dt}\right)dx_o \Rightarrow \frac{\partial v}{\partial x_o}dx_o \Rightarrow \frac{\partial v}{\partial x}\frac{\partial x}{\partial x_o}dx_o \Rightarrow \frac{\partial v}{\partial x}dx \tag{A3.5}$$

Therefore, we may write the following relations given in 1D and generalized to 3D versions as follows:

$1D \Rightarrow 3D$

$$\frac{\partial}{\partial x} \Rightarrow \nabla \tag{A3.6a}$$

$$v \Rightarrow \mathbf{v} \tag{A3.6b}$$

$$dx \Rightarrow d\Omega \tag{A3.6c}$$

$$\frac{\partial v}{\partial x}dx \Rightarrow (\nabla\cdot\mathbf{v})d\Omega \tag{A3.6d}$$

Appendix 4

The Gauss elimination method and its implementation

The Gauss elimination method is one of the commonly used methods for solving simultaneous equation systems in numerical methods such as the finite element method. First, the fundamental algorithm of the Gauss elimination method is explained.

The formulation of governing equations by the finite element method results in the following simultaneous equation system:

$$[K]\{U\}=\{F\} \tag{A4.1}$$

The number of unknowns is n and the size of the matrix [K} is n by n.

To explain the algorithm of the Gauss elimination method, let us consider a simultaneous equation system of 3×3.

$$\begin{bmatrix} k_{11} & k_{12} & k_{13} \\ k_{21} & k_{22} & k_{23} \\ k_{31} & k_{32} & k_{33} \end{bmatrix} \begin{Bmatrix} u_1 \\ u_2 \\ u_3 \end{Bmatrix} = \begin{Bmatrix} f_1 \\ f_2 \\ f_3 \end{Bmatrix} \tag{A4.2}$$

In the Gauss elimination method, the lower half of the matrix [K] of coefficients is reduced to zero through some manipulations, which is known as the decomposition procedure. Once the unknown u_n is determined, the unknowns $u_{n-1}, \cdots, u_1$ are obtained step-wise through the back-substitution procedure.

The coefficient matrix and vector on the right-hand side of Eq. (C.2) are rewritten as follows:

$$\begin{bmatrix} k_{11} & k_{12} & k_{13} & f_1 \\ k_{21} & k_{22} & k_{23} & f_2 \\ k_{31} & k_{32} & k_{33} & f_3 \end{bmatrix} \tag{A4.3}$$

There are two different approaches to obtain the upper triangular form of the coefficient matrix and the methods are explained in the following sections.

A4.1 Method 1

A4.1.1 Step 1

The second and third rows are multiplied by k_{11}/k_{21}, k_{11}/k_{31}, respectively, and subtracted from the first row. This operation results in the first components at second and third rows becoming 0 as given below:

$$\begin{bmatrix} k_{11}^* & k_{12}^* & k_{31}^* & f_1^* \\ 0 & k_{22}^* & k_{23}^* & f_2^* \\ 0 & k_{32}^* & k_{33}^* & f_3^* \end{bmatrix} \tag{A4.4}$$

where

$$k_{11}^* = k_{11},\ k_{12}^* = k_{12},\ k_{13}^* = k_{13},\ f_1^* = f_1$$

$$k_{22}^* = k_{12} - \frac{k_{11}}{k_{21}} \cdot k_{22},\ k_{23}^* = k_{13} - \frac{k_{11}}{k_{21}} \cdot k_{23},\ f_2^* = f_1 - \frac{k_{11}}{k_{21}} \cdot f_2$$

$$k_{32}^* = k_{12} - \frac{k_{11}}{k_{31}} \cdot k_{32},\ k_{33}^* = k_{13} - \frac{k_{11}}{k_{31}} \cdot k_{33},\ f_3^* = f_1 - \frac{k_{11}}{k_{31}} \cdot f_3$$

A4.1.2 Step 2

Next, the third row is multiplied by k_{22}^*/k_{32}^* and it is subtracted from the second row so that the second component of the third row becomes 0 as shown below:

$$\begin{bmatrix} k_{11}^{**} & k_{12}^{**} & k_{31}^{**} & f_1^{**} \\ 0 & k_{22}^{**} & k_{23}^{**} & f_2^{**} \\ 0 & 0 & k_{33}^{**} & f_3^{**} \end{bmatrix} \tag{A4.5}$$

where

$$k_{11}^{**} = k_{11}^*,\ k_{12}^{**} = k_{12}^*,\ k_{13}^{**} = k_{13}^*,\ f_1^{**} = f_1^*$$

$$k_{22}^{**} = k_{22}^*,\ k_{23}^{**} = k_{23}^*,\ f_2^{**} = f_2^*$$

$$k_{33}^{**} = k_{23} - \frac{k_{22}^*}{k_{32}^*} \cdot k_{23}^*,\ f_3^{**} = f_2 - \frac{k_{22}^*}{k_{32}^*} \cdot f_{23}^*$$

A4.1.3 Step 3

Once the lower half of the coefficient matrix is transformed into components with a value of 0, the back-substitution operation is implemented and unknowns are obtained one by one as given below:

$$u_3 = \frac{1}{k_{33}^{**}} \cdot f_3^{**},\ u_2 = \frac{1}{k_{22}^{**}}\left(f_2^{**} - k_{23}^{**} \cdot u_3\right),\ u_1 = \frac{1}{k_{11}^*}\left(f_1^{**} - k_{12}^{**} \cdot u_2 - k_{13}^{**} \cdot u_3\right) \tag{A4.6}$$

A4.2 Method 2

A4.2.1 Step 1

The first row is multiplied by $k_{21}/k_{11}, k_{31}/k_{11}$, respectively, and subtracted from the second row and third row successively. Thus, we get the following:

$$\begin{bmatrix} k_{11} & k_{12} & k_{13} & f_1 \\ 0 & k_{22}^* & k_{23}^* & f_2^* \\ 0 & k_{32}^* & k_{33}^* & f_3^* \end{bmatrix} \tag{A4.7}$$

where $k_{22}^* = k_{22} - \dfrac{k_{21}}{k_{11}} \cdot k_{12},\ k_{23}^* = k_{23} - \dfrac{k_{21}}{k_{11}} \cdot k_{13},\ f_2^* = f_2 - \dfrac{k_{21}}{k_{11}} \cdot f_1$

$$k_{32}^* = k_{32} - \frac{k_{31}}{k_{11}} \cdot k_{12},\ k_{33}^* = k_{33} - \frac{k_{31}}{k_{11}} \cdot k_{13},\ f_3^* = f_3 - \frac{k_{31}}{k_{11}} \cdot f_1$$

A4.2.2 Step 2

The second row is multiplied by k_{32}^*/k_{22}^* and is subtracted from the third row. This results in the following relation:

$$\begin{bmatrix} k_{11} & k_{12} & k_{13} & f_1 \\ 0 & k_{22}^* & k_{23}^* & f_2^* \\ 0 & 0 & k_{33}^{**} & f_3^{**} \end{bmatrix} \quad k_{33}^{*+} = k_{33}^* - \frac{k_{32}^*}{k_{22}^*} \cdot k_{23}^*,\ f_3^{**} = f_3^* - \frac{k_{32}^*}{k_{22}^*} \cdot f_2^* \tag{A4.8}$$

A4.2.3 Step 3

Once the lower half of the coefficient matrix is transformed into components with a value of 0, the back-substitution operation is implemented and unknowns are obtained one by one as given below:

$$u_3 = \frac{1}{k_{33}^{**}} \cdot f_3^{**},\ u_2 = \frac{1}{k_{22}^{**}} \cdot \left(f_2^* - k_{23}^* \cdot u_3\right),\ u_1 = \frac{1}{k_{11}} \cdot \left(f_1 - k_{12} \cdot u_2 - k_{13} \cdot u_3\right) \tag{A4.9}$$

Appendix 5

Constitutive modeling of discontinuities and interfaces

A5.1 Multi-response theory

Discontinuities and interfaces are encountered in many engineering problems such as excavations in rock mass, dam constructions, piling and bolting and so on. Relative sliding or separation movements in such localized zones present an extremely difficult problem in mechanical modeling. The discontinuities have been generally regarded as planar bodies that can only sustain and transfer the normal and tangential tractions between two adjacent bodies in geomechanics. The same type of approach is also followed in the mechanical engineering field in dealing with contact problems. Although this approach seems to be appropriate at first glance, it presents extremely difficult problems mathematically, experimentally and analytically. In this type of consideration, such planes will be internal boundaries within the body at which several conditions of constraints will be required to be satisfied. For example, if the condition of non-penetration is required, it will simply require that the so-called normal stiffness must have a value of infinity. The actual geometry of interfaces or discontinuities is never smooth and has asperities of varying amplitude and wavelength (Figure A5.1). Therefore, assigning a thickness to discontinuities related to the asperity height would

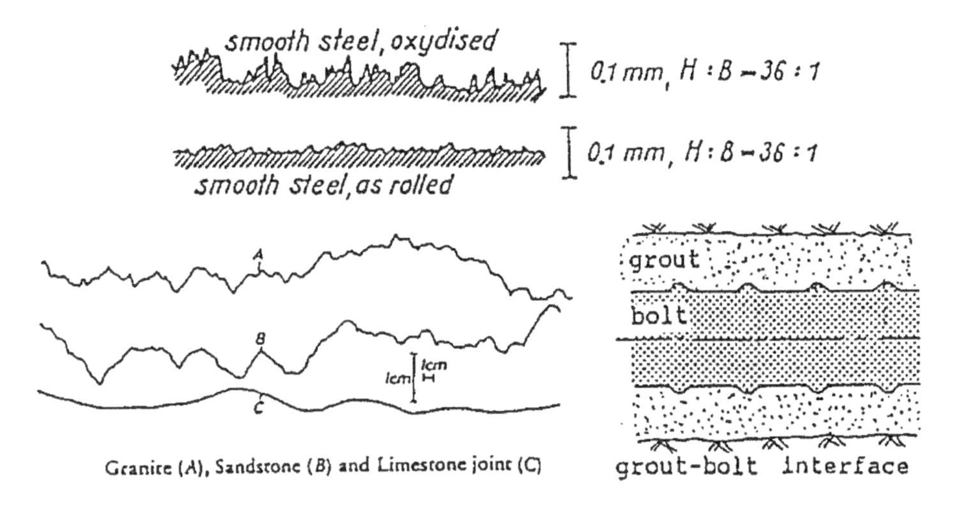

Figure A5.1 Surface configurations of discontinuities and interfaces.

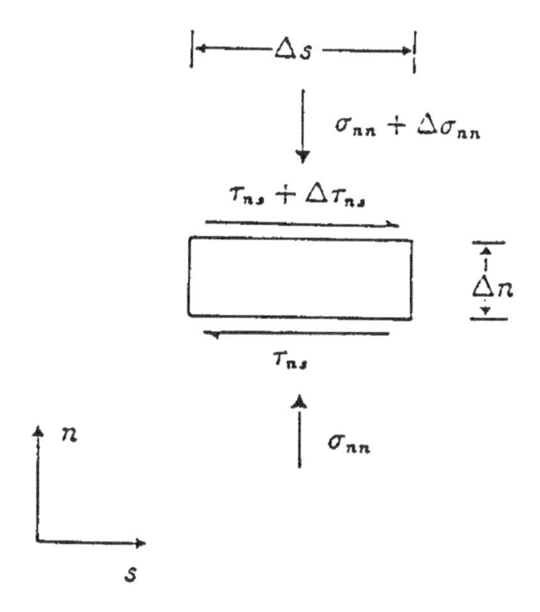

Figure A5.2 Mechanical model for discontinuities.

simplify the problem. Aydan et al. (1990a, b, 1996) and Aydan and Kawamoto (1990) suggested that the thickness of the discontinuity should be twice the average height of asperities, which can be easily determined from the surface morphology measurements (e.g. Aydan and Shimizu 1995).

Let us consider a thin two-dimensional tabular body in a Cartesian coordinate system (s,n) as shown in Figure A5.2. Within the thin tabular body of discontinuity/interface, stress components may be contemplated as σ_{nn}, σ_{ss} and τ_{ns} in two dimensions. If a discontinuity/interface is regarded as an internal thin body, the components of the stress tensor, which tend to remain on the discontinuity, may be σ_{nn} and τ_{ns} only. The relations among the strain and displacement components for discontinuities may be assumed to be of the following forms:

$$\gamma_{ns} = \frac{\partial u}{\partial n}, \quad \varepsilon_{ns} = \frac{\partial v}{\partial n} \tag{A5.1}$$

The behavior of discontinuities and interfaces is highly non-linear and they may show an elasto-plastic behavior from the very beginning of loading. Ichikawa (1985) proposed a multi-response theory for the behavior of materials, which separates the deviatoric and hydrostatic responses. This theory was adopted for interfaces and discontinuities by Aydan (1989, 2018) and Aydan et al. (1990a, b). The response functions of the discontinuities or interfaces for shear and normal stresses are given as

$$\begin{aligned} &\tau_i = \Phi_i^e(\gamma_i^e), \quad \tau_i = \Phi_i^p(\gamma_i^p, \varepsilon_i^p), \quad \tau_i = \tau_{ns} \quad \sigma_i = \sigma_{ns}, \\ &\sigma_i = \Psi_i^e(\varepsilon_i^e), \quad \sigma_i = \Psi_i^p(\gamma_i^p, \varepsilon_i^p). \quad \varepsilon_i = \varepsilon_{ns}, \quad \gamma_i = \gamma_{ns} \end{aligned} \tag{A5.2}$$

where γ_i, τ_i are the shear strain and stress, and ε_i, σ_i are the normal strain and stress. Superscripts e and p stand for elastic and plastic, respectively.

The incremental constitutive equations for the elasto-plastic behavior are given in the matrix form as:

$$\left\{\begin{array}{c} d\tau_i \\ d\sigma_i \end{array}\right\} = \left[\begin{array}{cc} \partial\Phi_i^e / \partial\gamma_i^e & 0 \\ 0 & \partial\Psi_i^e / \partial\varepsilon_i^e \end{array}\right] \left\{\begin{array}{c} d\gamma_i^e \\ d\varepsilon_i^e \end{array}\right\}, \quad \left\{\begin{array}{c} d\tau_i \\ d\sigma_i \end{array}\right\} = \left[\begin{array}{cc} \partial\Phi_i^p / \partial\gamma_i^p & \partial\Phi_i^p / \partial\varepsilon_i^p \\ \partial\Psi_i^p / \partial\gamma_i^p & \partial\Psi_i^p / \partial\varepsilon_i^p \end{array}\right] \left\{\begin{array}{c} d\gamma_i^p \\ d\varepsilon_i^p \end{array}\right\} \tag{A5.3}$$

Denoting

$$G_1^e = \frac{\partial\Phi_i^e}{\partial\gamma_i^e}, \quad E_1^e = \frac{\partial\Psi_i^e}{\partial\varepsilon_i^e}, \quad G_1^p = \frac{\partial\Phi_i^p}{\partial\gamma_i^p}, \quad G_2^p = \frac{\partial\Phi_i^p}{\partial\varepsilon_i^p}, \quad E_1^p = \frac{\partial\Psi_i^p}{\partial\varepsilon_i^p}, \quad E_2^p = \frac{\partial\Psi_i^p}{\partial\gamma_i^p}$$

and taking the inverse of the above expressions and after some rearrangements, we have

$$\left\{\begin{array}{c} d\gamma_i^e \\ d\varepsilon_i^e \end{array}\right\} = \left[\begin{array}{cc} 1/G_1^e & 0 \\ 0 & 1/E_1^e \end{array}\right] \left\{\begin{array}{c} d\tau_i \\ d\sigma_i \end{array}\right\}, \quad \left\{\begin{array}{c} d\gamma_i^p \\ d\varepsilon_i^p \end{array}\right\} = \left[\begin{array}{cc} 1/h_s & \mu/h_s \\ \beta/h_n & 1/h_n \end{array}\right] \left\{\begin{array}{c} d\tau_i \\ d\sigma_i \end{array}\right\} \tag{A5.4}$$

where

$$h_s = G_1^p - G_2^p \frac{E_2^p}{E_1^p}, \quad h_n = E_1^p - E_2^p \frac{G_2^p}{G_1^p}, \quad \mu = -\frac{G_2^p}{E_1^p}, \quad \beta = -\frac{E_2^p}{G_1^p}$$

In the above expression, parameters h_s and h_n are physically interpreted as the hardening moduli for the respective responses. The symbols μ and β denote the friction and the dilatancy factors, respectively. These terms are interrelated to each other in conventional plasticity. On the other hand, the present theory evaluates these terms independently from each other.

The elasto-plastic incremental constitutive law can be easily written by assuming that the total strains are a linear sum of elastic and plastic strains as

$$\left\{\begin{array}{c} d\gamma_i \\ d\varepsilon_i \end{array}\right\} = \left[\begin{array}{cc} \dfrac{1}{G_1^e} + \dfrac{1}{h_s} & \dfrac{\mu}{h_s} \\ \dfrac{\beta}{h_n} & \dfrac{1}{E_1^e} + \dfrac{1}{h_n} \end{array}\right] \left\{\begin{array}{c} d\tau_i \\ d\sigma_i \end{array}\right\} \tag{A5.5}$$

The yield function is vectorial and is written in the following form:

$$f\begin{Bmatrix} f_1 \\ f_2 \end{Bmatrix} = 0 \qquad \begin{matrix} f_1 = \tau_i - \Phi_i^p(\gamma_i^p, \varepsilon_i^p) \\ f_2 = \sigma_i - \Psi_i^p(\gamma_i^p, \varepsilon_i^p) \end{matrix} \tag{A5.6}$$

If the relation between stresses and strains are assumed to be linear elastic, then we have the following conventional form:

$$\begin{Bmatrix} \tau_{ns} \\ \sigma_{nn} \end{Bmatrix} = \begin{bmatrix} G_i & 0 \\ 0 & E_i \end{bmatrix} \begin{Bmatrix} \gamma_{ns} \\ \varepsilon_{nn} \end{Bmatrix} \quad G_1^e = G_i, \quad E_1^e = E_i \tag{A5.7}$$

where E_i and G_i denote the Young's modulus and shear modulus of the discontinuity/interface material.

References

Aydan, Ö. 1989. The stabilisation of rock engineering structures by rockbolts. Doctorate Thesis, Nagoya University, 204 p.

Aydan, Ö. 2018. *Rock Reinforcement and Rock Support*, CRC Press, Taylor & Francis Group, London, 486 p.

Aydan, Ö. and T. Kawamoto 1990. Discontinuities and their effect on rock masses. *Proceedings of International Conference on Rock Joints, ISRM*, Loen, 149–156.

Aydan, Ö. and Y. Shimizu 1995. Surface morphology characteristics of rock discontinuities with particular reference to their genesis. Fractography, Geological Society Special Publication No. 92, 11–26.

Aydan, Ö., Y. Ichikawa, S. Ebisu, S. Komura and A. Watanabe 1990b. Studies on interfaces and discontinuities and an incremental elasto-plastic constitutive law. *International Conference on Rock Joints*, ISRM, 595–602.

Aydan, Ö., Y. Ichikawa and T. Kawamoto 1990a. Numerical modelling of discontinuities and interfaces in rock mass. *The 4th Symposium on Computational Mechanics of Japan*, 254–261.

Aydan, Ö., I.H.P. Mamaghani and T. Kawamoto 1996. Application of discrete finite element method (DFEM) to rock engineering structures. *NARMS'96*, 2039–2046.

Ichikawa, Y. 1985. Fundamentals of the incremental elasto-plastic theory for rock-like materials (in Japanese), Dr. Thesis, Faculty of Engineering, Nagoya University, Nagoya, Japan, 132 p.

Appendix 6

Thin band element for modeling discontinuities and interfaces in numerical analyses

A6.1 Derivation of stiffness matrices

The equilibrium equations without body force for the thin tabular element shown in Figure A6.1 are written as (Aydan et al. 1990)

$$\frac{\partial \tau_{ns}}{\partial n} = 0 \quad \text{for } s \text{ direction} \quad \frac{\partial \sigma_{nn}}{\partial n} = 0 \quad \text{for } n \text{ direction} \tag{A6.1}$$

The relations among the strain and displacement components for discontinuities are assumed to be of the following forms:

$$\gamma_{ns} = \frac{\partial u}{\partial n}, \quad \varepsilon_{ns} = \frac{\partial v}{\partial n} \tag{A6.2}$$

The constitutive law for linear behavior is written as

$$\left\{ \begin{array}{c} \tau_i \\ \sigma_i \end{array} \right\} = \left[\begin{array}{cc} G_i & 0 \\ 0 & E_i \end{array} \right] \left\{ \begin{array}{c} \gamma_i \\ \varepsilon_i \end{array} \right\} \tag{A6.3}$$

Let us consider a body with a statically admissible stress field $\boldsymbol{\sigma}$ and a kinematically admissible displacement field $\mathbf{u}$, in which special zones are represented as subdomains Ω_i (Figure A6.1a). As the equilibrium Eq. (A6.1) is valid in the domain of interfaces, it should be also valid in each discretized interface element. Thus, the variational form of Eq. (A6.1) may be element-wise written by taking a variation on relative displacement $\Delta\mathbf{u}$ over a thin-band element in the following form:

$$\int_{s_e}^{s_{e+1}} \int_{n_e}^{n_{e+1}} \left(\delta(\Delta u) \frac{\partial \tau_{ns}}{\partial n} + \delta(\Delta v) \frac{\partial \sigma_{nn}}{\partial n} \right) ds dn = 0 \tag{A6.4}$$

Equivalently, carrying out integration by parts yields

$$\int_{s_e}^{s_{e+1}} \left(\delta(\Delta u) \tau_{ns} + \delta(\Delta v) \sigma_{nn} \right) ds \Big|_{n_e}^{n_{e+1}} = \int_{s_e}^{s_{e+1}} \int_{n_e}^{n_{e+1}} \left(\frac{\partial \delta(\Delta u)}{\partial n} \tau_{ns} + \frac{\partial \delta(\Delta u)}{\partial n} \sigma_{nn} \right) ds dn \tag{A6.5}$$

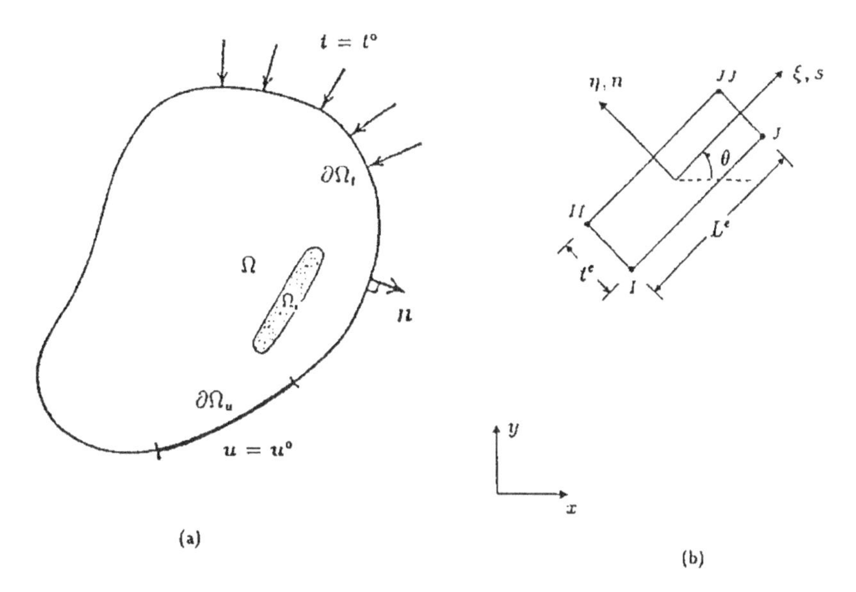

Figure A6.1 Notations for finite element modeling. (a) Illustration of discontinuity as a sub-domain in the body, (b) geometrical illustration of the band element.

Let us assume that the relative displacements in an element with four nodes (I, J, II, JJ) (Figure A6.1b) are approximated by the following expression with the use of linear type interpolation functions of s only in the local coordinate system (s, n):

$$\Delta\mathbf{u} = \mathbf{N}\Delta\mathbf{U} \quad \textit{or explicitly} \quad \begin{Bmatrix} \Delta u \\ \Delta v \end{Bmatrix} = \begin{bmatrix} N_{II} & 0 & N_{JJ} & 0 \\ 0 & N_{II} & 0 & N_{JJ} \end{bmatrix} \begin{Bmatrix} \Delta U_{II} \\ \Delta V_{II} \\ \Delta U_{JJ} \\ \Delta V_{JJ} \end{Bmatrix} \tag{A6.6}$$

where

$$\Delta U_{II} = U_{II} - U_I, \quad \Delta V_{II} = V_{II} - V_I, \quad \Delta U_{JJ} = U_{JJ} - U_J, \quad \Delta V_{JJ} = V_{JJ} - V_J,$$

$$N_{II} = \frac{1}{2}(1-\xi), \quad N_{JJ} = \frac{1}{2}(1+\xi), \quad \xi = \frac{2s}{L^e}, \quad \eta = \frac{2n}{t^e}$$

L^e and t^e are the length and thickness of the element, respectively.

The most crucial aspect of the formulation is the definition of strain components in terms of nodal displacements. Using the relations given by Eqs. (A6.2) and (A6.6), one can write

$$\gamma_{ns} = \frac{\partial(N_{II}\Delta U_{II})}{\partial n} + \frac{\partial(N_{JJ}\Delta U_{JJ})}{\partial n} = N_{II}\frac{\partial \Delta U_{II}}{\partial n} + N_{JJ}\frac{\partial \Delta U_{JJ}}{\partial n}$$

$$\varepsilon_{nn} = \frac{\partial(N_{II}\Delta V_{II})}{\partial n} + \frac{\partial(N_{JJ}\Delta V_{JJ})}{\partial n} = N_{II}\frac{\partial \Delta V_{II}}{\partial n} + N_{JJ}\frac{\partial \Delta V_{JJ}}{\partial n}$$

As the variation of strains in the direction of the axis n is assumed to be constant and the thickness t^e of the element is sufficiently small as compared with the length L^e of the element, the above relations are written as

$$\boldsymbol{\varepsilon} = \mathbf{B}\,\Delta\mathbf{U} \quad \text{or explicitly} \quad \begin{Bmatrix} \gamma_{ns} \\ \varepsilon_{nn} \end{Bmatrix} = \frac{1}{t^e}\begin{bmatrix} N_{II} & 0 & N_{JJ} & 0 \\ 0 & N_{II} & 0 & N_{JJ} \end{bmatrix}\begin{Bmatrix} \Delta U_{II} \\ \Delta V_{II} \\ \Delta U_{JJ} \\ \Delta V_{JJ} \end{Bmatrix} \tag{A6.7}$$

Introducing the above relation into Eq. (A6.4) yields the following

$$\Delta\mathbf{F} = \Delta\mathbf{K}\Delta\mathbf{U} \tag{A6.8}$$

where

$$\Delta\mathbf{F} = \left|\left(\int_{-1}^{1} \mathbf{N}^T \boldsymbol{\sigma}^i \frac{\partial s}{\partial \xi} d\xi\right)\eta\right|_{\eta=-1}^{\eta=1}, \quad \boldsymbol{\sigma}^{i\,T} = \{\tau_{ns} \quad \sigma_{nn}\},$$

$$\Delta\mathbf{K} = \int_{-1}^{1}\int_{-1}^{1} \mathbf{B}^T\,\mathbf{D}\,\mathbf{B}\left|\frac{\partial(s,n)}{\partial(\xi,\eta)}\right| d\xi d\eta$$

The explicit form of the matrix $\Delta\mathbf{K}$ can be obtained by carrying out the integration as

$$\Delta\mathbf{K}^e = \begin{bmatrix} 2K_g & 0 & K_g & 0 \\ 0 & 2K_e & 0 & K_e \\ K_g & 0 & 2K_g & 0 \\ 0 & K_e & 0 & 2K_e \end{bmatrix} \quad K_g = \frac{G_i L^e}{6t^e}, \quad K_e = \frac{E_i L^e}{6t^e} \tag{A6.9}$$

The relation between the relative nodal displacement vector and the nodal displacement vector can be written as

$$\Delta\mathbf{F} = \mathbf{R}\mathbf{U} \tag{A6.10}$$

where

$$\mathbf{R} = \begin{bmatrix} -1 & 0 & 0 & 0 & 1 & 0 & 0 & 0 \\ 0 & -1 & 0 & 0 & 0 & 1 & 0 & 0 \\ 0 & 0 & -1 & 0 & 0 & 0 & 1 & 0 \\ 0 & 0 & 0 & -1 & 0 & 0 & 0 & 1 \end{bmatrix},$$

$$\mathbf{U} = \{ U_I \quad V_I \quad U_J \quad V_J \quad U_{II} \quad V_{II} \quad U_{JJ} \quad V_{JJ} \}^T$$

Then, the stiffness matrix for the relative nodal displacements can be transformed to the stiffness matrix for nodal displacements in the local coordinate system (s,n) by the following relation:

$$[K]_l^e = [R]^T[\Delta K][R] \tag{A6.11}$$

or explicitly

$$[K]_l^e = \begin{bmatrix} 2K_g & 0 & K_g & 0 & -2K_g & 0 & -K_g & 0 \\ 0 & 2K_e & 0 & K_e & 0 & -2K_e & 0 & -K_e \\ K_g & 0 & 2K_g & 0 & -K_g & 0 & -2K_g & 0 \\ 0 & K_e & 0 & 2K_e & 0 & -K_e & 0 & -2K_e \\ -2K_g & 0 & -K_g & 0 & 2K_g & 0 & K_g & 0 \\ 0 & -2K_e & 0 & -K_e & 0 & 2K_e & 0 & K_e \\ -K_g & 0 & -2K_g & 0 & K_g & 0 & 2K_g & 0 \\ 0 & -K_e & 0 & -2K_e & 0 & K_e & 0 & 2K_e \end{bmatrix}$$

It should be noted that the above expression has the same form as those given by Ghaboussi et al. (1973), Heuze and Barbour (1982) and Desai et al. (1984). It is also interesting to note that the division of the components of the stiffness matrix proposed by Goodman et al. (1968) by the thickness t^e also yields the same result. The same expression for the stiffness matrix can be also derived using a four-noded rectangular solid element together with components of constitutive relations among σ_{ss} and $\varepsilon_{ss},\varepsilon_{nn}$ and σ_{nn} and ε_{ss} are nil as proposed by Zienkiewicz and Pande (1977).

The above stiffness matrix derived in the local coordinate system (s,n) should have to be transformed into that in the global coordinate system (x,y). This is done using the transformation law given as

$$[K]_g^e = [T]^T[K]_l^e[T] \tag{A6.12}$$

where

$$[T] = \begin{bmatrix} [T]_s & [0] & [0] & [0] \\ [0] & [T]_s & [0] & [0] \\ [0] & [0] & [T]_s & [0] \\ [0] & [0] & [0] & [T]_s \end{bmatrix}, \quad [T]_s = \begin{bmatrix} l_1 & m_1 \\ l_2 & m_2 \end{bmatrix}, \quad [0] = \begin{bmatrix} 0 & 0 \\ 0 & 0 \end{bmatrix},$$

$$l_1 = m_2 = \cos\theta, \quad -l_2 = m_1 = \sin\theta, \quad \theta = \tan^{-1}\left(\frac{\Delta x}{\Delta y}\right), \quad \Delta x = x_{JJ} - x_{II}, \quad \Delta y = y_{JJ} - y_{II}$$

The above formulation can be easily reduced to that for a contact element by restricting the number of nodes of the element to 2 instead of 4 (Ngo and Scordelis 1967; Aydan et al. 1996).

A6.2 A numerical example to compare various joint/interface elements

The material properties and the finite element mesh used in the analyses to be reported herein are given in Table A6.1 and shown in Figure A6.2, respectively. The model was subjected to normal and shear loading ($t_n = t_n = 10$ MPa) respectively at given nodes under a plane-strain condition. US and LS stand for upper and lower sides of the thin-band element with a thickness of 50 mm under a plane-strain condition. The thin-band element is geometrically represented in the finite element mesh. The constitutive parameters of the Goodman-type joint element and the stiffness coefficients are assumed to be the same as those of the band element but the coordinates of nodes at both sides of discontinuity are same in the Goodman joint-element, which is the common procedure in practice. C1 and C2 stand for the representation of discontinuity

Table A6.1 Material and geometrical parameters

E_r (MPa)	ν_r	E_i (MPa)	G_i (MPa)	t (m)
10,000	.25	10,000	100	0.05

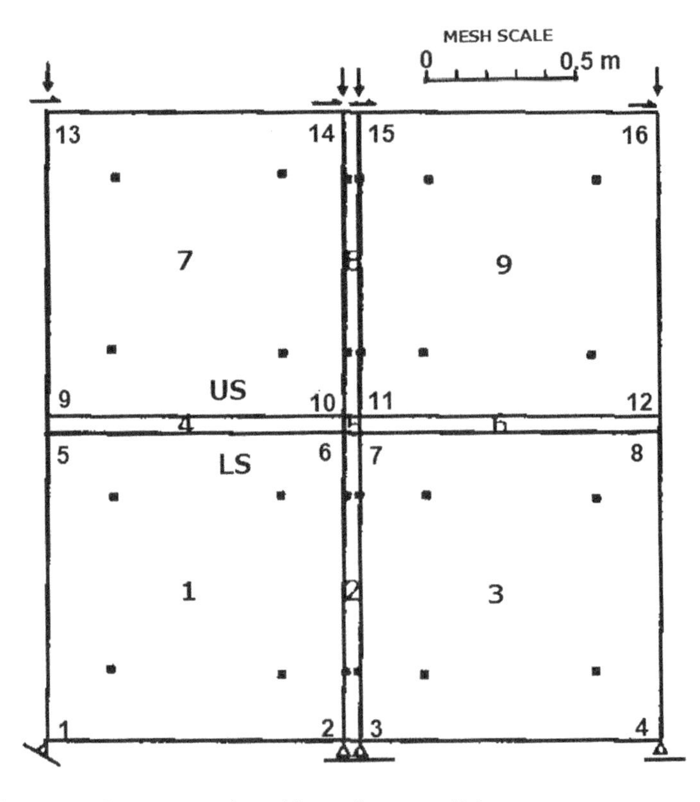

Figure A6.2 Finite element mesh and boundary conditions.

Table A6.2 Comparison of displacement responses at selected nodes

	Node no.	Case no.	$u\ (m)$	$v\ (m)$
LS	5	C1	7.536E-03	2.718E-03
		C2	7.536E-03	2.718E-03
	6	C1	6.556E-03	−1.471E-03
		C2	6.556E-03	−1.471E-03
	7	C1	6.590E-03	−1.527E-03
		C2	6.590E-03	−1.527E-03
	8	C1	7.569E-03	−4.403E-03
		C2	7.569E-03	−4.403E-03
US	9	C1	1.857E-02	2.778E-03
		C2	1.857E-02	2.778E-03
	10	C1	7.040E-03	−1.417E-03
		C2	6.818E-03	−1.440E-03
	11	C1	7.049E-03	−1.525E-03
		C2	6.836E-03	−1.528E-03
	12	C1	7.714E-03	−4.778E-03
		C2	7.512E-03	−4.649E-03

C1, thin-band element; C2, Goodman joint element; US, upper side; LS, lower side.

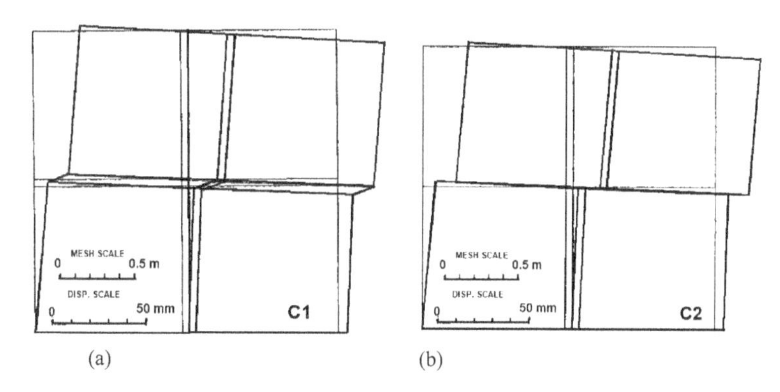

Figure A6.3 Comparisons of displacement responses. (a) Deformed configuration of finite element model with band element with a given thickness, (b) deformed configuration of finite element model for Goodman type joint element.

by the thin-band element and Goodman-type joint element in respective analyses. Table A6.2 and Figure A6.3 compare the computed components of displacements of selected nodes. As noted from the results given in Table A6.2 and Figure A6.3, the calculated results are exactly the same as the resulting stiffness matrices for the finite element analysis using thin-band element and Goodman joint element are exactly the same. However, there will be penetration of the sides of the block, which would appear as an overlap in plotted deformed configurations in the graphical outputs of computational results.

References

Aydan, Ö., Y. Ichikawa and T. Kawamoto 1990. Numerical modelling of discontinuities and interfaces in rock mass. *The 4th Symposium on Computational Mechanics of Japan*, 254–261.

Aydan, Ö., I.H.P. Mamaghani and T. Kawamoto 1996. Application of discrete finite element method (DFEM) to rock engineering structures. *NARMS'96*, 2039–2046.

Desai, C.S., M.M. Zaman, J.G. Lightner and H.J. Siriwardane 1984. Thin layer element for interfaces and joints. *International Journal for Numerical and Analytical Methods in Geomechanics*, 8, 19–43.

Ghaboussi, J., E.L. Wilson and J. Isenberg 1973. Finite element for rock joints and interfaces. *Journal of the Soil Mechanics and Foundations Division, ASCE, SM10*, 99, 833–848.

Goodman, R.E., R. Taylor and T.L. Brekke 1968. A model for the mechanics of jointed rock. *Journal of the Soil Mechanics and Foundations Division, ASCE, SM3*, 94, 637–659.

Heuze, F.E. and T.G. Babour 1982. New models for rock joints and interfaces. *Journal of the Geotechnical Engineering Division, ASCE, GT5*, 108, 757–775.

Ngo, D. and A.C. Scordelis 1967. Finite element analysis of reinforced concrete beams. *American Concrete Institute Journal*, 152–163.

Zienkiewicz, O.C. and G.N. Pande 1977. Time-dependent multi-laminate model of rocks*a numerical study of deformation and failure of rock masses. *International Journal for Numerical and Analytical Methods in Geomechanics*, 1, 219–247.

Index

action 11, 82, 246, 257
  arching 257
  springs 246
analytical solution 101, 108, 132, 134, 137, 150, 155, 159
  circular/spherical cavity 137, 159
  diffusion 155
  falling head tests 108
  fluid flow 155
  heat flow 155
atmospheric 113, 174, 180
  pressure 113, 174
  temperature 180
axisymmetric 114, 252–278
  cylindrical cavern 272
  finite element 253, 257, 260, 263
  model 114
  problem 263
  rock sample 83
  simulation 257, 278
  structure 253

bar 101–119, 132, 145, 244, 257
  divided 101, 102, 119
  embedded 132
  radius 145
  steel 244, 257
Barton, N. R. 92, 93, 118
behavior 72, 74, 85, 87, 125, 126, 137, 138, 148, 155, 177, 189, 264, 281, 282, 283, 284, 287
  brittle 85, 87
  creep 101
  ductile 85, 87
  elastic 134, 154, 257
  elasto-plastic 49–50, 242, 243, 247, 250, 282, 311, 312
  interface 287, 311
  linear 314
  material 43, 49, 58, 59, 62, 155, 236, 311
  mechanical 43, 251
  non-linear 59, 65, 85, 155
  plastic 61, 134, 154
  post-failure 85
  seismic 281
  softening 137, 236
  time dependent 150, 189, 261
  triaxial 85
  visco-elastic 149
  visco-elasto-plastic 262, 263
  visco-plastic 53
Bieniawski, Z. T. 148, 189–190
blasting 158
body force 27, 29, 36, 40, 148, 203, 211, 219, 314
borehole 30, 145, 158, 159, 274–279, 300
breakout 158, 159, 274, 275
Brekke, T. L. 249, 320
Brown, E. T. 65–70, 78, 80, 119, 140, 148, 155–170, 252, 301

cavern 74, 261–273
cavity 134, 137, 153, 155–156, 190
  circular 134, 137, 155, 156, 190
  spherical 137, 153, 155, 259
condition 32–34, 50–63, 68, 82–85, 89, 97, 105, 109–121, 130–150, 164–166, 170–182, 185, 192–198, 201–204, 211, 218, 226–227, 233–235, 261, 278–286, 318
  boundary 82, 84, 110, 116–119, 121, 130, 131, 135, 136, 139, 141, 146, 150, 164, 166, 172, 175, 185, 192–194, 198, 201–204, 211, 218, 226–227, 233–235, 261–266, 278–283, 318
  compatibility 137
  consistency 50, 54, 59, 63
  continuity 142, 146
  creeping 179
  drained 33
  dynamic 34, 286
  environmental 80, 93, 115, 182
  initial 109, 112, 116, 117, 121, 124, 131, 132, 155, 170, 173, 182, 198, 204, 210, 211, 218, 226

condition (*cont.*)
laboratory 97
loading 89, 149, 266
saturated 99, 100
steady-state 59
stress 122, 149, 159, 265
undrained 32
unstrained 105
three-point bending 85
yield 68
Cook, N.G.W. 43, 59, 68, 78, 80, 83, 119, 121, 122, 153–158, 190
creep 46–53, 97–101, 119, 148–149, 190
Brazilian 99
compression 99, 100
device 99
experiment 62, 100
failure 119, 148, 275, 277
impression 99, 100, 119, 190
secondary 149
shear 89, 101
steady-state 46, 49
strength 149
tertiary 149
test 53, 97–101, 149, 190
transient 47, 48, 148

dam 261, 310
construction 310
gravity 261
rockfill 261
damping 132, 236, 247, 248
degradation 250, 261
diffusion 26, 35, 42–43, 57, 80, 101, 114, 116, 119, 155, 180, 182, 183, 203, 208, 242
coefficient 43, 57, 183
constant 182
experiment 80, 114, 116, 208
equation 185
flux vector 35
phenomenon 35
problem 43, 180, 208, 242
process 182
discontinuity 74, 95–98, 164–167, 174–176, 243–246, 284–287, 301, 311–315
aperture 166, 167
permeability 167, 174–176
porosity 167
saw-cut 96–98
set 74, 243, 301
displacement transducer 96, 182

earthquake 77, 118, 177, 180, 188, 250, 265–282, 290, 297–302
fault 280, 282
Great East Japan earthquake 266
2018 Iburi 271
inland 265
loading 266
2014 Nagano Hokubu 280–282
Nankai-Tonankai-Tokai 267
prediction 250
2018 Soya 271
2000 Tottori 279–280
effect 65, 82, 90, 125–128, 144, 147–148, 153, 157
body forces 148
corner 65
elasticity 126
frictional 82
loading rate 125–127
pressure 147–148
reinforcement 144, 300
rotational 90
sample height 126, 128
temperature 157
tensile yielding 153
viscosity 125, 126
element 41–42, 83–85, 118, 121, 159, 177, 185–188, 192–196, 202–203, 207–209, 218, 226, 242–263, 271–302, 307, 313, 315, 318–320
boundary 121, 192, 249
brick 241
contact 242, 244
cubic 26, 30
finite 41–42, 83–85, 118, 121, 159, 177, 185–188, 192–196, 202–203, 218, 226, 207–209, 242–263, 271–302, 307, 313, 315, 318–320
Hookean 47, 48
infinitesmall 194, 251, 254
interface 242, 244
iso-parametric 240, 241
joint 242, 244
Kelvin 47, 48
linear 238, 240
Maxwell 48
quadratic 241
representative 73
triangular 238
equation of motion 132, 198, 226, 236, 242
Eringen, A. C. 1, 10, 11, 17, 18, 25, 26, 42–43, 78, 121, 127, 129, 190
experiment 62, 79–80, 82, 84, 91–94, 97, 100, 120
bending 80
Brazilian 80
compression 79–80, 82, 84, 120
creep 62, 100
drying 182
impression 153, 257, 259
shear 91–94
tilting 92, 94, 97

fault 70, 74, 91, 177–180, 191, 246, 265, 279–283, 299–302
Fick's law 43, 44, 57, 182
flow rule 50, 54, 55, 59, 63
foundation 78, 153, 249, 250, 257, 320
friction angle 65, 91–96, 118, 138, 283–285, 289

Gauss divergence theorem 303, 304
geoengineering 42, 118, 189, 300
Goodier, J. N. 121, 153, 155, 191, 259, 302

Hinton, E. 66, 79, 192, 236, 249
Hoek, E. 65–70, 78, 138, 148, 155, 157–158, 190, 252, 301
homogenization technique 76–78
hydraulic conductivity 34, 45, 57, 163

Ichikawa, Y. 78, 118, 248, 301, 311, 313, 320
Inglis, C. E. 155, 190
iteration Scheme 243
Ito, T. 77, 97, 99, 118–120, 189–190, 263, 300–302

Jaeger, J. C. 43, 59, 68, 78, 80, 83, 101, 116, 119, 121, 122, 153–158, 190

Kirsch, G. 121, 155, 159, 190
Kreyszig, E. 111, 119, 121, 122, 129, 185, 190, 192, 249
Kumsar, H. 188, 190, 294–297, 300–301

Ladanyi, B. 149, 190
laser displacement transducer 96, 182
law 28–36, 40, 43–79, 80, 89, 118, 121, 129, 132, 135–137, 162–167, 175, 182, 202, 205, 211, 218–219, 226–227, 236, 248, 254, 263, 280, 312–314
  constitutive 30, 43–79, 80, 89, 118, 121, 129, 132, 135–137, 202, 205, 211, 218–219, 226–227, 236, 248, 254, 263, 280, 312–314
  Darcy 162–167, 175
  effective stress 33, 34, 219, 227
  energy conservation 28, 36, 40, 121
  Fick 43–44, 57, 182
  Fourier 30, 43–44, 57, 163, 177
  mass conservation 26–31, 35–40, 181
  momentum conservation 27, 33, 36–41, 181
  transformation 14, 317
lining 188, 190, 248, 252, 289

measurement 40–41, 85, 101, 105, 118, 120, 145, 180, 182, 189, 202, 252, 272, 296
  deformation 85
  field 40, 145
  permeability 41, 189
  stress 252
  temperature 105, 180
  thermal properties 101, 118, 120
  wave velocity 182
method 59, 101, 168, 224, 241, 251, 257–260, 265, 270, 272, 276, 285, 293, 298–299
  discrete finite element (DFEM) 242, 245, 249, 282, 300, 302, 313, 320
  finite difference 192, 193, 265
  finite element 42, 83, 159, 177, 185, 192, 194, 242–245, 249, 250, 253, 257, 261, 282, 300, 302, 307, 313, 320
  İnitial stiffness 236, 237
  secant 236, 238
micro-structure theory 74–76
  GPLSM 74–76
  GSLPM 74–76
modulus 84, 122, 127, 135, 137, 145, 153, 182, 187, 192, 217, 250, 259, 282, 313
  deformation 187, 192
  elastic 84, 122, 135, 137, 145, 153, 182, 217, 250, 259, 282
  shear 127, 145, 187, 313
mudstone 299

Nawrocki, P. 54, 62, 63, 77, 149, 189, 261, 300
numerical analysis 71, 249, 267

overburden 142, 252, 263, 278, 290, 292, 293
Owen, D. R. J. 66, 79, 192, 236, 249
Oya tuff 66, 79, 91, 114–116, 187

permeability 34, 35, 41, 45, 66, 107–113, 118–119, 165, 167–170, 174, 176, 189, 191
Poisson's ratio 58, 82, 108, 119, 138, 153, 250, 259

relaxation 148
Reynolds transport theorem 37, 38, 306
rock anchor 118, 132, 144, 189, 261
rockbolt 119, 132, 144–148. 155, 160, 188, 252, 254–258, 287–289, 299–301, 313
Rock reinforcement 189, 248, 313
rock salt 113, 114. 189
rock support 313
Ryukyu limestone 85, 87, 91, 92, 96, 98, 297, 299

sandstone 66, 67, 176, 277, 278
shale 254
shape function 192, 196, 220, 221, 225, 228, 229, 237–241
shear 15–17, 84, 93, 127, 145, 156, 163, 165, 178, 186, 187, 191, 260, 263–264, 273, 275, 287, 295, 312, 313
  modulus 127, 145, 187, 313
  strain 129, 165, 175, 178, 186, 312
  stress 15–17, 84, 93, 156, 163, 165, 178, 186, 187, 191, 260, 263–264, 273, 275, 287, 295

simultaneous equation system 194, 197, 202, 208, 233, 307
softening 137, 152, 187, 236, 249, 295, 297, 298
specific heat coefficient 41, 101–106
strength 65–70, 89, 92–94, 118, 138, 149, 150, 153, 157, 242, 245, 278
  shear 89, 92–94, 118, 245
  tensile 153, 242, 278
  triaxial 66, 68, 157
  uniaxial compressive 65–70, 138, 149, 150, 153
stress tensor
  Cauchy 11–13, 73. 75
  net 74

Taylor expansion 26, 27, 29, 165, 174, 193, 201, 202, 204, 207, 214, 224, 232, 305
tensor 4, 9, 10–12, 14–23, 36, 40, 55, 59–63, 73–74, 76, 78, 156, 311
  crack 73–74, 78
  damage 73–74, 78
  elasticity 55, 58, 60, 63, 76
  operation 6
  second order 4, 9, 10, 73
  strain 18–23, 55, 59, 60, 63, 73–74
  stress 11, 12, 14–17, 36, 40, 55, 60–63, 73–74, 78, 156, 311
  viscosity 55, 58, 63
Terzaghi, K. 33, 127, 128, 131, 155, 189, 191, 219
test 70, 81–82, 89–98, 100, 118–119, 123, 182
  direct shear 89–93, 97, 100, 118
  stick-slip 118
  tilting 92–98, 118
  triaxial compression 70
  uniaxial compression 81, 82, 119, 123, 182
thermal conductivity 101–107, 119–120
Timoshenko, S. P. 121, 153, 155, 191, 259, 302
toppling 282, 284–289, 300
tunnel 191, 252–254, 301
  circular 253
  face effect 252
  support 191, 254, 301

Ulusay, R. 80, 94, 118–120, 182, 189, 190, 262, 263, 300–302
updated Lagrangian scheme 245

yield criterion 65–69, 135, 138, 149, 155–160, 263
  Aydan, Ö 66–70, 157
  Drucker-Prager 65–69, 157, 263
  Hoek-Brown 65–69, 138, 158
  Mohr-Coulomb 65–69, 135, 138, 149, 155–160, 263
yield zone 158–161, 265, 268, 270

ISRM Book Series

Series editor: Xia-Ting Feng and Reşat Ulusay

ISSN: 2326–6872

Publisher: CRC Press/Balkema, Taylor & Francis Group

1 **Rock Engineering Risk**
Authors: John A. Hudson & Xia-Ting Feng 2015
ISBN: 978-1-138-02701-5 (Hbk)

2 **Time-Dependency in Rock Mechanics and Rock Engineering**
Author: Ömer Aydan 2016
ISBN: 978-1-138-02863-0 (Hbk)

3 **Rock Dynamics**
Author: Ömer Aydan 2017
ISBN: 978-1-138-03228-6 (Hbk)

4 **Back Analysis in Rock Engineering**
Author: Shunsuke Sakurai 2017
ISBN: 978-1-138-02862-3 (Hbk)

5 **Discontinuous Deformation Analysis in Rock Mechanics Practice**
Author: Yossef H. Hatzor, Guowei Ma & Gen-hua Shi 2017
ISBN: 978-1-138-02768-8 (Hbk)

6 **Rock Reinforcement and Rock Support**
Author: Ömer Aydan 2018
ISBN 978-1-138-09583-0 (Hbk)

7 **Continuum and Computational Mechanics for Geomechanical Engineers**
Author: Ömer Aydan 2021
ISBN 978-1-138-68053-4 (Hbk)